D0688016

Field Natural History

A GUIDE TO ECOLOGY

Field Natural History

written and illustrated by

LONDON: G. BELL AND SONS LTD

A GUIDE TO ECOLOGY

ALFRED LEUTSCHER

First published 1969

G. Bell and Sons Ltd.,
York House, Portugal Street,
London W.C. 2

Printed in Great Britain by
T. and A. Constable Ltd., Edinburgh

Contents

PART 2: MAJOR HABITATS IN BRITAIN

PART 3: FIELD WORK

Acknowledgements

To acknowledge the help and understanding of all those who have made the writing of this book possible would involve the listing of a lifetime circle of friends and acquaintances who have shared with me the pleasures and rewards of natural history pursuits.

In particular I have in mind the encouragement given by my mother during the difficult days of student studies, the earthy wisdom of a New Forest gypsy and Epping Forest keeper, and the love of the outdoors inspired by my old scoutmaster.

For the immediate present I have to thank, in particular for their generous help and guidance Dr Ernest Neal of badger fame for his kindly advice and criticism of the manuscript, and my wife for her patience and forbearance in sharing her home with a temperamental author, a bush-baby, python, plague rat, jackdaw, and a host of other individualists.

Meanwhile, for any inaccuracies which still appear in the present work, the author is entirely to blame.

ALFRED LEUTSCHER

Wanstead
Essex
1968

Introduction

When the famous Darwin-Wallace concept of Evolution burst upon a startled world in the mid-nineteenth century, a new era in natural history was to begin. Up till 1858 pursuit of living nature was mainly in the hands of a few critical, and usually isolated, devotees, of which Darwin himself was an outstanding example. The foundations for classification laid down a hundred years before by Linnaeus were still in a somewhat dogmatic state, in that each species was looked upon as immutable. In other words, once a dog, always a dog, and once a man, always a man. From here on a new band of students of the evolutionary faith began to apply themselves to examining the startling possibility that a constant flow and change of life had been going on all the time, and that this was governed by natural laws. Darwin's *Origin of Species* was on trial.

To test the validity of a classification which was linked by blood relationship, many more specimens were required as evidence. To carry out this herculean task the naturalist turned into a collector, and with his trophies then went into hiding, so to speak. The museum and laboratory became his meeting-places, and the pickled animal and herbarium specimen his tools. The fact that these were all dead seemed of little importance. Collecting, dissecting and describing each fresh species became the order of the day. Meanwhile the animal and plant as a living and breathing thing was largely ignored.

This had a somewhat unfortunate influence on biological research and education, in that the student seemed to lose contact, or indeed never made any, with the living world outside his classroom and laboratory. His became an academic pursuit, and natural history largely an 'indoor' science. To coin a word, the name Mortology might well have applied to his activities—the study of death.

By the turn of the century a 'new look' began to show. A growing band of amateurs, many of them joined into societies and clubs, was showing an increasing interest in living nature of the countryside. By searching for and recording wildlife and its activities, these amateurs revived the pursuits of the field naturalists begun by such pioneers as Gilbert White. But there was still a gap. The academic scholar with his scalpel and classification key tended to scorn the useful pursuits of the naturalist with his binoculars and net. He in his turn largely ignored the important and necessary spade-work of the museum worker.

Today this rift is rapidly closing, and there was really no need for it in the first place. In thinking over Darwin's classical words, the 'survival of the fittest', one has only to ask 'fitted for what?' and the significance of an organism linked with its surroundings becomes apparent.

This relationship between organism and environment has been the spur to a relatively new and exciting branch of the biological sciences, in which the biologist and field naturalist can work as a team. This discipline, called Ecology (from the Greek *oikos*—a home), is concerned with the study of living plants and animals in their living surroundings, and in their relations with each other. It attempts to seek out and explain the problems of life's struggle with its environment.

One thing becomes clear from all this. No organism, even man, is entirely independent and a law to itself. Each is influenced by natural forces around it, and other organisms with which it comes into contact. Among the latter man has now become the strongest biotic factor and his presence or passing is felt in all corners of the earth, especially in heavily populated areas such as Britain. With rare exceptions, such as remote mountain tops and occasional fragments of shore-line being won from the sea, the whole of these islands are now in a semi-natural state, if not entirely artificial.

This influence may complicate the work of the ecologist but should not deter him. Indeed, it adds interest, for a twofold struggle is going on among animals and plants, one with the environment and the other against man. In some cases they have lost the struggle, and in other cases have come to terms. There are even situations in which, through man's activity, new and artificial habitats have been made which are actually exploited by nature. This is especially so in built-up areas.

This book is an attempt to explain how nature is coping with the problem of living, what is actually going on at the present, and what the future is likely to bring about in Britain. Part 1 gives a theoretical background to the ecological scene, and discusses the relationship between soil, plant and animal. From this there emerges a set of factors in which a certain type of habitat and its peculiar flora and fauna can be recognised. Part 2 deals with those major types of habitat which make up the British scene. Part 3 offers a practical guide from the author's own experience, to the kind of field work which can be carried out by the ecologist, in order to examine, record and confirm the workings of nature.

It is hoped that this book will be of interest and guidance to senior school children, college students, their teachers and instructors, to amateur naturalists, and to all who have pride and concern for our countryside and its wildlife, and who appreciate the urgent need for its proper understanding and conservation.

Part One

The Organism and its Environment

Part 1: Section A

Rocks and soils

To a field naturalist some understanding of the nature of rocks and soils in any given locality is necessary because this will have an influence on the types of plants and animals which it supports. Life depends upon the soil. It is in the nature of living things that animals owe their lives to plants, and plants in turn are dependent upon substances in the soil. Without the inorganic 'foods' supplied by soils life could hardly exist. The subjects of Geology (study of rocks) and Pedology (study of soils) are closely allied, and both materials are necessary to plants. The product of one becomes the substance of the other.

ROCKS

Rocks crumble into dust, and from dust new rocks are made. As with life and death one follows the other, and this has been going on ever since the earth's molten surface hardened into a crust. In a geological sense rocks are the mineral substances which go to make this crust. Some, like clay, are soft and break up readily. Others, like granite, are hard and enduring. In the beginning all rocks originated from the molten magma in the earth's interior. There are three kinds:

Igneous Rocks

In Palaeozoic times, and at intervals ever since, molten matter has reached the surface through volcanic activity, flowing out of long cracks in the crust to spread as lava-fields over the land, or has built up as mountains from chimney-like openings. Active fissure and cone eruptions are still going on today. In time the erupted material cools and solidifies, and shows a characteristic crystalline structure. Such igneous or 'fire' rocks make up some of the oldest parts of the earth's crust, and Britain rests on a foundation of igneous origin, the so-called Palaeozoic platform (Fig. 1). Igneous rocks in Britain are seen at the surface mainly in the west and north where the more rugged and mountainous landscapes are to be seen.

Sedimentary Rocks

Countless ages of exposure and weathering by the elements have worn loose the crystals of igneous rocks, and so provided soils with much of their mineral content. In time a whole mountain may disappear. Most of the worn material is carried away by rivers (or wind) and deposited in the sea (or lake) (Fig. 2). The coarser and heavier material comes to rest near the shore, and the finer material further out to sea, in a graded series of deposits. It is sedimented, or laid down, on the sea-bed. Many times in the past seas have filled up or sea-beds have

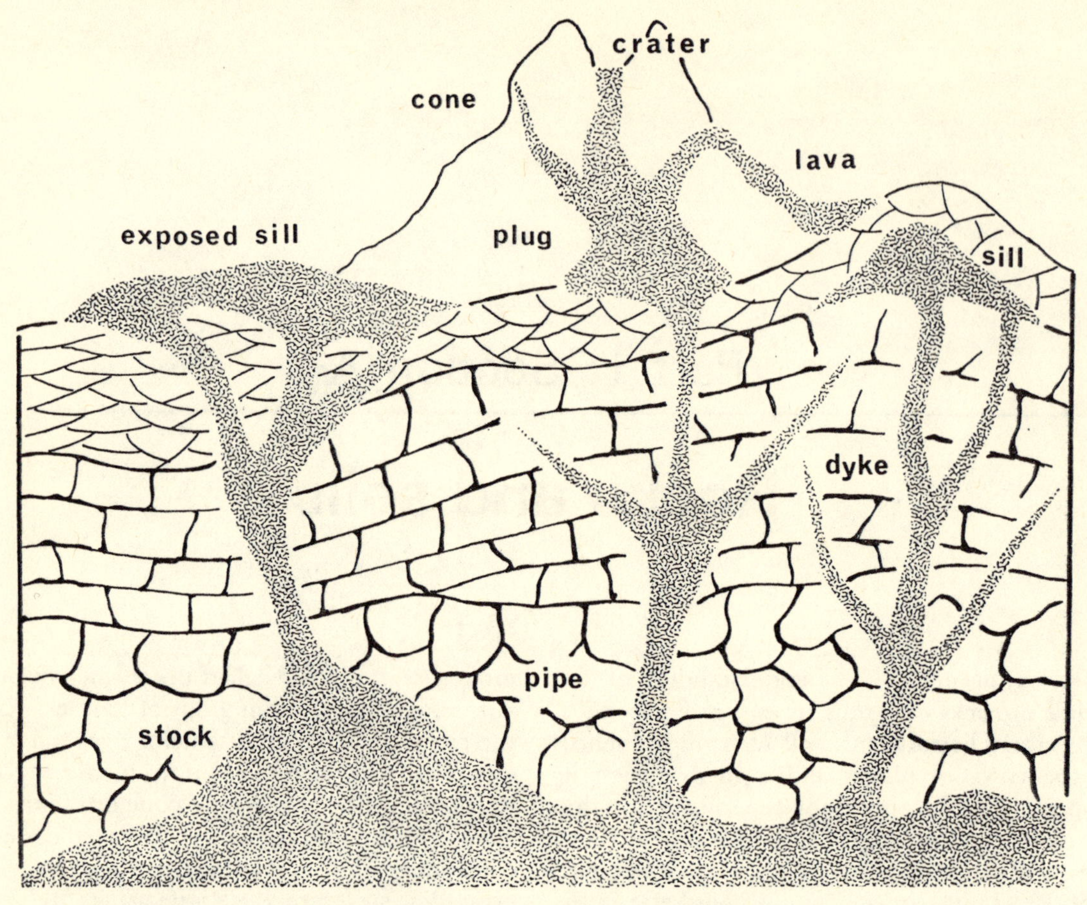

FIG. 1. Section of Earth's crust showing layers of sedimentary rocks and volcanic intrusions of igneous material (shaded).

consolidated and hardened into rock where earth movements have caused uplifts of the crust, and the water has receded. In this manner, by alternate sedimentation and erosion, such rocks have formed and then disappeared many times. Some of Britain's sedimentary rocks have originated in the sea itself. An example is the familiar white chalk found over much of south-east England. Sediments deposited in old, shallow seas have accumulated where they were made. Fine particles of carbonate of lime, made in part from the shells and skeletons of dead sea creatures, and sometimes from sea-weeds, rain down on the sea-bed to form a soft ooze which may one day become a calcareous rock. The process of erosion, river transport and deposition in the sea goes on continually, so that with further consolidation and uplift new sea-beds become rock layers on top of the older ones. Ideally, the earth's crust should look like the layers of a sponge cake. However, due to changes in pressure in the earth's crust, which are fortunately very slow processes in most cases, irregularities in these layers are usually found. The rock layers are compressed into folds, or are fractured by faults (Fig. 3). Sometimes these movements have been so violent that older rocks are forced on top of the younger ones. This happened during the Alpine 'storm' in the Miocene Period when the mountain chains of south Europe were formed. Shock waves from this upheaval reached Britain to produce the more gentle folding of rocks in south-east England. A combination of earth movement and rock erosion may produce a complicated jumble of rock layers which can set a problem for the geologist

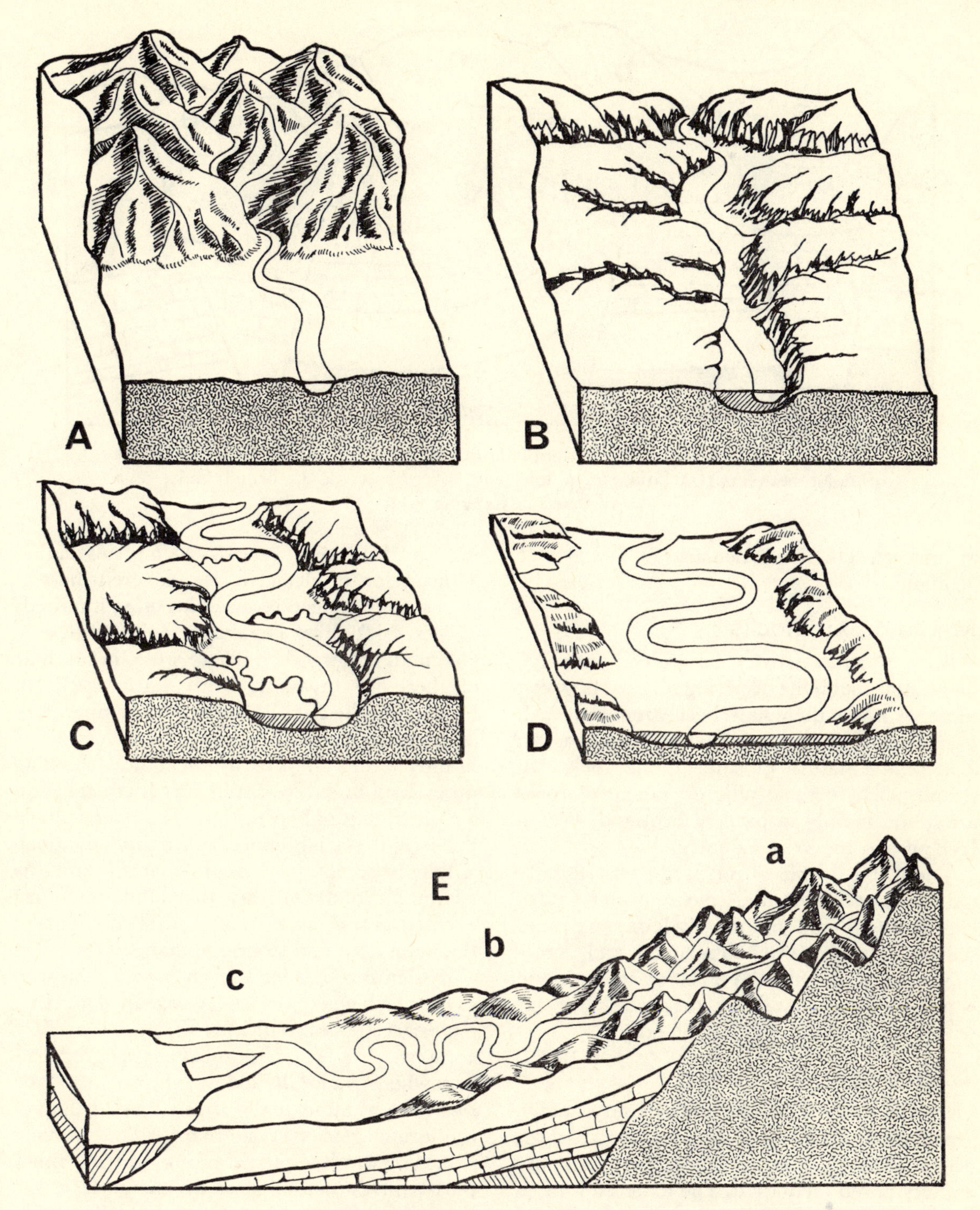

FIG. 2. River action. A-D Stages in river erosion. E. View of river showing all stages. a, the young stage (mainly erosive); b, mid or transport stage, forming meanders; c, old or deposition stage, forming a delta.

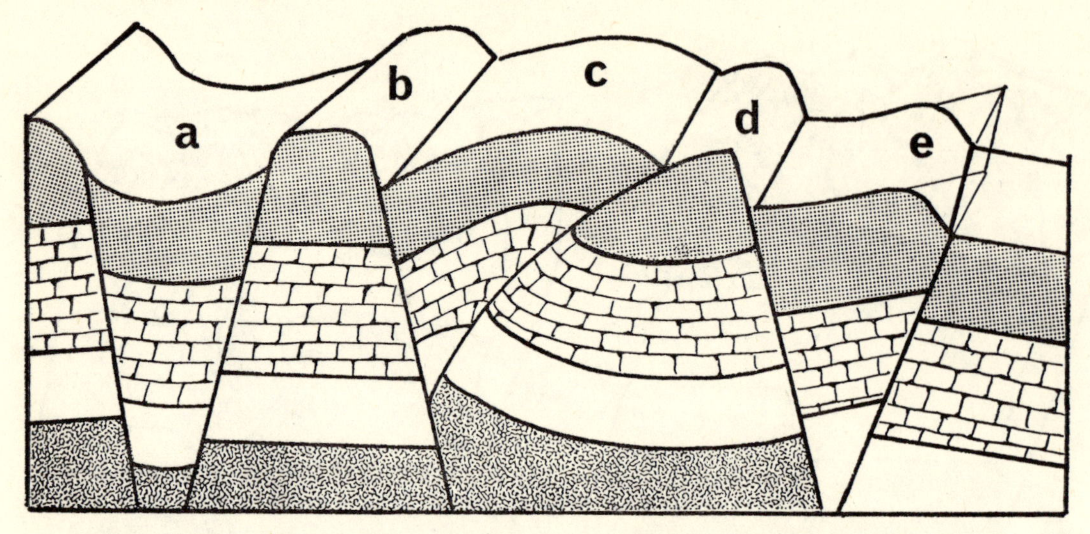

FIG. 3. Folding and Faulting. a, a double fault making a rift valley; b, a raised block or horst between two faults; c, a fold with overthrust; d, a normal fault; e, a reversed or hanging fault.

to unravel. This is particularly so in parts of Scotland.

Metamorphic Rocks

Where volcanic activity has forced intrusive material through the overlying rocks or between them, or where there have been strenuous earth movements, the resultant heat and pressure can change the nature of surrounding rocks. In Britain such metamorphic or 'changed' rocks are found mainly in parts of Scotland, Wales, Ireland and the West Country.

These three types of rock—igneous (usually old, hard and crystalline, metamorphic (hard, chemically altered and usually laminated) and sedimentary (grained, compacted and sometimes soft), provide the bulk of the mineral contents of soils, and are the foundation on which plants and animals live and grow.

The Ice Age

Here and there in Britain are deposits of clays, gravels and sand, even large boulders, whose origin cannot be explained through igneous activity or sedimentation. The material is often foreign in nature to the locality, and must have been brought in from other areas. Such drift material is largely the result of the Ice Age which affected much of Britain's landscape during the Pleistocene Period. During this Period a succession of glacial phases covered most of the country in deep layers of ice often hundreds of feet thick. The landscape looked much as Greenland does today. There were, in all, four of these ice invasions, with warm spells in between, called interglacials. As temperature fell and a glacial phase built up, rivers of ice formed in the valleys to the north, slowly linking up to form broad ice-sheets which advanced in a general southerly direction. The major phase, the second glacial, ended approximately along a line between the Bristol Channel and the Thames. South of this the land remained exposed as a bleak strip of tundra, the home of the mammoth and stone-age man.

Evidence of the ice movement may be seen in the north where the ice-rivers, called glaciers, have carved out a typical U-shape in the valley contours (Fig. 4). The so-called hanging valleys where waterfalls today spill out from the side streams indicate the approximate height of the former glacier in the main valley. Moving slowly forward it carried along with it, either buried in the ice or loose on the surface, the rock material torn away from the valley sides. These erratics were later deposited as moraines where the glacier ended, sometimes damming up the valley and trapping the melting ice. This

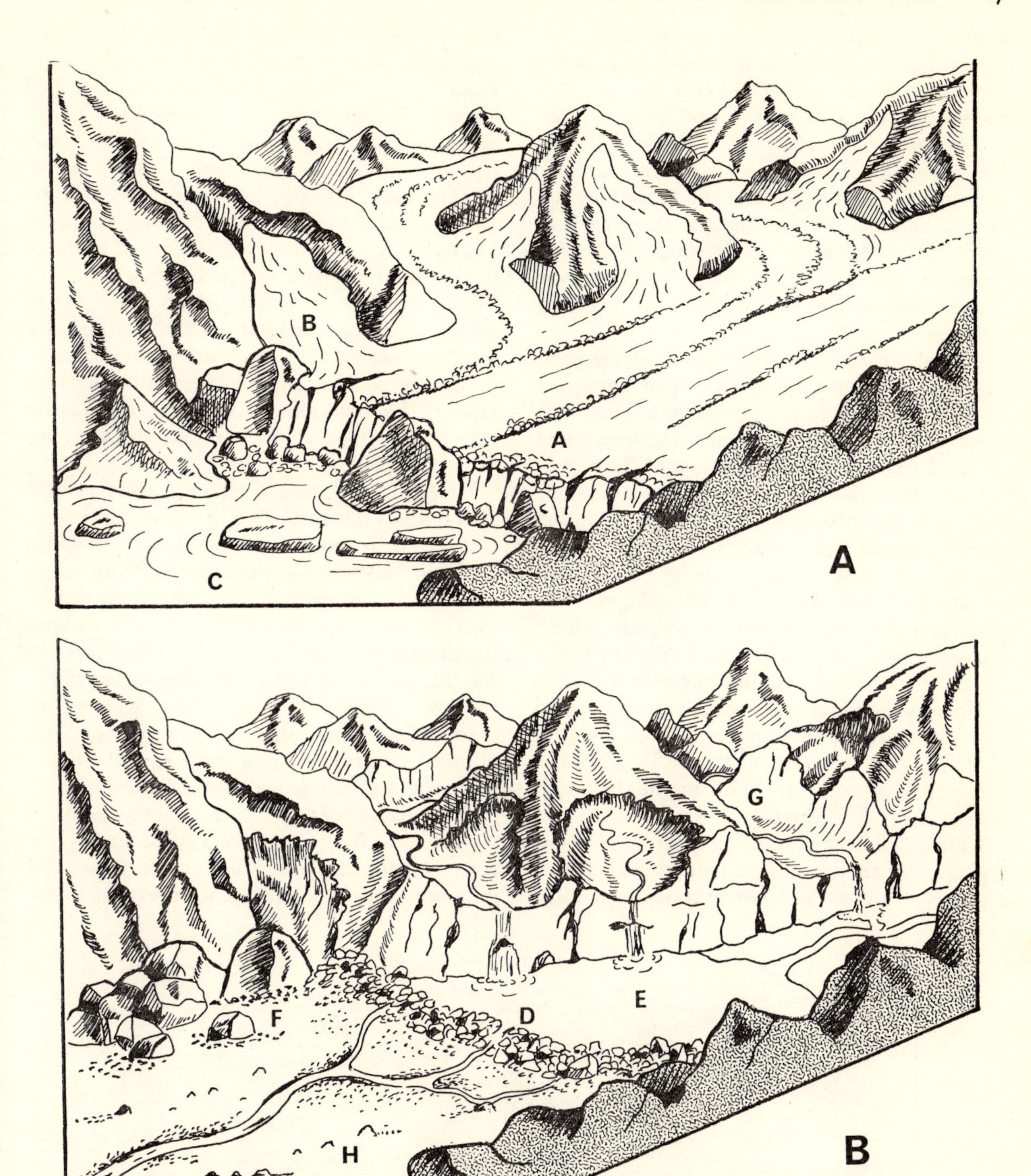

FIG. 4. A, a glacier at the time of the Ice Age; B, the same glacier valley today. Note the vertical sides giving the valley a U-shape. A, end of glacier; B, snowfield; C, melting ice; D, terminal moraine; E, glacial lake; F, erratics; G, hanging valley; H, boulder clay drift.

explains the origin of many of the elongated glacial lakes, called lochs or meres, in Scotland, the Lake District, Wales and Ireland.

Moraines mark the end of a glacier. A further deposit, called boulder clay, was laid down by the expanding ice-sheet as it ground its way over more level ground. Streams of melting ice flowed beneath the ice-sheet, to spread out a layer of finer glacial drift which fanned out over a wide area. Such drift material is a common feature over much of the low-lying parts of East Anglia.

In some places parallel scratch marks may be seen on exposed bedrock which has been scoured by the passing ice, whose movement and direction is recorded by the orientation of these marks. Large boulders left exposed by the receding ice have become smooth and rounded. Resembling the backs of resting sheep they are known as *roches moutonnées*.

The extent and depth of the ice-sheets may be judged from the presence and whereabouts of erratics scattered over the countryside. These are rock fragments once carried in the ice. Some found just north of London have been traced to Norway, and material found over a thousand feet up in Snowdonia originated from the bed of the Irish Sea. One can imagine a Britain once drowned in a sea of ice with mountain peaks jutting above like islands.

Violent winds are a common feature of polar regions, and are due to warm air currents meeting a cold atmosphere over the ice-cap. Fine dust particles are carried by these winds and deposited over a wide area (see also Sand Dunes p. 98). This superficial deposit is called loess. In glacial times a similar fine dust spread along the southern border of the ice-sheet, and the brick-earths in the lower Thames valley originated in this way. During the bleak arctic spells a tundra vegetation covered southern England. Much of the arctic flora and fauna of this southern region has now become extinct in Britain, or has retreated to the north. Such arctic 'relicts' once lived at sea-level in the south, but have been driven into our northern mountains. One such example is the mountain avens (*Dryas*). Among animals the flatworm, *Planaria alpina*, occurs only in mountain streams (see pp. 107 and 119).

River gravels are a common feature of drift soils in southern England. The rivers here were unaffected by ice action, and show the typical V-shape valley sides which result from water action. Former levels of the valley floor can be seen as jutting terraces at various levels above the present river course (Fig. 5). Old rivers like the Thames have wide curves or meanders in the lower reaches, passing over a wide valley floor, called a flood-plain. This receives deposits of mud and gravel during flooding. In the past when a glacial phase ended, the flood-waters of melting ice carried to the sea caused a rise of sea-level. This had the effect of damming the river mouth, slowing the current, so that material carried down was dropped on to the

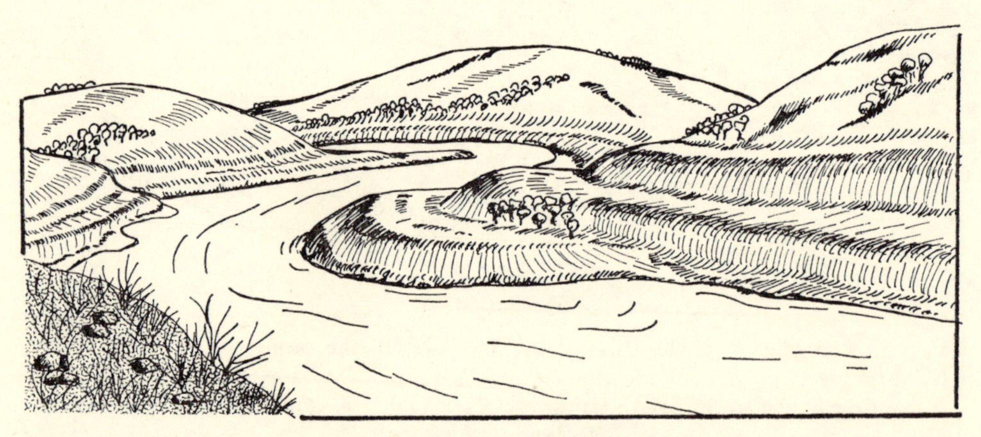

FIG. 5. Terraces along a mature river formed during the Ice Age (see text). The V-shaped valley slopes are due to water action.

flood-plain. During the next glacial the river's flow was once again unimpeded, and it continued its work in cutting a new channel into the old flood-plain. This left behind a shoulder, or terrace, along each valley slope. In succession the Thames has cut itself a new flood-plain with each successive glacial phase. The terraces serve as reminders of former river levels. It should be noted that with sea-deposits (i.e. the beginning of sedimentary rock formation) the younger beds are on the top of the older ones. With gravel terraces the highest one is the oldest.

Types of Rock

Names which are given to different kinds of rocks may give some idea of their composition, but not necessarily of their origin. Some examples are given here:

Igneous Rocks

BASALT. This common larval rock of fine-grained texture and dark colouring is seen in many conspicuous landmarks. The Giant's Causeway in Northern Ireland is part of an ancient larval field which contracted on cooling to produce the striking hexagonal columns. Volcanic dykes, that is, vertical intrusions into the earth's crust, are found on the Isle of Arran in Scotland. Being more resistant to weathering, the basalt stacks stand out from the softer, surrounding rocks. The Whin Sill in the Pennines is a horizontal intrusion between layers of Carboniferous rocks. Where it has been uncovered this bluish-grey rock stands out as an escarpment. The old Roman wall passes over it at the Border, and waterfalls are features in places like Teesdale.

GRANITE. This more crystalline rock is composed of such minerals as quartz, felspar and mica. It is of plutonic origin, that is, the igneous material has solidified below the earth's surface under heavy loads of overlying sediments. Crystals are often large and visible to the naked eye, the result of a slow cooling process. These are the rocks from which sediments such as beach sand are composed. Grey and pinkish granites cover much of Scotland and N. England.

GABBRO. This dark rock is also of plutonic origin and contains large crystals. It occurs in places like the Isle of Skye.

Metamorphic Rocks

As explained above, these are the rocks which have become chemically altered by heat and pressure due to volcanic activity or earth movement. As a result sedimentary sandstones are changed into quartzites, shales into schists, limestones into a crystalline marble, and so on.

GNEISS is a common metamorphic rock resembling granite, but which has a banded structure in grey, pink and dark layers.

SCHISTS are usually fine, crystalline rocks of transformed sediments which tend to split into layers.

Sedimentary Rocks

SANDSTONES AND GRITS. These rocks consist chiefly of sand particles, usually of silica crystals, which are derived from granite and then cemented together, e.g. Devonian Sandstone. Some of the sandy rocks have retained their loose character, and these form a base to many sandy regions. Examples occur in the Hampshire Basin (Bagshot Sands) and in East Anglia (Breckland). The harder and coarser sandstones are well represented in Derbyshire by the Millstone Grit, whose name betrays its former use. On hilltops in the Peak District can still be found the discarded millstones which were carved out of the rock by the old craftsmen. Sandstones usually occur in layers a few feet thick, and suggest how the material was deposited at intervals in estuaries and shallow seas. Such rock can often be split along horizontal planes of weakness, called bedding planes. Vertical joints may also occur, so that regular squarish blocks can be quarried as building material. The reddish nature of many sandstones and grits is due to an oxidising process.

CONGLOMERATES. Rounded and water-worn pebbles of old shingle beaches which have become cemented together by water action indicate the former presence of a shore-line. Their discovery has helped to map out the position of former sea coasts. One such conglomerate has the fanciful name of pudding-stone, after its appearance to a plum pudding.

Breccias. Angular rock fragments may fall away from cliffs and mountain crags to gather in heaps at the foot of slopes, or they may be swept into dry river beds and gullies by heavy torrents. These do not show the rounded, smooth surfaces of water-worn material such as one finds in rivers or along the sea-shore. Some breccias are volcanic in origin and are the weathered remains of lava flows. Screes at the foot of mountain slopes may be described as breccias in the making.

Clays, Shales and Slates. Fine particles which are weathered from rock by physical or chemical erosion are washed away and deposited in lakes and seas as layers of mud. Successive layers become consolidated with uplift, to become rocks which may be soft or hard. Clay is a term usually reserved for the softer material, such as the London Clay, Oxford Clay, Kimmeridge Clay and Gault. Shales are harder and tend to split along their bedding planes. Slates are the hardest sedimentary rocks, and often exploited as roofing material. They can originate from sedimentary rock particles washed away by rivers, or from volcanic ash blown into the sea by wind, or possibly from subterranean eruptions over the sea-bed. Under heavy pressure these slaty rocks may become semi-plastic, so that the particles arrange themselves at right angles to the pressure plane. This facilitates the splitting of slate when it is quarried. Slate is a common builder's term and its origin may be from calcareous or clay deposits.

Limestone and Chalk. Whereas sands and clays are mostly weathered from land sources, then transported by rivers and so deposited in the sea, calcareous sediments which produce chalk and limestones have usually been laid down in the sea itself. The sands and clays are deposited near the coast, the coarser material inshore and the finer particles further out. Beyond this, outside the reach of river-borne muds and sands the chalky deposits occur. These are either carried out from the land, or come from the shells and skeletons of dead sea animals, such as the minute protozoans, called Foramenifera. Some limestones originate in freshwater, such as stalactite and stalagmite growths in caves, and as a stone called travertine which comes from evaporated spring water rich in carbonate of lime. Much of south-east England is covered in chalk (see p. 82).

Oolite. Some limestone deposits have a peculiar structure of tiny particles welded together and resembling the roes of fish (Greek *oon*—an egg). Concentric layers of carbonate of lime have covered fragments of sand or shell to produce these spheres which are visible on the rock surface (see p. 86).

Marble. As a builder's term this refers to a rock which takes a polish and is used for decorative purposes. The rock may be of igneous or sedimentary origin, and the fossils contained in the latter reveal its age and give it character.

Rocks of vegetable origin

Coal is an example of this. It is composed largely of plant remains in the better-class coal (up to 90 per cent carbon) or as hardened mud from the ancient Coal Age forest, which is mixed with sand and carbon. In some beds a grey or blackish slate often alternates with the true coal seams, and may be rich in fossil remains. Peat (see Soils, p. 25) is coal in the making.

Pebbles, Gravels and Sands

Fragments of rock variously known by these common names are classified according to their size (see p. 26). They have broken away during weathering. Some are angular in shape, others appear smooth and rounded. The latter have been worn smooth by friction caused by river and tidal action, and are a common sight along rivers and coasts. Their presence below ground which is uncovered during digging operations, often far from the sea, reveals the position of former estuaries, and helps in mapping out the ancient river beds of prehistoric ages.

SOILS

Soils are composed, in varying amounts, of humus, air, water, minerals, and various animal and plant organisms. Each of these components will give a certain texture to the soil, as well as fertility, and the resulting factors will determine the type of vegetation which it supports. This can range from a desert flora to that of a tropical rain forest.

Humus

This is derived from dead organic materials in process of decay. Dead animals and plants, also animal droppings, contribute towards soil richness, as every farmer knows, and no gardener should be without his compost heap. Decay is an important phase in the Nitrogen cycle, and the value of the humus will depend to some extent on the activities of the soil organisms, and to what degree they break down the organic compounds into the mineral salts required by plants (Fig. 6). Animal droppings make a rich soil nutrient, and may affect localised areas on cliffs and shores (sea-bird guano), fields (farm animals), latrine areas (rabbit, deer, badger and otter) and some tropical caves (bats). In the last example this may provide the only food supply to the cave community. Bat guano supports the invertebrate population inside the cave. This is turn provides food for the cave amphibians and reptiles. Dead bats become a feast for scavengers, and nothing is wasted. The speed of decay is to some extent controlled by the degree of light, temperature and soil chemicals, also the work of the soil animals and plants, such as the litter-fauna, fungi and bacteria, sometimes called 'dustmen' of nature.

Air

The amount of air in the soil will depend upon the available space between the soil particles. A clay soil will hold less air than an equal

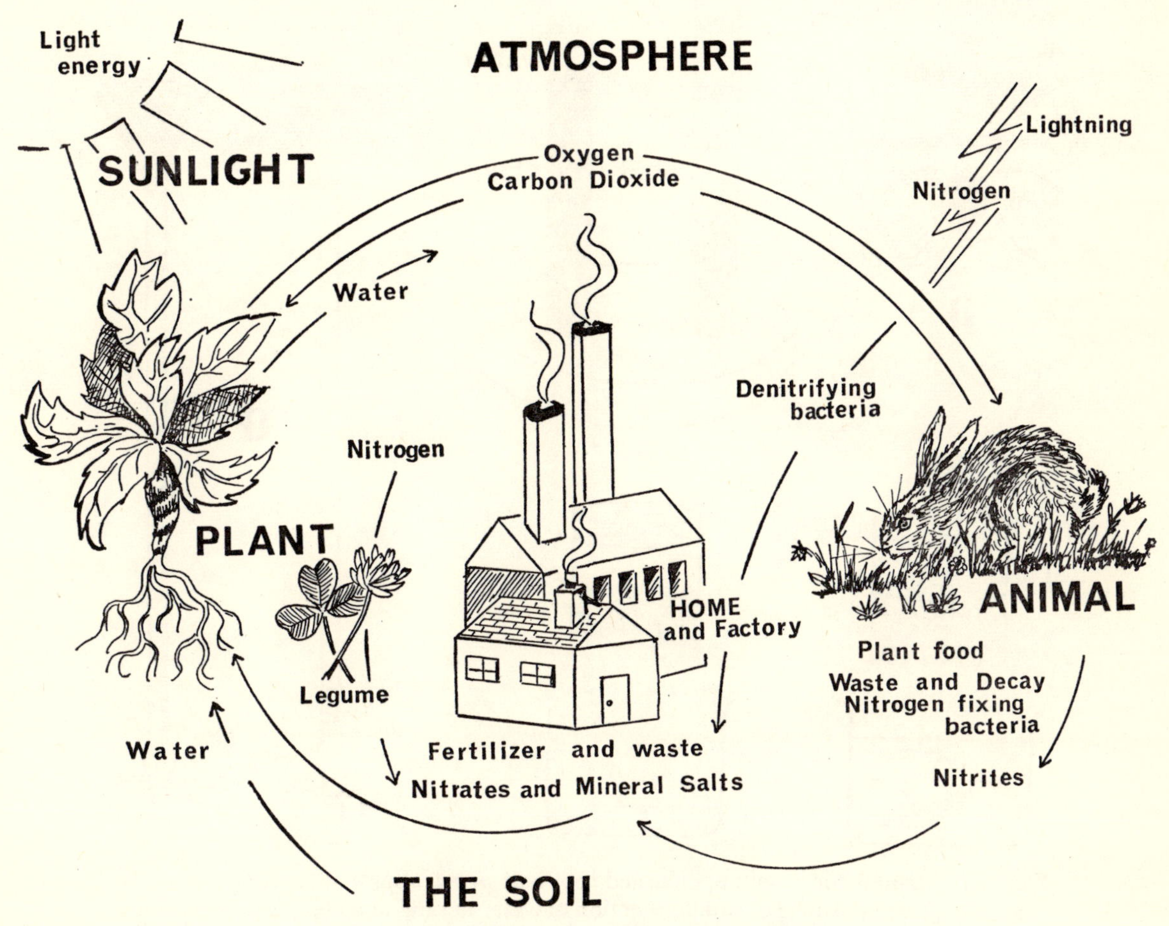

FIG. 6. The Carbon and Nitrogen cycles.

volume of sand. Air finds its way in from the surface, so the more this is broken up the more easily air will penetrate. Plant roots, burrowing animals, freezing and drying are some of the agencies which make the soil more friable.

Water

Water in the soil originates in most cases from rain or melting snow at the surface, or flooding and underground seepage. Should the underground water level reach near to the surface, then the ground becomes waterlogged, and may even support an area of standing water such as a marsh or lake. How long the water is retained by the soil, or how quickly it is lost, depends on the soil itself. With air, the larger the particles, the more space in between can be filled. With water this works in reverse. Most of it is retained by surface tension, that is, it clings to the face of each particle. Consequently, the smaller these are, the larger becomes the surface area to the volume. A fine soil which is more tightly packed also lessens evaporation and drainage (Fig. 7). Such a 'heavy' soil, like clay, may even become waterlogged. A 'light' soil, like sand, which consists of larger particles and more air space, allows for better drainage and more evaporation. The gardener's practice of breaking up the top-soil with a hoe helps to introduce air, and also draws up the hidden water by capillary action. A mulch of grass cuttings or peat on the surface helps to cut down evaporation.

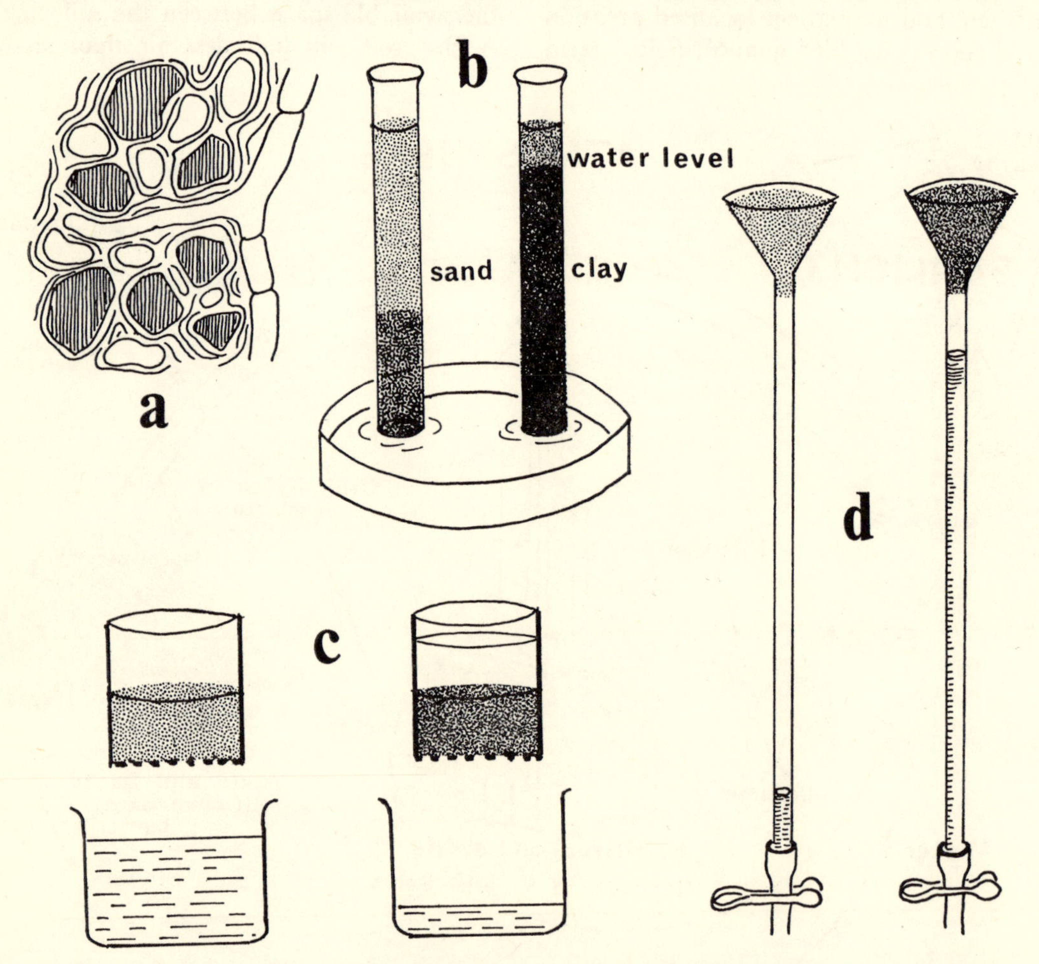

Fig. 7. Water and Air in soil: a, enlarged root hair penetrating soil particles (shaded) containing air and water; b, capillary action of water in sand and clay; c, permeability of sand and clay to water; d, permeability of sand and clay to air.

Minerals

These are derived from the underlying bedrock, from rock material brought in from outside, or as products of decay. A slow breakdown process, called weathering, releases the mineral salts. In water solution these can be taken in by osmotic pressure through the roots of plants (or through the surface in submerged aquatics).

Animal and Plant Organisms

A large range of animals and plants, collectively called the litter fauna and flora, play an important role in the break-up of soil and its dead organic contents (see p. 64).

Weathering

This process, which consists of the breakdown of rocks to form the non-organic ingredients of soils, can function in various ways, and the agencies at work can be physical, mechanical, chemical or biotic:

A. *Physical*. *1*. Heating and cooling. Daily and seasonal changes in temperature cause expansion and contraction of exposed rock faces which in turn break up into mineral fragments. Frost action in cracks will also break up rock. Water expands by 10 per cent of its volume into ice. *2*. Wetting and drying. Rain and dew-water on rocks together with alternating spells of hot sun and cold air will produce a slow physical breakdown of rock.

B. *Mechanical*. *3*. Moving water. Percolation of water through cracks in rocks, or dripping from ledges, cave roofs, bushes and trees, helps in the mechanical destruction of rock. On a more impressive scale rivers, waves, waterfalls and glaciers wear away bedrock by scouring out river beds or beating against the cliffs. *4*. Wind. In sandy areas particles of sand carried by the wind have an abrasive action on the rocks, and will weather them however slowly. This can happen to buildings. *5*. Gravity. Falling rocks and stones will help to break up material which has broken loose, especially in mountains and along steep slopes and cliffs, sometimes building up screes of considerable size. As a guide to this gravitational force, a sandy soil will begin to slide at about 50° of slope. *6*. Pressure and friction. Weathering of this kind is due to movement. Water action of rivers and tides will wear rock fragments into smooth and rounded pebbles and stones. Moving glaciers and ice sheets once played an important role in eroding much of the surface of Britain, leaving behind large sheets of drift material originally rubbed off the valley sides.

C. *Chemical*. *7*. Oxidation. The addition of oxygen to a mineral will alter its nature (e.g. the rusting of iron exposed to water), and this will contribute to the weathering of rocks. *8*. Reduction. This is a reverse process, in which new minerals form from rocks deficient in oxygen, as in waterlogged areas and places rich in humus. *9*. Hydration. The addition of water may form a new compound by chemical action, and lead to rock disintegration. *10*. Carbonation. Carbon dioxide from the air or soil forms carbonic acid when dissolved in water. This has a solvent action on some minerals, especially calcareous rocks such as limestone. Ultimately a whole cave may be weathered out of such rock. In reverse, should the water evaporate, it leaves behind the calcium carbonate which is familiar as stalagmite and stalactite formations where water has dripped into caves, or covered the walls and floors.

D. *Biotic* (weathering caused by living organisms). *11*. Chemical. As weathering agents certain plants and animals may attack rocks and soil particles. The release of carbon dioxide will help in carbonation. Chemicals released by such rock plants as lichens, also acids from root hairs, will slowly dissolve the rock material. This can be seen on grave stones and on old buildings or walls. Soil fertility is added to by the action of nitrogen-fixing bacteria during decay. In some plants (Pea family) these bacteria live in the root nodules. To enrich his farm soil a farmer will grow a leguminous crop, then plough it back before planting a cereal crop. *12*. Mechanical. Disturbance due to activities of soil animals helps to break up and mix the soil ingredients, so allowing for entry of air, and also improves drainage. From centipede to badger, not overlooking Darwin's 'tiller of the soil' the earthworm, all assist in the breakdown of soil. The worm speeds the process of decay by pulling dead leaves into its burrow. Bacteria and fungi, aided by innumerable soil members of the litter-fauna, also take part (see p. 64).

Topography and Water Supply

The surface contours of present-day landscapes are the result, in part, of the weathering process on the type of bedrock which they contain. The original features will have been laid down in the past, due perhaps to some earth disturbance of volcanic activity. Over this the soil will form as weathering proceeds. In this, topography has a strong bearing. In high places and against steep slopes the rains and winds will tend to inhibit soil formation by steady erosion. The leaching process of its chemical content, and the general downward movement of the soil and rock material due to gravity and water movement, will tend to produce deeper and more fertile ground at the lower levels. Also, valleys, hollows and low ground will attract water, so that the water level comes nearer to the surface (Fig. 8).

On slope sof more than 35° a continuous grassy turf does not usually hold, so that the ground is exposed in parts. Once a turf community is firmly established, however, it acts as a blanket against frost and over-dryness, even more so if there are bushes or trees. The plant roots help to bind the soil, and the ground vegetation acts as a sponge to retain moisture. The plants, in effect, protect the soil which gives nourishment in return. Without cover erosion may set in as the sun and wind combine to dry it out, and the rainwater carries it away. One of the tragic lessons learned in many parts of the world is the result of selfish removal of trees. Once the trees are down and none planted in their place, the soil is lost through erosion, especially on sloping ground, and a man-made desert is formed.

Some soils are produced on the spot as on bare mountain crags, and in sheltered places such as earthworks and on bared ground (see Bombed Sites, p. 123), also on land reclaimed by the sea. Other soils are derived from outside, and may accumulate under the action of gravity

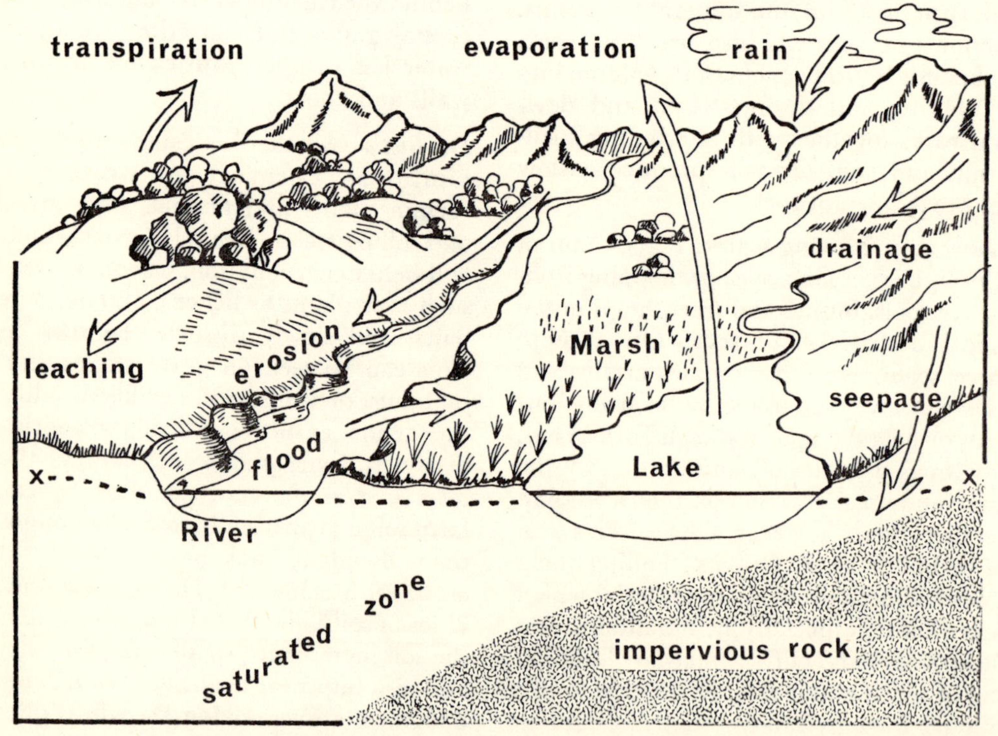

Fig. 8. Topography of land in relation to the circulation of water (x-----x water table.)

(screes on mountain slopes), wind action (sand-dunes) or by water transport (alluvial muds and sands in estuaries). Height above sea-level, latitude, and its position relative to the sun can also affect soils and their vegetation. Colder ground where there is a shorter growth period occurs in the north and on mountains with a short summer season (see Mountains, p. 104). Slopes facing south are warmer than those facing north, and there may even be a marked contrast in vegetation on opposite sides of the same valley or mountain.

Soil Types

In examining a profile of soil and its underlying bedrock (e.g. along the edge of a cliff top, quarry or railway cutting), certain types of soil can be recognised. Each has its characteristic colouring, chemical nature and water content. The following main soil types occur in Britain (Fig. 80):

A. *Brown forest earth* (see Woodland, p. 73).
Colour: uniform brown to grey, with a dark, humus-rich topsoil.
Drainage: normal to impeded.
pH: normal to acid, and never base saturated.
Base Rock: varies, but typically on clays.
Habitat: deciduous woodland.

B. *Podsol* (see Heath and Moorland, p. 87).
Colour: topsoil a dark humus, then a greyish horizon of leached soil. Below this a dark brown to reddish band containing iron compounds which may form a hard and impervious 'pan', finally grading into the base rock.
Drainage: excessive.
pH: strongly acid (below 7). For explanation of pH see Appendix 15, p. 233.
Base Rock: usually a sand or porous sandstone.
Habitat: heathland.

C. *Chalk soil* or *rendzina* (see Chalkland, p. 82).
Colour: uniform greyish to white, sometimes with white chalk particles showing on surface, and a dark, usually thin and loamy top-soil.
Drainage: free.
pH: 7 or above, and often base saturated.
Base Rock: chalk, oolite or limestone.
Habitat: cliffs, hills and downs, depending on locality and bed-rock.

In the formation of a podsol condition there is excessive drainage with leaching. The upper layer loses its base chemicals, and iron compounds are drained into the lower and darker horizon. This does not normally happen with a brown earth or a rendzina. In the former seepage is usually too slow since the rich humus layer holds back the moisture. In a rendzina any leaching is usually countered by a replacement of base material from the parent rock. Podsols which are leached usually occur on sandy soils under heavy rainfall, also on exposed mountain tops. In both places one or other kind of heath is found.

D. *Peat soil.*

Peat is plant material in the process of decay which may ultimately end up as a form of coal. It can be regarded as a soil where about 65 per cent. or more of the content is composed of plant material. There are some plants which can grow in peat soils even of different chemical composition. In fen peat which is alkaline the base chemicals come from the parent or neighbouring rock. In acid peat these base materials are missing (see Heath and Moorland, p. 87).
Colour: very dark brown, almost black in places, and of varying thickness.
Drainage: usually impeded, the peat acting as a sponge.
pH: alkaline (fen peat) or acid (moorland peat).
Base Rock: variable—usually chalk under fenland, or base deficient rock such as granite on moorland.
Habitat: fenland or moorland.

E. *Glei soil.*

This occurs in places where waterlogged land is fairly permanent.
Colour: topsoil a mottled, rusty brown due to presence of iron compounds which overlie a saturated horizon of greyish-green to bluish soil (see Pond and rivers, p. 111).
Drainage: permanently water-logged, or almost so, to just below the surface.
pH: usually acid, but depending on bedrock.

Base Rock: variable.
Habitat: marshland and fen in low-lying areas, borders of lakes and rivers, moorland bogs, etc.

Models of the above soil profiles can be made as a record (see Project 16).

Soil Texture

For study purposes soils are described and identified by their feel and appearance. The texture will depend upon the amounts of gravel, sand, silt and clay which they may contain. These commonly used terms for loose rock material are linked with the relative sizes of the particles. One standard of classification gives the measurements:

Gravel: from 2 mm. down to 1 mm. in size. Larger gravels are called pebbles or shingle.
Sand: grains ranging from 1 mm. to 0·005 mm.
Silt: from 0·005 mm. to 0·002 mm.
Clay: particles below 0·002 mm.

The smaller the particles the more surface area is exposed to water, so that finer soils tend to hold more, and to bind more easily. Where sand grains form the bulk of the soil the feel is gritty and friable. Where silt content is high a silky texture is noticed. A dominantly clay soil feels sticky and smooth to the touch. In testing soils by hand a degree of wetness should be allowed for. If possible the soil should be judged when slightly moistened. Wetness or dryness of the humus content can affect the judgement.

Mixtures of the above named particles are called loams, and are named according to the commonest ingredient e.g. a sandy loam, clay loam or silt loam. A balanced loam is given as equal parts of silt and sand. It is neither gritty nor silky, and does not stick.

Part 1: Section B

The living organism

LIFE'S FUNCTIONS

Life, like other active processes, requires energy to function, and it is the manner in which this energy is obtained and utilised that distinguishes a living from a non-living thing. Energy for boiling a kettle of water comes from the continuous supply of heat from a stove, and the stored energy in a clock from the mechanical action of winding up a spring. Energy in a firework or piece of radium is released by the disintegration of part of itself.

Ultimately such energies give out, whereas that which is used by a living thing acts in such a way that it can preserve, renew or add to its parts, and even appear to direct and increase itself in a purposeful manner. This activity goes on within the living material of which it is composed, the protoplasm.

Molecules that compose the substance of protoplasm include carbon, hydrogen and oxygen. They form complex and unstable materials which are rich in energy, and can be broken up to form much simpler and more stable molecules which are low in energy content. Energy which is thereby freed can be utilised for the business of living. This involves two metabolic functions—a building-up process for the storage of energy, called anabolism, and a breaking-down process which liberates energy, called katabolism.

In performing these activities plants and animals are complementary in that they depend more or less on one another. However, it is plants which are mainly engaged on a building process, and animals which then break down these plant products. In other words, plants make their own 'foods' or energy supplies, whereas animals get this ready made. It is this which brings out the differences between the two living organisms.

Plant and Animal Distinctions

In comparing plants with animals certain obvious differences can usually be noticed. A plant, we say, is static and anchored to one spot all its life, whereas an animal moves about. Plants are mainly of a green colouring, which is rarely found in animals. Plants appear to absorb nourishment in liquid and gaseous form, which occurs all around them, through broad surfaces. Animals take in solid food through an opening into a food canal. Animals have senses—plants appear to be nerveless. These distinctions are only generalities to which exceptions can be made. Among the lower organisms they are not so readily apparent, since the simpler animals and plants have more in common.

They share the same aquatic environment, in which many of the tiny plants behave as actively as their animal neighbours do.

Taken as a whole, however, all these broad distinctions which separate the two living kingdoms are due to one basic difference. This is the method of nutrition, and where the food and consequent energy supply comes from.

Photosynthesis

It is the plants which are the food builders in nature, a function denied to animals. This is due to the presence in plants of a chemical called chlorophyll, or leaf-green (the Fungi are a peculiar exception to this—see p. 155). In some animals, like the green hydra (*H. viridis*), chlorophyll is visible, but this is due to the presence of minute green plant bodies, called zoochlorellae, which are embedded in the hydra's tissues.

Chlorophyll in the leaf acts as a screen to filter off certain wave bands of sunlight. Light is the energy source which sets in motion a chemical reaction during which certain elements natural to air and water, called the inorganic foodstuffs, interact with the chlorophyll to form into the manufactured or organic foods. These are then utilised by the plants or eaten by animals (Fig. 6).

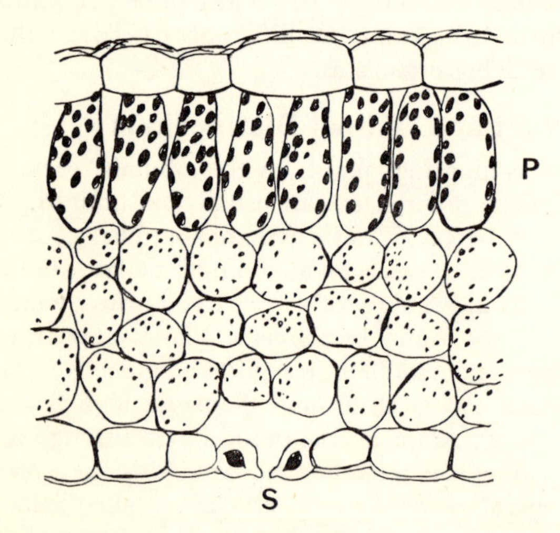

Fig. 9. Cross section of a leaf to show chloroplasts in the palisade layer of cells (P), also a stoma (S) through which gases including water vapour can pass.

The building process in the plant is called Carbon Assimilation, in which the element carbon takes an essential part. The alternative expression of Photosynthesis indicates a need for light energy to make it work (Greek *photos*, light, and *synthesis*, putting together). In this process carbon dioxide (CO_2) and water (H_2O) combine to form oxygen O_2 and a soluble carbohydrate glucose, as shown in this equation —$6CO_2 + 6H_2O$ and light energy $= C_6H_{12}O_6$ (glucose) $+ 6\ O_2$. The chlorophyll which acts as catalyst during this conversion can be seen in leaves within bodies called chloroplasts. These are mainly found in the layer of palisade cells just below the upper epidermis. They are capable of movement so that their position in the leaf changes with the intensity of light (Fig. 9).

The oxygen which is liberated can be seen as bubbles rising from a water plant. By counting the rate of bubbles per minute it can be demonstrated that photosynthesis is slowed down or speeded up according to the available light, or to a rise or fall in temperature. This constant supply of oxygen into the atmosphere is of great importance, both to plants and animals, when it comes to utilising the energy stored up in the above process (see below).

In addition to carbon, hydrogen and oxygen obtained from air and water, a plant also uses further chemicals, collectively called mineral salts, which come from the rocks and as products of decay. These include nitrogen, sulphur, calcium, magnesium, phosphorus and iron. All these elements appear to be necessary for healthy plant growth, and occur in the organic foods and as part of protoplasm.

Organic or manufactured foodstuffs. Carbon, a vital ingredient of living material, occurs in nature in its pure form as diamond or charcoal. In living matter the carbon atoms can join up with hydrogen, oxygen, nitrogen, sulphur and phosphorus to produce compounds which include the three classes of foodstuffs made by plants and eaten by animals—the carbohydrates, proteins and fats.

Carbohydrates. These are the energy providing compounds, the so-called 'starchy' foods which are manufactured during photosynthesis. They contain the elements carbon, hydrogen and oxygen, the last two in the same ratio of 2:1 as in water (H_2O). Some carbohydrates occur as sugars, such as glucose (grape sugar) and sucrose (cane sugar). These crystalline solids are soluble in water and have a sweet taste. They occur in cell sap and can be transported to storage organs and transferred into insoluble starch which is composed of gigantic sugar molecules and visible as starch grains (e.g. in a potato tuber). Carbohydrate molecules may also combine to form cellulose, the material of cell walls, or into certain pigments which give colour to flowers.

Proteins. These are the so-called 'meaty' foods or body builders, composed, in addition to carbon, hydrogen and oxygen, of the soil elements of nitrogen, sulphur and sometimes phosphorus. Enzymes, such as the digestive juices in animals, are protein substances acting as catalysts in plant synthesis and animal digestion. Alkaloids in plants are the result of protein synthesis. Their poisonous quality to animals may serve as protection. Others like oils and resins can safeguard a plant from extreme cold, or act as antiseptics in sealing an injury.

Fats. These also contain carbon, hydrogen and oxygen and are concentrated as food reserves in seeds and fruits. Fat in animals can also act as a heat insulator, especially when concentrated below the skin in warm-blooded animals (e.g. whales). Fats also provide energy, more than twice as much as do the carbohydrates and proteins. Whereas one gram of either of the two latter will give only 4·1 kilocalories, this amounts to 9·3 in fats.

Vitamins. In addition to all of the above, plants also produce the complex substances called vitamins, which are so vital to a balanced diet and good health in animals, including ourselves. To a limited degree an animal can also rearrange organic molecules into its own vitamins, such as the vitamin A found in the liver oil of some fishes, and the vitamin D formed through sunlight, i.e. the 'sunshine' vitamin. This D vitamin helps in the body's metabolism of calcium and phosphorous, two ingredients necessary in the formation of bone and teeth. Some proteins, called nucleic acids, occur in the cell nuclei, and are arranged in strings along the chromosomes. They are believed to form the basis of the hereditary material which regulates the activities of an animal or plant. One of these is the deoxyribose nucleic acid, or DNA.

With all these food-building activities going on in the leaves of plants we may regard them as the great chemical factories of nature. During synthesis of these foods, which is a way of storing up the sun's energy, more is usually produced than is immediately required. Reserves can be laid up for the renewed growth of buds and flowers each season, or to give a start in life to an embryo seedling.

Respiration

Photosynthesis is but half the story of life's activities. Having made its food a plant can then break it down by a katabolic process so that energy is released. This goes on all the time, whereas photosynthesis ceases before it gets dark. By an oxidising process, for which free oxygen is needed, a plant can obtain energy from its carbohydrate stores. This is called respiration, during which carbon dioxide and water are liberated. In animals 'respiration' is a term which includes its breathing activity, during which there is an exchange of the gases oxygen and carbon dioxide through its gills or lungs. It also includes the oxidising of its foods with the liberation of energy as in plants.

Animal Feeding

Since animals cannot make food they must search for it. This requires movement and awareness of their surroundings, quite apart from the dangers of being eaten themselves. Food is taken in, or ingested, through a mouth, then broken down or digested within a food canal. It can then be rebuilt into animal tissue. Waste products are passed out in the process. The search for food is performed in many ways, according to whether the animal is a carnivore, herbivore, scavenger or parasite. This is one of the main concerns of animals, a competition for food, whereas plants mainly compete for living space. Digested food undergoes chemical changes to convert it into substances which compose the body of the feeding animal. This is necessary because the nature of flesh differs from one animal to the next. Among the most important are the protein foods. As colloidal substances they are converted into amino-acids. Starches become sugars and fats are converted into fatty acids and glycerol. In these simpler forms food is conveyed in the blood stream to the tissues, or to a storage organ such as the liver.

The following elements are found to occur in animal tissue, and are given by percentage weight: Oxygen 65; Carbon 18; Hydrogen 10; Nitrogen 3; Calcium 1·5; Phosphorus 1·0; Potassium 0·35; Sulphur 0·25; Sodium 0·15; Chlorine 0·15; Magnesium 0·005; Iron 0·004; Iodine 0·0004; also, traces of Copper, Manganese, Zinc and Fluorine.

The Carbon and Nitrogen Cycle

This remarkable process, in which the living protoplasm has the power to convert unlike materials into additional matter of its own composition, is one of the marvels of life. Continuous building up and breaking down of foods goes on all the time, and it is possible to trace the course of the various elements involved such as oxygen, nitrogen and carbon. This can be illustrated by means of well known diagrams (Fig. 6). They show how elements necessary to life fluctuate from plant to animal, via air, water and soil, and serve to emphasise the most important lesson that one can learn from a study of nature. There is no separate independence of plant or animal, and they, together with the soil, water and air on which they depend, form an intricate but co-operative pattern of life on this earth. This give-and-take is as important to life as is its competition to survive. The first helps to maintain life—the second merely changes it.

Perhaps the most remarkable thing about the living process is that it appears to have a direction and purpose which is aimed at self-preservation. There is, as Darwin pointed out, a struggle to keep alive. Each animal and plant responds to events in its environment, so that it may continue to survive. A non-living machine which is more rigid in its structure and performance, must be given a constant supply of energy and is, so to speak, controlled from without. A living thing can exist by its own volition, is self-energising and can reproduce itself. To illustrate the two different ways of life, of plant and animal, the highest groups are chosen for discussion in the following chapter—the flowering plants and the mammals.

THE FLOWERING PLANT

Flowering plants which stand on the topmost rung of plant evolution are well represented in Britain. They also tend to attract more attention than the lower plants, even among botanists, because of their size and bright colouring. Both in appearance and numbers they dominate the scenery. In particular the flowering tree which overshadows all, helps to preserve the deciduous climax woodland which is natural to our mild and maritime climate. Here also, beneath the oak, ash or beech, will be found many flowering plants forming the lower herb layer. In open country where grass and heather grow they may be found in profusion as herbal communities. Each of these, whether herb or tree, is adapted to its environment and can greatly vary in shape and size. On closer examination a number of features may be noted which are shared by flowering plants.

The shoot and root

Typically a flowering plant consists of a shoot and a root portion. The former usually grows above ground and is exposed to air and light (unless aquatic). It has a definite structural pattern. There is usually a vertical axis, such as the main trunk in a tree, bearing side branches which in turn may support supplementary branches. These arise in the angles made by the leaves and the stem from which they grow. Such a position on the plant is called a node, and the part in between two nodes, an internode (Fig. 10).

The great variety of shapes and designs among flowering plants is due to the proportionate size and growth of the various parts. Some plants have long internodes—in others

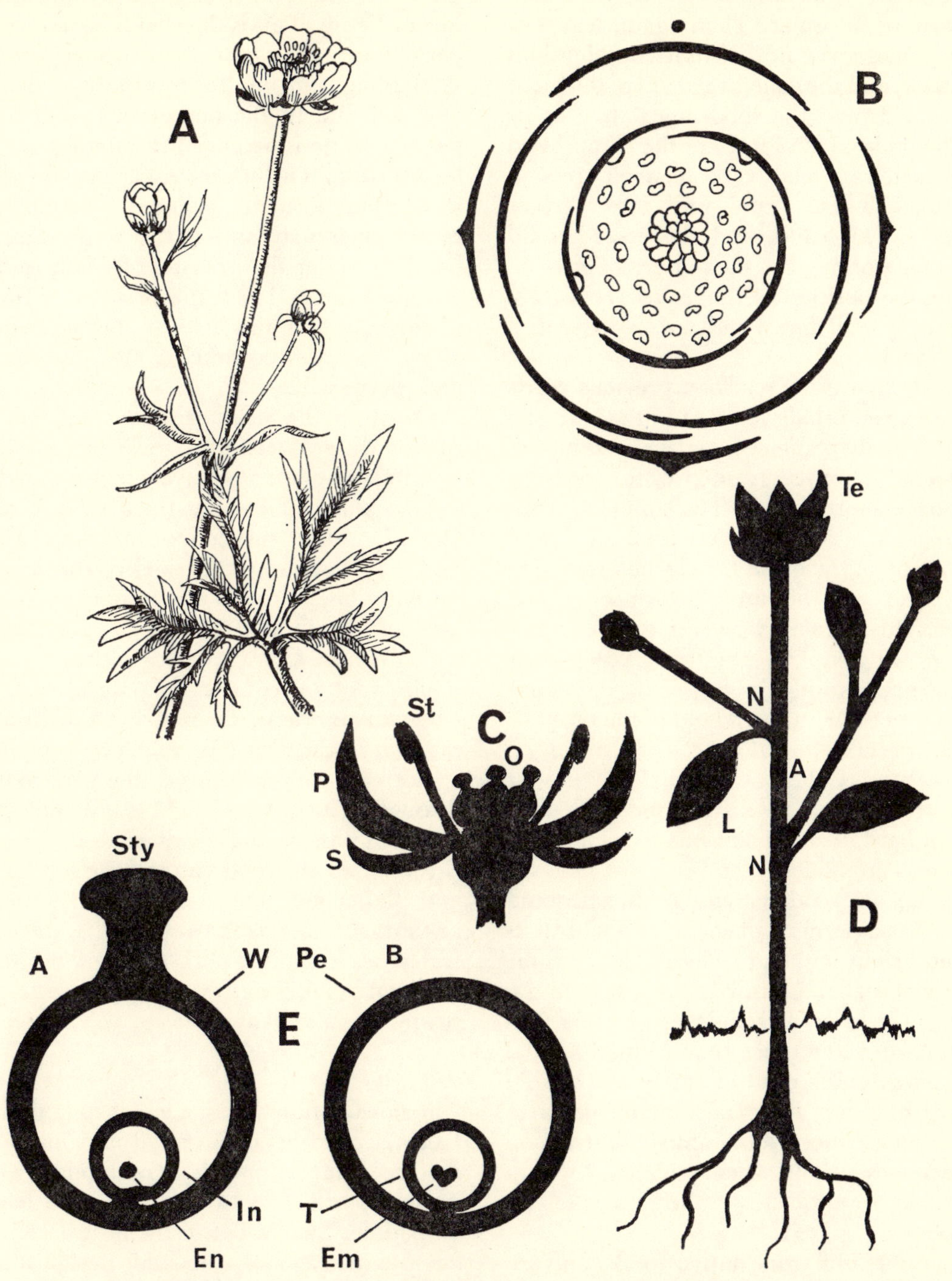

FIG. 10. A, The Meadow Buttercup (*Ranunculus acris*); B, floral diagram of a buttercup; C, parts of the flower; D, anatomy of a flowering plant; E, section through A, an ovary, and B, a fruit for comparison. A, axillary branch; EN, egg nucleus; EM, embryo; IN, integument; L, leaf; N, node; O, ovary; P, petal; PE, pericarp; S, sepal; ST, stamen; STY, style; T, testa; TE, terminal bud, and W, ovary wall.

they are short. Branching of side growths may be uniform or irregular. Then again, leaves of varying shapes grow in different combinations on the stems, and the general growth of the plant is more rapid near the shoot apex than at its base. The buds which form are the compressed shoots which can readily be seen on trees in winter-time. These grow out into further branches, or as leaves and flowers. A useful criterion for working out the pattern of growth, is to note the position of the leaves. It is from the axil of a leaf that a bud arises to form a lateral branch.

In woody plants the scar of a previous year's leaf position can usually be seen at the base of a new bud. Woody plants such as trees and bushes may exist for many years and grow to large size. Others have shorter lives. The annual germinates, grows up, sets seed and dies, all in one season. Many of these will rapidly colonise bare ground, and can be an annoyance to the gardener. Others take two years to live (biennials), growing up in the first year, then flowering and fruiting in the next. Perennial herbs take more permanent hold of the ground, dying down every winter to their roots or storage organs. Other perennials, the largest and long lived such as trees, develop a secondary growth of hard, woody elements to which successive rings are added each year, just beneath the bark. Such a woody perennial can withstand the winter as a permanent shoot, and add to its girth and height with age. There may be some danger in this ageing growth. An oak tree, for example may grow to hundreds of years, but is much harder to replace than a small annual which has a short life, early maturity, and a rapid turn-over of seed. Trees take many years to mature, and are more vulnerable to destruction as a species, especially when exploited by man in a careless or selfish way by wholesale felling (see Woodland, p. 73).

The shrubs and trees native to Britain are mainly flowering plants which shed their leaves in autumn (deciduous). The few native conifers such as the Scots Pine, retain their needle leaves, and are better suited to the cooler northern latitudes, or to mountains.

To give a plant support the aerial shoot has layers of cells with strengthened walls of cellulose or lignin. This is the woody structure of the permanant shoots of trees which acts as a scaffolding and for the transport of the water and cell sap. In the herbaceous plant the supporting tissue is arranged in bundles of cells to form a ring. This passes along the stems of the shoot in the form of a cylinder. There is enough resilience and strength in this to resist excessive bending under the pressure of wind, rain, or a passing animal. The water pressure or turgidity in the cells will also help to hold a herb erect. When there is too much water loss the plant and leaves wilt.

Covering the shoot is a layer of epidermal cells with a watertight cuticle on the outside, forming a plant's equivalent to a skin. In woody plants the much thicker layer of bark has a similar protective function. Gaseous exchange between the tissues of the shoot and the exterior is through the stomata in the leaves (see on) or through openings called lenticels which can be seen on woody stems.

The root by contrast is pale coloured and grows in darkness. It branches downwards at random to the root tips, each one protected by a root cap. Behind this are the tiny root hairs through which water and dissolved mineral salts can pass on their way up to the leaves. In this manner the roots can tap the soil for water and plant food, and at the same time provide anchorage. The skeletal elements are concentrated as a central core through this part of the plant, giving roots strength against a pulling strain (see also Water Plants, p. 112).

The leaf

The normal function of a leaf is to manufacture food. It contains chlorophyll and has a broad surface area on the blade or lamina which is normally attached to the stem by a leaf stalk or petiole. Spaces between the leaf cells allow for the passage of air, and the products of photosynthesis such as oxygen and water pass out through the tiny pores or stomata built into the under epidermis (Fig. 9). These pores are encircled by a pair of kidney-shape guard cells. These can swell up and contract with changes of light and humidity, opening by day and when the air is damp, and closing by night and in dry conditions. This mechanism helps to control the rate of water loss from evaporation. In some plants the leaves have reduced surfaces, or are rolled up to cut down the transpiration (see Xerophytes, p. 100). Other leaves are reduced to mere bracts whose function is to protect growing buds rather than to make food.

Storage organs

In many herbaceous flowering plants which die down for the winter, there is some portion below ground which survives and acts as a storage organ. It will grow out a new shoot the following spring. Many such organs are actually part of the shoot. For example, a bulb (e.g. bluebell) is a compressed shoot of fleshy leaves, a corm (e.g. wild arum) is a compact, upright stem base, and a rhyzome (e.g. iris and grasses) a horizontal underground stem. A tap-root (e.g. carrot and turnip) is a swollen root, whereas a tuber (e.g. potato) is a swollen part of an underground stem (Fig. 24).

The flower

The flower by which a flowering plant can be recognised is a highly specialised growth of modified leaves which function in reproduction. Leafy structures which operate in this way are more noticeable in organs called sporophylls (i.e. spore-bearing leaves) which occur in the lower plant-forms such as the club-mosses. The sporangia which contain the spores are borne on these leaf-like growths. In true flowers the different parts may be looked upon as specialised leaves within a compressed shoot apex (Fig. 10). A common feature in most flowers is their bisexual nature. At the centre is the female organ or ovary, composed of one or more carpels. These may occur separately, as in a buttercup (*apocarpous*), or are joined together (*syncarpous*). A carpel bears a style at its apex, ending in a stigma. This has a receptive surface which will receive the pollen grains. Surrounding the ovary are the male organs, called stamens, consisting of a pair of swollen pollen-laden anthers borne on a stalk or filament. Stamens on a buttercup are held erect, since this flower is insect-pollinated.

With wind-pollinated flowers the stamens usually hang loosely, as in grasses. In others, like the hazel, the male flowers grow separately in hanging clusters, called catkins (Fig. 11). Pollen can thus be transferred from anthers to stigmas by insects (*entomophilous*), by wind (*anemophilous*) and even by water (e.g. the Hornwort, p. 114). There is usually some mechanism which can prevent self-pollination, such as the maturing of anthers and ovary at different times. This will help to ensure cross-pollination.

In the case of insect-pollinated flowers there are some ingenious mechanisms and shapes among flowers which are closely connected with certain kinds of insects. In such a plant-animal relationship there is a mutual advantage, in that the flower is pollinated by the insect, and the latter receives nectar or pollen as food.

Since a buttercup has a regular shape and is unspecialised, it may be visited by numbers of flies, bees and butterflies. With flowers of the pea family, however, there is irregular shape. Below the large overhanging petal, called the standard, are two lower petals, called the keel, which together form a boat-shaped covering to the stamens inside. A heavy-bodied insect such as a bumble bee alights on this, depresses the keel with its weight, so that the stigma and stamens are pushed out at the tip. In this way pollen is dusted on to the bee, and any pollen it carries in from other flowers can be dusted onto the stigma (Fig. 11). In the case of the honeysuckle flower the petals form a deep tubular corolla which can only be reached by long-tongued insects. Pollination usually takes place at dusk when the air is still and heavy with scent. This attracts the large hawk moths then abroad, and they hover in front of the flowers in humming bird fashion to sip at the nectar.

Surrounding the ovary and stamens are the non-essential organs making a perianth. The outer layer or calyx is composed of the sepals, which are usually dull in colour and protect the growing flower bud; they also protect those flowers which show 'sleep' movements at night or during dull weather. Within the calyx is the corolla of petals, usually the brightest part of

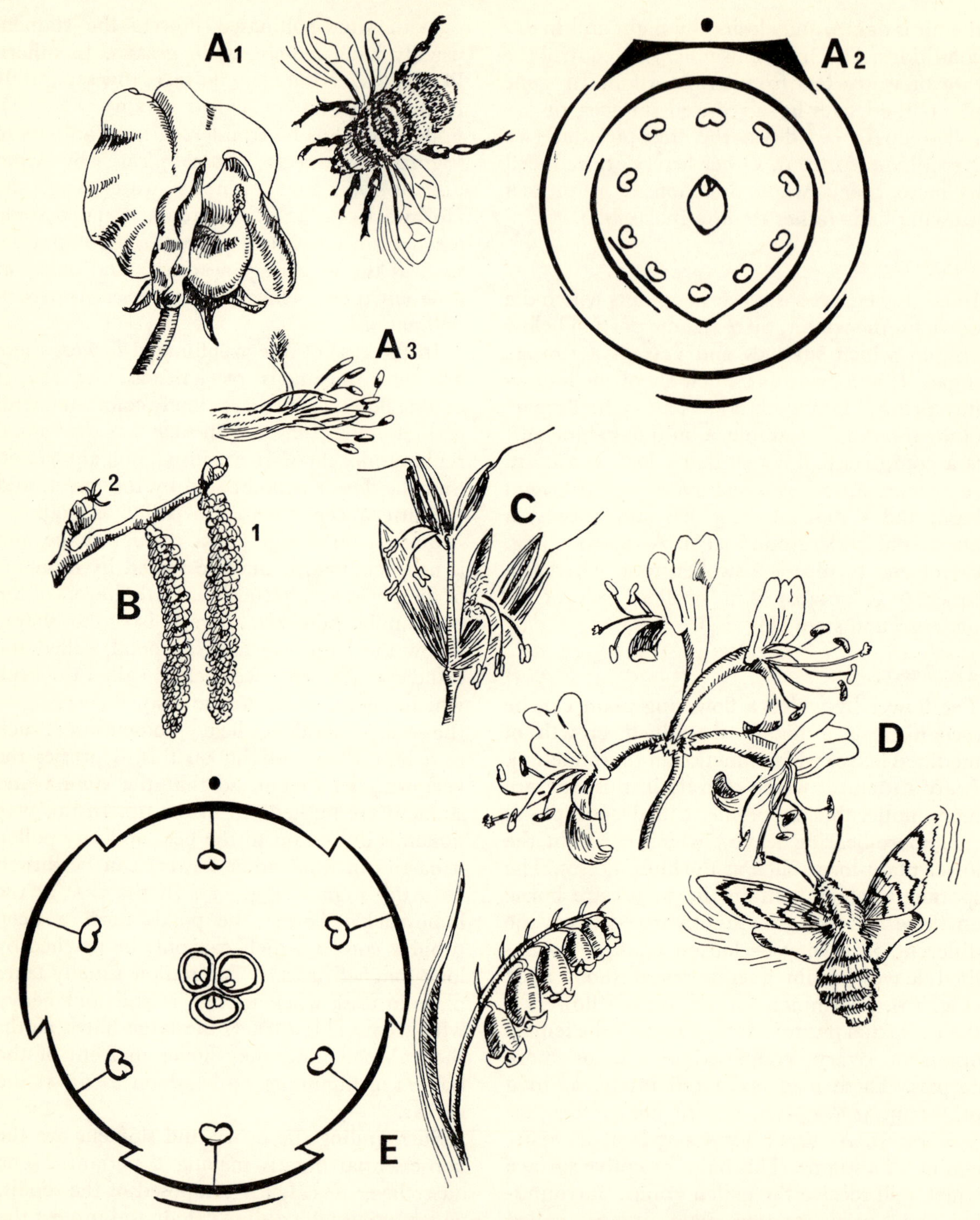

FIG. 11. A_1, bumble bee after visiting a sweet pea flower (*Lathyrus*). Note the emergence of the stigma; A_2, floral diagram of sweet pea; A_3, stamens and style of sweet pea; B, hazel flowers showing two male catkins (1) and a female flower (2); C, flowers of a grass with pendulous stamens; D, honeysuckle flowers (*Lonicera*) visited by a humming-bird hawk moth; E, bluebell (*Endymion*) with floral diagram.

the flower. Their colours help to attract insects. The buttercup has yellow petals and there is a nectary at the base of each for insects to visit. The honeysuckle in contrast attracts mainly by means of scent. To draw an insect in the right direction once it has alighted, many flowers have spots and lines on their petals, called honey guides.

The fruit

Once on the stigma each pollen grain can send out a pollen tube. This penetrates the style and enters the ovary chamber so as to reach one of the ovules. A male nucleus passes down the tube and unites with the egg-cell nucleus in the act of fertilisation. The embryo develops from this and the ovule becomes a seed. A seed is a complicated structure among flowering plants, consisting of the embryo which grows from the fertilised egg-cell, a mass of food material called endosperm surrounding it, and a protective covering, the seed-coat or testa (Fig. 10). Surrounding the seed is a further wall, the pericarp, and the whole struction comprises a fruit in the strict botanical sense.

Technically, a seed arises from an ovule, and a fruit from a carpel. Thus, each buttercup carpel forms a fruit with a single seed, whereas in the pea fruit, called a pod or legume, there are many seeds (i.e. the peas).

Fruits can either be dry or succulent, depending on the nature of the pericarp. Those with succulent covering, such as berries, attract the attention of birds and mammals which eat them. The seeds then pass undigested through the body, and may finally be dropped many miles from the parent plant. Hard fruits such as nuts (achenes in the buttercup) and acorns are collected, stored, then perhaps forgotten, by squirrels, mice and jays, or may be carried away on muddy feet, shoes or tyres (Fig. 55).

Some fruits have hook-like attachments and can adhere to fur and clothing (e.g. burdock and bur-marigold). Those with winged and plumed devices, such as sycamore and dandelion, are carried by wind. Yet other fruits float on water (e.g. yellow flag). Explosive mechanisms occur in those fruits which burst open, or dehisc, as in the pea pod, so that the seeds are scattered. Box-like fruits called capsules shed the seeds through holes, in pepper-pot fashion, as in the poppy (see also Bombed Sites, p. 124).

The great variety of such scattering and spreading devices of pollen and fruit in flowering plants has arisen in evolution as a struggle for living space, in organisms which otherwise would crowd each other out of existence. The very immobility of a plant may well have brought about the line of development already discussed, in which two entirely different kinds of organism, the flower and the insect, have perfected a mutual relationship of value to both. The one is made fertile and the other receives food. Renewed vigour is also bestowed by cross-fertilisation of the flowers, as the bee passes from one to the next. A greater mixing of genes from different parents brings about more variety in the offspring. Without this mutual help between plant and insect the world to our eyes would probably lack much of its beauty, for only such dull-coloured plants as the wind-pollinated grasses could exist.

Identifying a Flowering Plant

As explained elsewhere (p. 56), each animal and plant species is given a binomial scientific name, and each is recognised by possessing certain characteristics. In the case of flowering plants the structure of the flower, and the numbers of its parts, will enable a field-worker to identify a specimen, certainly down to its family, if not its genus and even species.

For example, in the family Ranunculaceae, the so-called buttercups, the flower is mostly of regular shape, bisexual or hermaphrodite, and all its parts are free or separate. Sepals and petals occur in fives, whereas the stamens and ovaries are numerous, the last growing in a superior position on the flower. This information can be shown in a floral diagram (Fig. 10) or written as a floral formula K5.C5.A $G\infty$, in which K stands for calyx (i.e. sepals), C for corolla (i.e. petals), A for the male stamens (androecium) and G for the ovaries (gynoecium). The fruit is an achene or follicle.

In contrast to the buttercup, the pea flower of the family Leguminoseae is irregular in shape. There are five joined sepals in the calyx. The corolla of five joined petals has a large overhanging 'standard', two lateral 'wings', and two lower petals forming a keel. Inside the last are ten stamens fused together by their filaments, surrounding a single monocarpous ovary. Pollination, as already mentioned, is performed by bees. In the sweet pea (*Lathyrus*) the posterior stamen is free from the others (Fig. 11). The formula reads K(5).C(5).A(9) + 1.G1. The dehiscent dry fruit is a legume.

Both buttercup and sweet pea are called dicotyledonous flowers. In their seedling stage two seed-leaves or cotyledons are present. Most flowering plants belong to this group, and many grow into trees with woody growth. The leaves have branching veins. In the other major division, the Monocotyledons, the far fewer species remain mostly as herbs without woody growth (e.g. bluebell). Storage organs are commonly present, and the long and narrow leaves have parallel venation (the wild arum is an exception in Britain). Floral parts grow in threes. In the bluebell, *Endymion* (= *Scilla*) *non-scripta*, the three petals and three sepals resemble one another closely, and collectively form a perianth. There are six stamens, in two whorls, and a trilocular ovary which is superior (Fig. 11). The formula is P3 + 3.A3 + 3.G(3). The fruit in the bluebell is a capsule. It will be remembered that the great systematist Linnaeus based much of his work on the classification of the flowering plants in his *Systema Naturae* by examining the structure of the flower (p. 56).

THE MAMMAL

Whereas flowering plants hold a dominant position in the Plant Kingdom, and are the most highly organised, the same applies to mammals in the Animal Kingdom. It is interesting to note that both these major groups of our modern world evolved side by side as successors to the reptiles and the evergreen seed-plants, called cycads. These flourished during the Secondary or Mesozoic Era. By the Tertiary, about seventy million years ago, many reptiles and cycads had vanished. The Age of Flowers and Mammals was beginning.

In the pursuit of field studies it is true to say that the mammals are not held in such high regard as the flowers. This position in animal popularity must be given to the birds. Where mammals perhaps fail to attract, through no fault of theirs, is in being too great a match for most people. Because of their quiet bearing, neutral colouring, secretive behaviour, and largely nocturnal movements, a study of our wild mammals can only be rewarded after much time, patience and inconvenience is taken. It is only the occasional naturalist, gamekeeper or poacher who knows them intimately. A constant intrusion into their lives has made these intelligent animals elusive. They have been hunted since earliest times, first for food and clothing, then for sport or venerie, but lately more for pleasure and study. Even so, many hours can be spent and wasted in uncomfortable positions when observing mammals.

With patience and a quiet approach, however, these fascinating creatures are worthy of our attention, making a rewarding pursuit in the field, as suggested by the activities and projects given in other parts of this book (i.e. projects on mice, deer and badger, studies of tracks and signs, and the keeping of small mammals).

Mammals in Britain are now recognised as deserving closer study, for there is still much to be learned. The recently formed Mammal Society, which has a flourishing membership, is now doing much to further this interest.

Mammals vary considerably in size, from the minute shrew to the hundred-foot Blue Whale. This and their pattern of behaviour is very much connected with different ways of food gathering. Since animals require made-up, that is, organic food, they must first find it and even pursue and catch it in some cases. It is then eaten or ingested, broken down or digested, and waste material discarded. Various body organs carry out these functions. Mammals are adapted to performing the act of food gathering according to where they live and what they feed on. The one is complementary to the other, and two parts of the body which are specialised, one for movement and the other for feeding, are the limbs and teeth.

Mammal Limbs

All mammalian limbs are derived from a basic plan which is found in their early land ancestors, the Amphibia. Each of the four limbs contained five digits. Known as the pentadactyl limb, this has since become modified in different ways, often with a reduction in the number of digits, but never more. Those limbs used for slow movements, as in digging, grasping or swimming, usually retain the full number, as in the mole, monkey and porpoise. In British mammals the badger makes full use of all five digits (Fig. 12). With palms and soles touching the ground as it moves, this powerful digger walks in a flat-footed fashion after the manner of humans. This is termed plantigrade. The same occurs in the otter, where the digits are joined by webs of skin to assist in swimming. In the dog, fox and cat the limbs are more extended to give better leverage and kick during running and leaping (Fig. 13). These faster-moving hunters travel on the tips of four toes, called digitigrade. The fifth toe, the equivalent of a thumb or big toe in man, is called the dew-claw, and lies clear of the ground. Claws in these mammals are used for digging (the fox) and for climbing and catching prey in the cat. A further reduction in the number of functional digits, with an even greater lengthening of the limbs, is found in the hoofed mammals or ungulates. These move on the tips of the horny coverings to the digits (unguligrade). Deer, antelopes, sheep and goats are examples of even-toed ungulates with two functional toes (cloven-hoofed). Extreme reduction of digits occurs in the one-toed mammals, such as horses and zebras which move on a single toe to each limb. Borne on such slender and elongated

FIG. 12. Badgers emerging from their sett after dark.

limbs these mammals can attain high speeds of movement.

Among the rodents which contain the most numerous of British mammals the hind feet are five-toed and the fore feet have four toes, the fifth being rudimentary. Limbs are to some extent adapted to food requirements. Climbing rodents such as squirrels and dormice use them for ascending the bushes and trees. The former make use of long toes and sharp claws for gripping rough bark, and the dormouse can hold on to branches with hand-like paws. In such situations the nuts, seeds, buds, bark and berries on which they feed are within easy reach. Other rodents such as mice, rats and voles tend to go underground by burrowing. Grass and other vegetation, fallen fruit, twigs and small animals come within their diet.

Rabbits and hares, placed in a separate Order (Lagomorpha) have elongated hind limbs for rapid flight. The former is also a tunneller.

An extreme case of adaptation occurs in bats. A skin membrane stretches over a framework consisting of greatly elongated digits of the 'hand'. Hence the name for the order of bats, the Cheiroptera (meaning 'hand-wing'). The equivalent of a thumb protrudes like a hook half-way along the wing, and is used in climbing or for attachment to a resting place.

Mammal Teeth

The above limb adaptations to different movements and speeds all help in the search for food, and in some cases the avoidance of capture. Specialisation in mammals is also found in teeth. From the earliest non-specialised teeth of the ancestral amphibians, used mainly for gripping the prey prior to swallowing it whole, those in mammals have become adapted to different uses. In gnawing mammals, or rodents, it is the front teeth, or incisors, which are in constant use. Prominent, curved and chisel

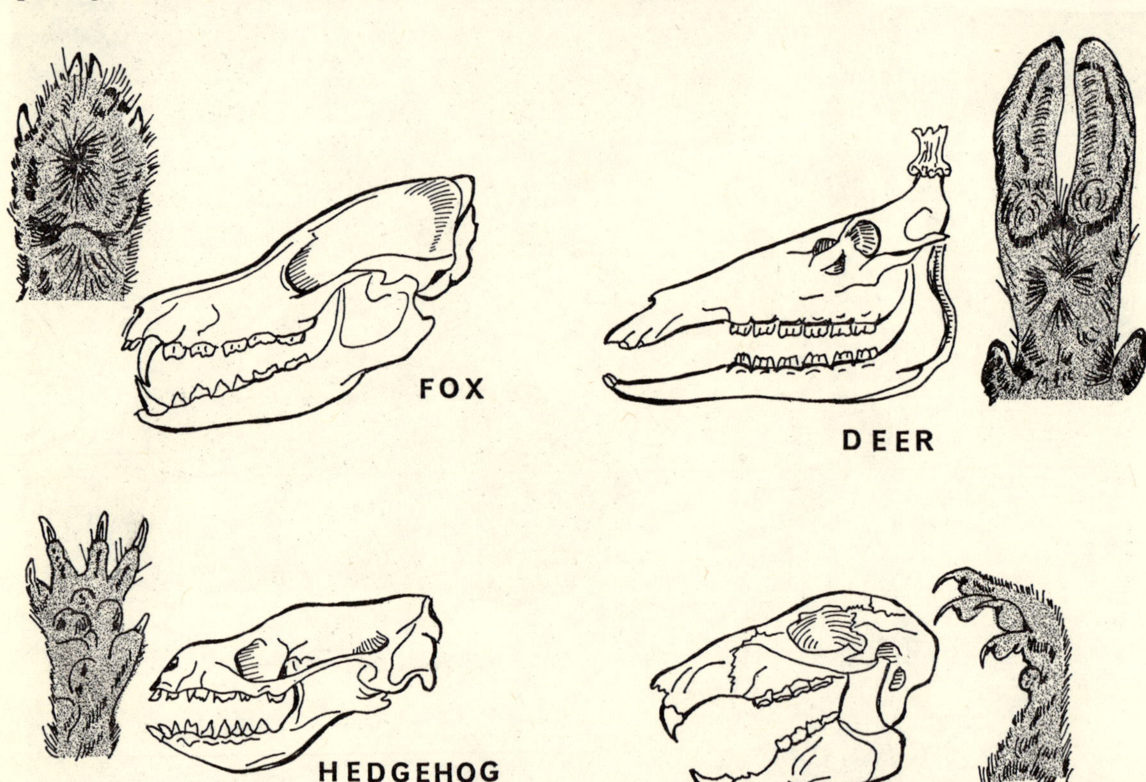

FIG. 13. Skulls, teeth and feet in some British mammals. Fox—a carnivore; Deer—a herbivore; Hedgehog—an insectivore and Squirrel—a rodent.

shaped, they make efficient tools for uncovering food which is protected by shell or bark. Canines are missing, and the gap where they should occur, called a diastema, makes a convenient outlet for inedible nut-shell and bark fragments which can be pushed outside the mouth. The cheek teeth or molars then take over to chew up the edible portions of food (Fig. 13).

Carnivores, as the name implies, are flesh-eaters whose teeth are used for dealing with animal food which must first be caught, then killed. Because of this they show a higher degree of intelligence than the usually more dull-witted plant eaters. Early training in the art of hunting which takes the form of play in the young makes the carnivores especially interesting and attractive as pets if reared from babyhood. This is particularly so with the more socially behaved mammals, such as the dog.

In the Carnivora the prominent teeth are the dog-teeth or canines, pointed and recurved for holding on to the prey and for tearing at flesh. In the more rigorous meat-eaters such as cats and dogs special premolars, called carnassials, are used to slice through meat and bone. They slide over one another in a shearing action. In this manner of feeding the food can be cut off and swallowed with a minimum of chewing.

Apart from these true carnivores, there are animal hunters from other groups which use their teeth for catching prey. Bats hunt insects in the air, hedgehog and shrew seek food in the leaf-litter and the mole searches below ground. Their teeth have pointed cusps for holding on to active prey and for biting through hard skins.

Ungulates such as deer are adapted to an almost exclusive plant diet. Usually the canines are missing, or very small. Strong and deep-rooted molars with grinding surfaces are built to withstand constant wear and tear during long spells of feeding or in chewing the cud. This interesting habit is found in other ruminants, such as cattle, and may have evolved as a means of breaking down the indigestible cellulose in plant tissues. This is carried out in places of comparative safety from predators. Feeding at dawn and dusk, in poor light, the shy deer nervously crop the grass or browse the bushes and low trees, then retire to hidden safety and rest. Meanwhile, in special stomach compartments, resident bacteria soften the food by breaking down cellulose. This can then be regurgitated at intervals as the 'cud', and thoroughly chewed up before being finally swallowed. While hiding, the deer are safe from enemies. Formerly these were the bear, wolf and lynx in Britain. Today it is largely man. A herd instinct in these animals helps to give warning as well as some protection to the individual and to the young when in danger (see Project 3, p. 157).

Predator and Prey

It will be seen that some mammals are vegetarian, but that they in turn fall victim to flesh-eating enemies. Consequently a serious matter of hide-and-seek has evolved in this life-and-death search for food. The senses of sight, hearing, touch and smell, all four in some cases, are developed to a high degree. Special adaptations to a particular food niche enables a species to live side by side with other species as members of the same community sharing the same territory. It is when two species compete for the same food that there may be an upset in the food-chain (see p. 63). As already mentioned, the plants in a given community will share out the space offered by the habitat, whereas the animals share out the food. Food is supplied in plant form to the herbivorous mammals, as animal prey to carnivores, and as both for omnivores. To this can be added the scavengers which feed on offal and left-over meals, and parasites which, in a sense, 'prey' on other animals without killing them (see p. 197). In the case of a prey-predator relationship there is a delicate balance of efficiency between the two. One is adjusted to ways of escape and the other to means of capture.

For example, a squirrel can disappear into the trees, a mouse can dart down a hole, and a deer escape in flight. How, then, do the predators manage to catch a meal? A cat will lie in ambush, then pounce on the emerging mouse, or a weasel will simply pursue it down a hole. The pine-marten, now our rarest mammal, can hunt and beat the squirrel in leaping through the trees. At one time the wolf hunted the deer by following the scent of its trail and wearing it down. A fox can adopt both cat and wolf tactics, and catch the mouse by surprise and follow the rabbit by scent. The extinction of the wolf and bear in Britain can partly explain the increase in deer, necessitating from time to time the organising of deer shoots. Possibly, too, an increase and reintroduction of pine martens might help to control the squirrel population.

Today there is a tendency to show more tolerance towards the carnivores, and to recognise their part in the food chain, helping in a balance of numbers. The fox, stoat and feral mink can do harm to farm stock, but so can the deer, rat and rabbit in the case of crops and stored grain.

Animal Behaviour

Ethology, the scientific study of animal behaviour, has attracted man's attention ever since he started hunting animals and needed to know about their ways and movements in order to survive. However, to observe and record an animal's actions may be easy to a huntsman or trained naturalist, but then to interpret and explain why an animal does just what it does is another matter. The difficulty here is that we are also animals with our own movements and behaviour patterns. Since we understand much of what we do among ourselves, it is easy to fall into what is called anthropomorphism. This happens when we project ourselves into an animal so as to humanise its actions. We see our own behaviour, or so we imagine, reflected in the way in which an animal looks or behaves, and we describe it as a sly fox, cheeky monkey, wise owl, vain peacock, greedy pig, and so on. These animals are simply behaving naturally, and it would be dishonest to attribute to them such human emotions and actions.

Animal behaviour, in simple definition, is 'the response which an animal makes to some kind of stimulus'. This may come, either from its surroundings, or from within, and it can be an automatic response or one which is controlled and directed by a brain.

Here are some examples. A *reflex action* is automatic. Some stimulus from outside, such as a smell, touch, movement or sound, is picked up by one or other sense organ, and a message or impulse transmitted through the nerves then reaches the spinal cord. From here, through further nerves, an outward impulse reaches a muscle or gland. As a result we salivate at the smell, jerk away at a touch, blink at a movement, and jump at a sound. This kind of 'closed circuit' works independently of the brain, responds immediately to an impulse, is of short duration, and works only for a certain set of muscles or organs.

Equally commonplace is the *instinctive* behaviour, also automatic, and which can be performed without any experience or training. A reflex works so long as the sense organ is being stimulated, whereas an instinct may involve a whole chain of reactions once it is set in motion. Reflexes are common to many species, but an instinct is more rigid and intraspecific. Each animal is born with the ability to perform an instinct. It is, so to speak, built into the individual and inherited. Some quite astonishing performances may take place in the life of a species, and a number have now been worked out by ethologists during the past fifty years, to explain what is going on, especially in following the complicated ritual of courtship, nesting and territorial behaviour. Two classic examples are the work of Niko Tinbergen on the stickleback, and David Lack on the robin.

In spite of the involved pattern of courtship and territorial rituals, it appears that these follow a fixed pattern in which one situation leads to the next. As each stimulus arises it sets in motion a certain drive which can then lead to the next act. In this the animal is brought into condition or put 'in the mood', by the release of chemical hormones which, in turn, are stimulated by some condition in the environment. For instance, the increased daylight hours in spring have some influence on the development of the reproductive glands in birds, probably under hormone stimuli, so that they come into what is called breeding condition. Perhaps for a similar reason it is during spring that 'a young man's fancy lightly turns to thoughts of love'.

Like us, animals can learn from experience,

or by imitation, and then perform acts which are deliberate rather than automatic, in which a brain is used. The knowledge then acquired can be modified so as to produce more efficient behaviour. For example, a dog will chase a rabbit automatically, by instinct. After some experience in this it may learn that the rabbit always bolts towards a certain hole. Now, by running straight for the hole, instead of after the rabbit, the dog gets there first and so catches the rabbit. This displays *intelligence*, a kind of behaviour which can be described as 'the ability to exploit a situation'. The degree of failure or success will depend on past experience, trial and error, and what memory is stored in the brain. This makes one wonder whether in this example of a dog's intelligence, this method of catching a rabbit might be just a lucky accident, or whether it came out of reasoning. Did the dog actually work out its plan for catching the rabbit? To answer this one would have to know the entire history of this dog, including that of its ancestors. Perhaps this hunting tactic was used in the past, and has simply been handed down as part of the hunting instinct of this dog.

It is the duty of the ethologist to sort out from his behaviour studies those actions such as reflexes and instincts which are automatic, from those which have been learned from experience and by imitation. This is not always easy, since an instinct which is at the bottom of an animal act can sometimes be modified through experience. This applies as much to animals as to humans. For instance, a loud bang makes us jump, but after repeated bangs we learn to control our reflex and cease to jump. In the classroom a child coughs repeatedly because it has a sore throat. Another child coughs deliberately in order to draw attention, and it takes an experienced teacher to spot the faker! In another way a reflex can become altered so that it responds to the wrong stimulus. In Pavlov's famous dog experiments, the animals were shown food, and salivated. At the same time a bell was rung. After some repeats the bell was rung by itself and no food produced, yet the dogs continued to salivate. This is a case of a *conditioned reflex*. Perhaps the reader can think of an example in his own experience.

With an instinct the stimulus which releases a drive and a particular act of behaviour may be blocked or withheld. It is *inhibited*. For example, when acutely embarrassed or alarmed, the 'fear' drive within us stimulates the adrenal gland which then pumps energy-giving sugar into the blood. Even the mouth may taste sweet. The impulse is then to run away or to attack the enemy. Civilised upbringing, or some code of conduct, inhibits this, but the instinct is set in motion and has to run its course. The 'bottled energy' must be released. In this case we seek relief in some harmless act, such as blowing one's nose or lighting a cigarette. These are *displacement activities* which commonly occur in us and animals, and may sometimes appear ridiculous, if not dangerous. A cornered mouse will sit up and wash its face in the very shadow of a cat. Birds often preen their feathers under similar situations of stress.

The behaviour, especially of socially behaved animals which are much akin to us, has become a fruitful study in ethology. Much of this behaviour is intended to help in binding the individual to his group or society. The whole group becomes involved, in what is termed a 'pecking order', seen in barnyard fowls. In the hierarchy of dominance and submission understood by all the individuals, there is a leader who acts as boss and holds the highest social rank. This could be a male or female individual. Next comes the second in command who has authority over all the rest, but is subordinate to the leader. In this way, right down the scale to the lowest rank, the pecking order holds good, and helps to keep discipline. In this system there are certain gestures and calls which are recognised from birth, and instantly acted upon for the common good of the community. Much of this social behaviour, and the required response, is acted upon from birth, in which an inexperienced baby can even recognise its own mother from a crowd of, say, sheep, seals, penguins or jackdaws.

Fascinating work has been carried out in the study of family and society behaviour, notably by Konrad Lorenz on his famous jackdaws and geese. There is fruitful work in this field of study for anyone who cares to keep a pet animal, in particular one which has a social background. It can often happen in such cases that the young animal, having no contact with its own kind, but at the same time being born with strong social instinct, may become attached or *imprinted* on its master or mistress. The human becomes, so to speak, a leader jackdaw, sheep, monkey, and so on. The author well remembers the raucous cries set up by a baby jackdaw abandoned by its rightful parents and reared as a pet. The garden resounded with its calls each time it lost sight of its 'mother'. At one stage during adult life it fell 'in love' with me, and since there was nothing doing turned its attentions to my wife. It was named Mary after the well known 'little lamb' in the nursery rhyme, which follows its young mistress, surely a case of imprinting which must be a common occurrence on any farm.

Camouflage, Warning Colouration and Mimicry

The value of camouflage to an animal, either as a life-saver for the hunted or as a trap for the unwary, is constantly brought to our notice during field-work. Even so, one wonders how many creatures are actually passed by and overlooked, even by a trained observer. Camouflage is the common experience of a naturalist, game-keeper, poacher and soldier, each of whom applies concealment techniques to his own use. Its importance to ourselves has now been recognised since the First World War, when the word entered our language in 1917. Today it is the hidden enemy who must be sought, rather than the foe in scarlet uniform.

Large numbers of animals can be cited which profit by camouflage in its many guises, and some understanding of the principles which are involved will help in penetrating the deception. That camouflage is indeed a form of deceit is true in that the 'hidden' animal remains in full view. It is 'seen' yet not recognised, and in this deadly game of hide-and-seek a trained observer can pick out from its surroundings the animal he is seeking, if he knows what to look for. Each aid to camouflage may be listed in this way under two groups.

Hidden or cryptic camouflage. In this deception the animal is made to blend with its surroundings so that it becomes part of the background. To make this effective certain revealing characters need to be eliminated.

a. Colour harmony. A similar body colour to that of the background will make an animal less conspicuous. It might be argued that this is unnecessary, since many animals are colour-blind (e.g. mammals). However, when reduced to tones of black and white, as in a monochrome photograph, the result can look just as effective. The colour only enhances the illusion. Colour harmony may be permanent, as in a green insect against leaves or a rabbit against the earth, seasonal (e.g. stoat to ermine) or variable (e.g. frog, chameleon, cuttlefish).

b. Disruptive colouration. An animal's outline is a means of identification. To confuse and draw the observer's eye away from this, a set of dots or stripes over the animal's body will tend to break up the outline, so that it appears to merge into the background (e.g. a tiger's stripes against tall grass, a moth against tree-bark or a trout against a pebble river-bed). The stripes or dots become part of the shadows and highlights which surround the wearer. This type of 'dazzle' camouflage is in common use by the armed forces and helps to conceal their vehicles, buildings and equipment (Fig. 14).

c. Counter shading. An animal's body is three-dimensional and has solidity, causing it to stand out from the background. This is because it is illuminated by the sky from above, but in shade below. This has the effect of giving a pale upper surface and darker belly, even in an animal of uniform colouring. If, however, the animal's colouring is graded, from dark above to pale below, i.e. in the opposite way, this has the effect of counteracting the light and shade effect from the sun. The body tends to flatten and recede into its background. By contrast, an illusion of solidity can be produced quite effectively by drawing a circle, then carefully shading over one side with a pencil or paint-brush.

d. Shadow elimination. The crouching attitude of a sleeping or alerted animal is a common stance during times of danger or when at rest. This is important in camouflage, since the lowered body tends to lessen the size of its shadow. A shadow is conspicuous, and since it cannot be camouflaged, may spoil the whole illusion and reveal the presence of an otherwise 'hidden' animal.

e. Stillness. Movement can mean a living presence, which in turn may mean food or danger. Consequently, stillness, or at most slow and cautious movement, is highly important to an animal. Hunters will frequently pause when searching for prey, and the hunted 'freeze' at any sign of danger. Warning cries from parent animals produce freezing attitudes in their young, especially those reared on the ground where most danger lurks (e.g. baby partridge, wader, leveret or deer). See the Deer Project 3.

Much useful information can be learned about camouflage, and with advantage put to good use by a field naturalist. Bearing the above principles in mind, one can be excused for turning up at a field meeting in old and faded, even mud-stained clothing and behaving in an anti-social, monosyllabic fashion. The dress and noisy small-talk expected of a lively cocktail party is hardly conducive to successful field observation, where silence and slow movement are essential to success.

Make-belief camouflage. In this form of deception it is the animal's appearance, and sometimes behaviour rather than its cryptic pattern which causes it to be overlooked or ignored. As with Cryptic Camouflage, it works both ways—for

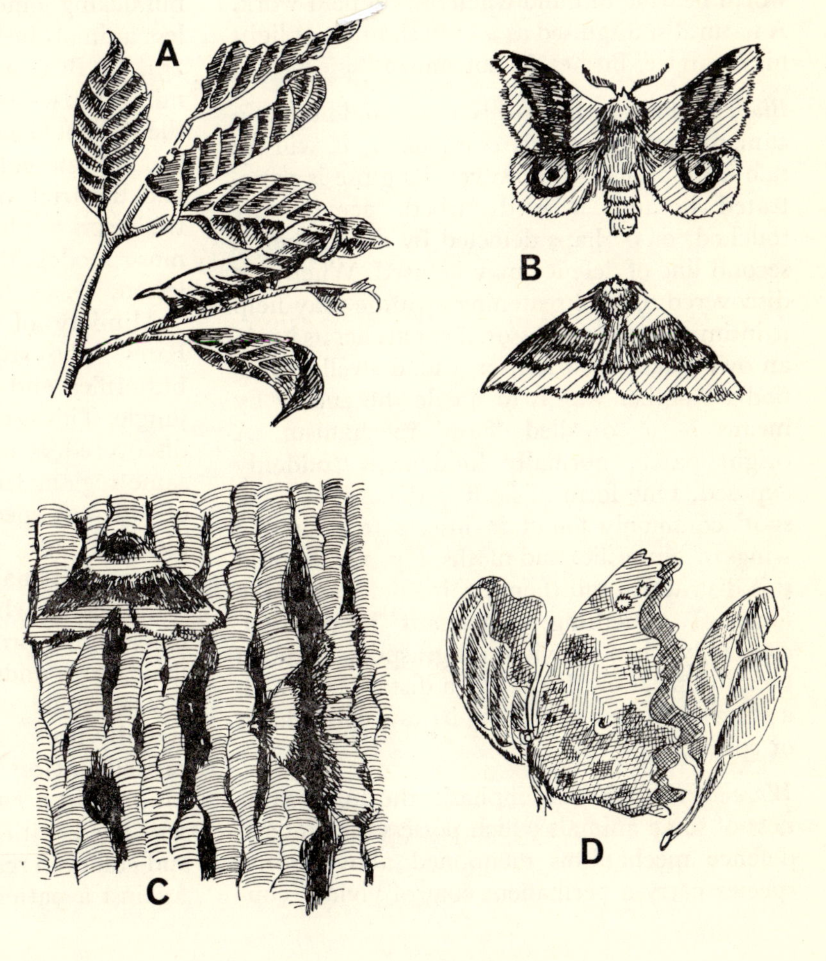

FIG. 14. Camouflage. A, a hawk moth larva in its normal upside-down resting position is well countershaded, but stands out when in the unnatural position (below); B, a moth in normal resting position displays a disruptive pattern of dark stripes. A flash mechanism comes into play when it is disturbed and raises its fore-wings; C, the same moth (right) will position itself on the bark in order to match up with the dark shadows between cracks; D. the comma butterfly resembles a dead leaf in shape and colour when resting on the ground.

the hunter as well as hunted. In the former case this is called Aggressive Camouflage and in the latter Protective Camouflage, and the object which is resembled is either harmless or inedible. Thus, a harmless animal might resemble a twig (e.g. stick insect), a leaf (e.g. butterfly) and so is overlooked (Fig. 14). However, a harmless animal might easily come within range of its unnoticed enemy (e.g. a toad resembling a clod of earth, a praying mantis a leaf, and an angler fish a piece of rock). Notice in the case of the mantis that it has two advantages, being aggressively camouflaged against smaller animals, and protectively camouflaged against larger ones.

A whole range of objects are exploited in this way, such as a leaf, twig, stone, animal dropping, piece of sea-weed, and so on. However, like good actors these animal masqueraders must act the part as well as look it, a point worth bearing in mind when out on field-work. A naturalist disguised as a bush should not light his cigarette. Bushes do not smoke.

Bluff. It follows from this that, so long as a camouflaged animal is overlooked, it will be safe from attack. If, however, its guise is penetrated because it is disturbed, accidentally touched, or perhaps detected by scent, then a second line of defence may be used. When it is discovered some threatening attitude may help to intimidate the aggressor. A cat arches its back, an owl makes itself tall, or a toad swells up its body. Another way is to startle the enemy by means of a so-called 'flash' mechanism. A bright patch, normally hidden, is suddenly exposed. One form of flash is the bright 'eye-spot' commonly found in insects (e.g. on the wings of butterflies and moths, Fig. 14). Should this distraction fail then a third defence in the form of something unpleasant is released, especially when the victim is grasped. This can be an unpleasant discharge, a distasteful smell, a sting or bite, or something irritating like hairs or spines.

Warning colouration. To emphasise the unpleasantness of some animals which possess the kinds of defence mechanisms mentioned above, some species carry a permanent coat of vividly contrasting colours in both sexes. This has nothing to do with recognition in the social or mating sense, but serves as a warning to any aggressor 'not to touch'. Like vivid poster colours used in advertisements, some mixture of black, white, yellow and red is adopted. Each inexperienced aggressor probably makes a first mistake, and sharply learns the lesson from the sting, objectionable smell or taste which it then experiences. Controlled experiments on domestic chicks have shown that two or three mistakes in selecting the wrong meal are enough to associate this with the warning colours. A wasp (sting), cinnabar moth (taste), badger (bite), skunk (odour) and salamander (poisonous skin) are examples of warning colouration.

Mimicry. There are also other examples of brightly-coloured animals which, if caught, are quite edible, but are usually left alone. By mimicking some unpleasant species, such harmless animals become immune from attack (Fig. 15). In Britain there are harmless flies which mimic the wasps or bees, also two moths such as the Hornet Clearwing and the Bee Hawk Moth. Taking into consideration the law of averages, and the trial and error method by which this trick works, it follows that there must always be more models than mimics to drive home the lesson.

Mimicry of this kind was discovered by Bates, who studied the brilliantly coloured butterflies and their models in the Brazilian jungle. This is called Batesian mimicry. Müller discovered a more complex situation in the same region. In Müllerian mimicry a number of different species of butterflies, all with unpleasant taste, use a similar kind of warning pattern. An analogy to this in business practice, is the way in which different firms manufacturing a similar product, will amalgamate and sell their wares under a common slogan.

The value of camouflage in survival has actually been seen to work in a brilliant study carried out on a British insect, the Peppered Moth (*Biston betularia*). This common moth is associated with the wooded areas of Britain, and comes to rest by day on the trunks of trees. Against a pattern of grey bark and patches of lichen, the pale, spotted wings of this moth are a perfect match. It is hidden from birds (Fig. 16). In 1848 a black specimen of this moth was discovered near Manchester, possibly the result of a mutation. By the turn of the century this dark form was still a rarity, about one melanic or *carbonaria* to every ninety-nine normal or *typica*. Since then the Industrial Revolution has polluted much of the countryside, and many trees in industrial areas have lost their lichens and turned black with sooty deposits. The result has been a large increase in dark moths.

Dr Kettlewell, the entomologist, performed a field experiment in which he set free large numbers of marked moths, both *carbonaria* and *typica*, on the trees of a clean wood in Dorset, then observed the work of the birds in search of food. On recovery with light traps it was found that a ratio of some ninety-nine white to only one black had survived. Obviously the *typica* forms were too well camouflaged among the lichens. A similar experiment was carried out in a polluted area near Birmingham. In this case most of the recoveries were the *carbonaria* form which were concealed against the sooty bark.

Most dark moths today are found in the Midlands and eastern England, whereas the cleaner west and north are largely unaffected. The spread of pollution is due to the prevailing winds which carry the dirt in an easterly direction, apart from a concentration in manufacturing areas. One wonders whether, eventually, all the dark moths will disappear, if the authorities are able to establish a supply of clean fuel in the future, so that pollution becomes a thing of the past. It is estimated that some 4 million tons of solid matter and 1 million tons of smoke particles are deposited each year on the surface of Britain at the present time.

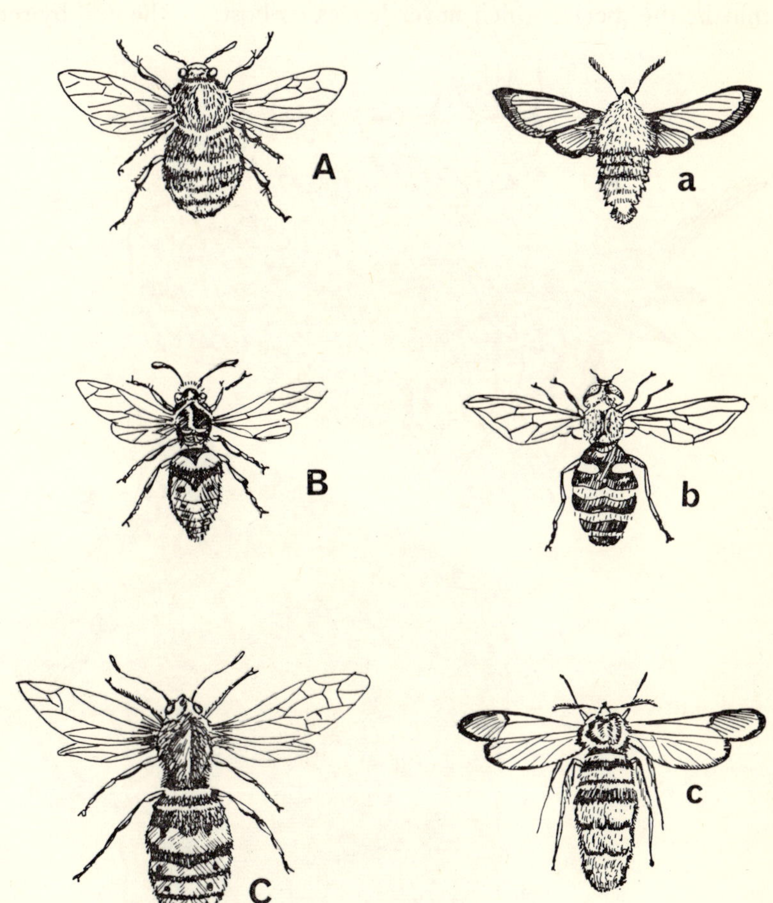

FIG. 15. Examples of Warning Colouration and Mimicry. Models are on the left and mimics on the right. A, a bumble bee and a, the bee hawk moth; B, a wasp and b, a hover fly; C, a hornet and c, the hornet clearwing moth.

Industrial melanism, as it is called, has occurred in a large number of macrolepidoptera, and not infrequently turns up in different animal groups. Melanism, in which there is a concentration of pigment in the skin and its products, is said to bestow greater vigour on the individual. Possibly this may have helped the spread of the *carbonaria* Peppered Moth, since the larvae are said to thrive more readily on polluted leaves than the *typica* individuals.

Parasitism

Within both plant and animal kingdoms there exist a number of species which live a highly specialised and, to some extent, dangerous kind of life. These are called parasites, and can be described as 'organisms which live and feed in or on other organisms'. The victim of the parasite, called the host, becomes its environment. In the case of the obligative parasite, that is, the species which never leaves its host, the latter is also a permanent feeding-ground. This is the case with many of those minute parasites, often referred to as microbes or germs belonging to the bacteria and one-celled protozoa, also the worm-like groups such as the flatworms (Platyhelminthes) and many threadworms (Nematoda). Without the security of a host many such parasites could not survive. In other cases there is a looser contact between parasite and host, in which the latter is only victimised at intervals. A flea or mite, for instance, becomes gorged on the blood of its host, then falls off and may remain free-living for long intervals before rejoining another host.

In some cases both individuals are free-living. One takes advantage of the other for the rearing of its young. The cuckoo mother parasitises another species which then must act as foster parent to the ravenous child which takes over the nest by removing all the rightful occupants.

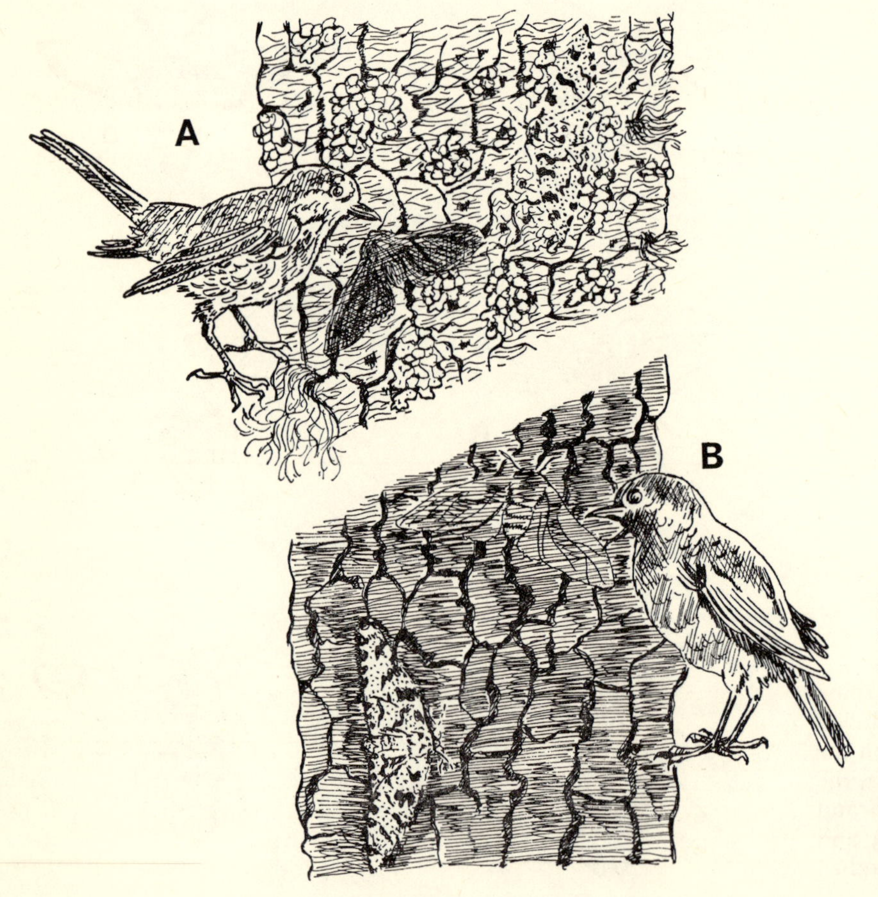

FIG. 16. Industrial Melanism in the Peppered Moth. A, a robin spots a black (*carbonaria*) on a clean tree trunk but overlooks the normal (*typica*) individual which is well camouflaged; B, a redstart eyes the *typica* form on the dirty tree trunk, but misses the *carbonaria* one.

The ichneumon uses some unfortunate caterpillar in which to lay an egg, so as to ensure an ample food supply for the growing larva.

Such loose intimacy has produced little change in cuckoo and ichneumon, and the one can be recognised as a bird, the other as a wasp. On the other hand the obligative parasite, permanently attached to its host, often in total darkness, and hanging on for dear life, may take on some bizarre shapes and abnormal functions. Pigment is lacking, and hooks or suckers take the place of limbs. Movement is minimal, and the body reduced to the bare essentials for feeding and reproducing. Little else is required of it.

Parasitism may provide a secure and comfortable existence once it is attained, but there are enormous obstacles to overcome in order to reach this security. Eggs of the nematode worm which leave the body of a pig or human must somehow be transferred to the mouth of another pig or human, and swallowed so that another generation may hatch out. A leech in the swamp, a tick larva on some meadow grass, or a lamprey in the river, must all wait until a suitable host brushes past or swims by, so as to become attached once more to a mobile feeding-ground. The right host may never turn up.

It is for this reason that the human species is the only primate to own a flea. Monkeys and apes which roam about in groups would give fleas little chance of returning, once they leave the body of their host. Humans tend to occupy territories and more or less permanent living-quarters, thereby giving the human flea, bed bug and louse an opportunity for re-attachment. Even so there is a break in the cycle during which the host is abandoned for a while, and the parasite must face a hostile outer world before re-entry into the host environment. This is the case, for example, of the egg and larval stage of many mites who may never find a home.

This hazard has been overcome by many parasites which adopt another host as a temporary home for the larval stage. The interesting, but not so surprising, thing is that this so-called intermediate host is the normal prey of the main host. Thus, the tape worm of the fox has a larval stage in the rabbit. Man's tapeworm has a larva in the pig, ox and freshwater fish, according to the species. In other words, advantage is taken of an existing food-chain so that the parasite can be passed on from one generation to the next. Notice how the larva of the tapeworm settles down in the muscle tissue of the intermediate host. It can then be devoured by the main host (man or fox). The adult parasite then lives in the intestine of man so that eggs may readily be passed out with the excreta.

It is obvious from the above example how precautionary measures may be taken against man's tapeworm, from the knowledge gained about its life history. Proper sanitation will prevent eggs from passing to pig or ox, and regular meat inspection will weed out contaminated beef or pork, so that larval tapeworms cannot enter man.

For economic and medical reasons considerable study has been made of those parasites which can cause harm, even death, to man and his domesticated animals and plants. Disease which at one time was attributed to the evil presence or bad air (e.g. malaria) is now known to be the result of some parasite, once invisible because of its minute size, but now seen through the eye of the microscope. It is now even possible to uncover the smallest of living things, the viruses, with the aid of the electron microscope.

This emphasis placed on hygiene and preventive medicine where parasitic diseases are involved may give us a false impression of the true role of a parasite. Like any normal species it is adapted to an environment, albeit a peculiar one, and is subjected to the natural laws even more rigidly than its free-living counterparts. For survival there has to be a delicate balance between parasite and host. While the former can enjoy security, protection and regular meals, it must not impair its host for fear of killing the goose. A dead host is of no use to a parasite, for this would only seal its own fate. This presupposes that there is something unnatural at work, and that the injurious parasite may be attached to the wrong host, and consequently is not in tune with it. This could be the situation in the case of the human species, a restless and wandering animal which comes into contact with a wide range of other animals, including parasites. For instance, when the white man settled in Africa both he and his cattle became the target of the tse-tse fly, an insect vector which transmits the trypanosome parasite, the cause of sleeping sickness in man and nagana in cattle. Yet the same protozoan parasites are found in a number of native game which can tolerate it. Game is part of the African scene, whereas the white man is a comparative newcomer. In time, maybe, his body may adjust itself to the parasite so that there is harmony between the two. As it is, the human body is perhaps the most abused and neglected of living machines. In the face of all its ills and ailments the miracle of human living would seem to be in staying alive, not just being alive.

By far the majority of parasites pass unnoticed, even though some may have to be tolerated to extremes by the host. Thousands of nematodes may crowd the interior of an apparently healthy animal. The author remembers helping to count over 2,000 nematodes from the intestine of a python which actually died from a blood disease. The study of parasites apart from its value to medicine and human welfare, is of interest to the ecologist where it may upset a food chain, or diminish populations. The widespread decrease in the rabbit population due to myxomatosis, and its consequent effect on the predators such as fox and buzzard, also the vegetation, is a case which cannot be ignored. Another parasite, the foot-and-mouth virus which raged in England during 1967-8, resulted in tragic losses of cattle, but ironically may have benefited parts of the countryside and wildlife. The standstill order restricted movement, and such sporting activities as hunting and fishing.

These are two outstanding cases of parasitic upset in Britain, but normally during field-work one's attention is not drawn towards parasites unless this kind of thing happens. Only if some deliberate search is made does a parasite receive attention, especially if it lives inside its host. The author can recall two instances in which parasites were accidentally found, and helped to solve a problem. These are discussed in Project 19, page 197.

When parasites are included in a piece of field-study or recording it should be borne in mind that, since the parasite is smaller than its host, the pyramid of numbers in a food chain is reversed. Parasites in turn may have even smaller and more numerous parasites to bear. The top of the pyramid consists of large numbers of small animals, in contrast to the single apex species, such as the sparrow hawk mentioned on page 62.

Part 1: Section C

The evolution theory

On July 1st, in the year 1858, a scientific paper was presented to the Linnean Society in London, under the title 'On the tendency of species to form varieties; and on the perpetuation of varieties and species by natural means of selection'. This was published in the year's *Proceedings* under the joint authorship of the two great nineteenth-century naturalists, Charles Darwin and Alfred Russel Wallace. It was the birth of the Evolution theory. In the following year Darwin elaborated his ideas in his famous work—*The Origin of Species*.

Until his voyage as a young man on the survey vessel, H.M.S. *Beagle*, in 1831, Darwin had no cause to disagree with the current belief of the day that species are permanently fixed and never change. It was during his studies on the Galapagos Islands that he began to doubt the idea of immutability. The birds which he observed on these islands, in particular his famous finches, differ slightly from one island to the next. They differ even more markedly from those on the Cape Verde Islands, where there are also slight variations. In spite of this the two island groups are physically and climatically similar, one in the Pacific and the other in the Atlantic. Also, those birds on the Galapagos resemble those on the mainland of South America, whereas the Cape Verde birds are more like those in Africa.

During his stay in South America Darwin saw the giant fossil remains of animals found on the Pampas which resemble the present-day armadillos in their peculiar armour covering. These and other discoveries stimulated him in finding an explanation. The obvious answer would seem to be that the finches of the Galapagos and those of South America have a common ancestor, as do those on the Cape Verde Islands and Africa. The same idea would apply to the fossil and modern armadillos. If this were so, then fossil discoveries would make a valuable contribution to this idea of change. Here one might find the missing links in the chain of evolution. As we know today the fossil record is a valuable form of evidence to support this idea. In some cases, as with ammonites and horses, two beautiful series of fossils clearly demonstrate this process of change.

On his return from his travels Darwin recorded his thoughts in a Sketch in 1842, which he then enlarged as an Essay in 1844, both of which were unpublished. He argued that if species always remained the same there was no answer to his problem. But, if species did change in time, then they could well have descended from a common ancestor.

Darwin then set to work to assemble all the evidence in support of change, and to find an explanation of how it works. He noticed the variations which occur within a species, and how these can be put to good use in the selective breeding of domestic plants and animals derived from wild ancestors, such as the dahlia and rockdove. He called this Artificial Selection, the work of man. But how did this operate without man's interference, and was there a form of Natural Selection? He gave the answer in his famous theory, a natural but automatic process governed by nature's laws.

Meanwhile, in Malaya, Alfred Russel Wallace hit on the same idea, and wrote, 'It is evidently possible that two or three distinct species may have had a common antitype, and that each of these may again have become the antitypes from which other closely allied species were created'. The two men finally pooled their ideas in 1858, and today their theory is accepted as fact, and that life is in a continual process of change.

Since such changes take place slowly, and cannot be observed in a man's lifetime, much of the evidence is circumstantial. First, there is the fossil record. Secondly, there is the significance of the Classification system, the *Systema Naturae* of Linnaeus, in which plants and animals are divided into groups according to their structure and blood relationship, and these into smaller groups, right down to the individual species (see p. 56). This is a natural classification which reflects the progress of Evolution. Thirdly, the close resemblance in the embryonic stage of a fish, reptile and mammal, to take one example, i.e. vertebrates, suggests a common form of ancestor. Fourthly, the presence of vestiges, those apparently useless parts of the body, can be explained on the grounds that at one time they had a function in the ancestor, which has since been lost (e.g. the traces of a pelvic girdle in snakes and whales, and the appendix and ear muscles in man). Then there is the discovery of those valuable 'missing links' such as the *Archaeopteryx* and *Proconsul*, the first a transition stage between reptiles and birds, and the other between ape-like primates and human primates. The evidence of biochemistry and physiology, in such things as blood-grouping and a weakness to certain diseases, also suggests a common link. In comparative anatomy the many shapes and forms in an organ common to a group, such as the flower and the mammal's limb, can be traced back to a common ancestral plan. At one time some confusion arose in anatomy, because of certain resemblances which are only superficial. A bat's wing and a whale's flipper may look quite different, but they are truly homologous and derived from the same plan, the so-called pentadactyl limb common to mammals (see p. 37). On the other hand a bird's wing and a butterfly's wing are on quite different plans and from separate origin, the one from vertebrate stock and the other from invertebrate. They are analogous.

Finally, geographical distribution. The range of present-day species indicates how they have spread from common ancestors, as in the case of Darwin's finches. All this evidence shows that the members of a particular branch of evolution share similar characters in their anatomy, chemistry, even in behaviour, because they came from a common ancestor.

Darwin's *Origin* puts a convincing case for Evolution. By reasoned and logical argument he enumerates the steps by which this theory can work. It is based on natural laws which can be listed in the following way:

1. *There is more life being produced than can possibly survive.* On the assumption that all offspring and subsequent generations live and breed into old age, then, starting with two original parents, an enormous progeny would result among the descendants. In the case of mice it is calculated that some 200,000 individuals could result from successive matings over a period of 2 years. This same astonishing multiplication could result in plants if, for example, every acorn from a single oak were to germinate and grow into a mature tree. Darwin chose the elephant. He calculated that if all the progeny from a single pair survived to breed at the rate of six elephants in the course of one century, then in 750 years there would be 19 million elephants alive. Wallace

chose a pair of birds. If these and their successive offspring produced four young four times a year, then in 15 years the numbers would reach ten million. This has not happened, and does not happen.

2. *A species remains fairly constant in number, and there must be a high rate of mortality.* On occasions what is called a population explosion takes place, and there is a geometrical increase in numbers of individuals, as is happening among humans today. With animals and plants this does not last, and some factor such as disease or sterility, or storage of food and living space, brings it to a halt.

3. *There is a constant struggle for existence.* Living things, as opposed to the non-living, have a direction and purpose which is aimed at self-preservation. Life competes for what it requires in order to keep alive.

4. *There is a survival of the fittest.* Because there is a limit in the supply of life's necessities, those species which are best fitted or adapted to use them will tend to survive and breed. The word fittest was not intended by Darwin to imply just a state of good health and well-being, as we use it in the popular sense. The 'fitness' is like a key which is made to fit into a certain lock. An earthworm is as much fitted to its surroundings, as a lion is to its habitat. Darwin chose the woodpecker as an example of this adaptation, a bird fitted for life in the trees (Fig. 17).

5. *Life can vary.* Again, by comparison, a damaged key or tampered lock will no longer operate. A new lock will only work if another key is made to fit. If the environment and conditions to which a species is adapted should happen to change, then it must re-adapt to the new way of life in order to remain fitted. This is precisely what is going on all the time. Climates alter, land is replaced by sea and vice-versa, food supply changes and life changes with it. This is shown from the studies of the earth's crust (geology) and the fossils, which they contain (palaeontology). In this 'story in the rocks' there are many examples of failures to adapt, and whole groups of animals and

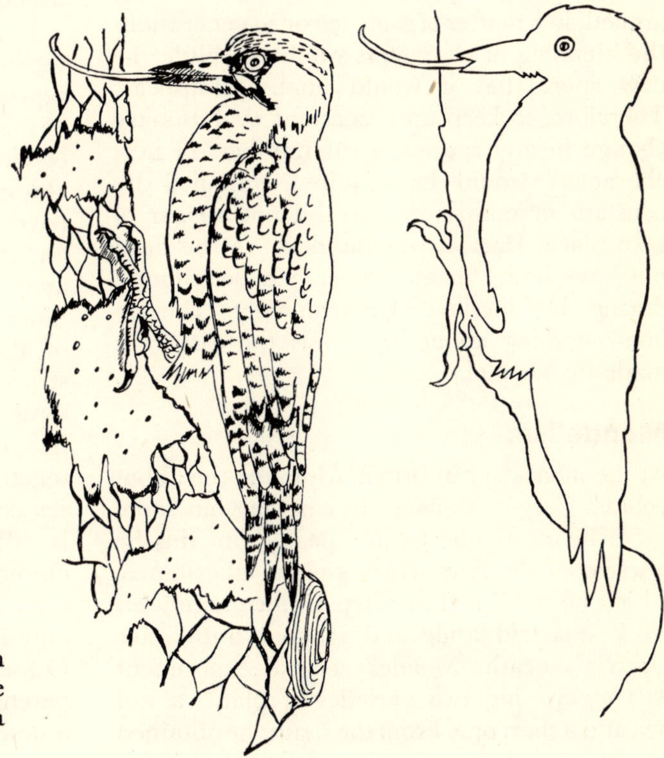

FIG. 17. The Green Woodpecker feeding on a tree trunk. The silhouette shows the specialised beak, tongue, feet and tail which fit this bird into a woodland habitat.

plants have become extinct, including some of the strongest and largest—the dinosaurs.

Since species show variation, however, those characters which might be of benefit would be passed on by those individuals which possessed them, giving a better chance of survival. Those which lacked them would be at a disadvantage, being less fitted, and would tend to die out.

6. *Variations can be inherited.* This is so commonplace that its significance may be overlooked. Although convinced that his theory worked, Darwin could not clearly explain to his satisfaction its actual mechanism. How did variations originate, and how were these passed on? He assumed that particles carried in the blood from the animals' own body entered the reproductive organs so as to alter the hereditary make-up, with the result that new characters arose in the offspring. He called these changes sports. Today we call them mutations. A mutation would mix with the other characters, so that the offspring showed some kind of balance between the characters contributed by the two parents. If this were so, then, Darwin argued, in a matter of some ten or so generations the blending of characters would so dilute the new sport that it would finally disappear. Therefore, to keep up a constant evolutionary change in any species, great numbers of new characters would be required, supplied by constant mutations. As this did not appear to take place, Darwin was puzzled. Yet he need not have been, for an answer had been found during his lifetime. Unfortunately, Darwin knew nothing about it. It was the discovery made by Mendel.

Mendelism

At the monastery in Brünn, Moravia, the abbot Johann Gregor Mendel was carrying out some experiments on the garden pea. From this he discovered the laws which govern inheritance, which he published in 1865. Unfortunately his work was laid aside and ignored until after Darwin's death. Mendel's classic experiment was in crossing two varieties of plant, a tall pea and a short one. From the results he obtained he deduced that, in the transmission of the characters from each parent to the offspring there is first a segregation of the factors (i.e. genes) which control these characters, followed by a random recombination but in a definite numerical ratio. Segregation occurs in the formation of the reproductive cells, called the gametes, each containing only half the number of chromosomes, and recombination on the fusion of gametes to form a zygote with full number of chromosomes.

Putting it in this language would have been puzzling to Darwin and Mendel, since little was known in their time about the mechanism of inheritance. Great strides have since been made, both by geneticists in their breeding experiments, and by cytologists who study the structure and working of the living cell. It is now clear that the nucleus of a cell takes part in cell division as an orderly process, in which the nuclear material assembles into structures called chromosomes. These occur in double sets in the normal cell. The chromosome number remains constant for any given species. In man it is 48. In the famous fruitfly, *Drosophila*, much used in Mendelian work, it is 8.

In normal cell division, as part of growth and repair called *mitosis*, the chromosome pairs reproduce themselves equally, so that the nuclei of the two daughter cells retain the full number of chromosome pairs, or N. On each of the chromosome pairs are situated the molecular particles which represent the so-called genes, and it is these which represent the characters of an individual, such as its colour, shape, size, etc. Since at mitosis each daughter cell is a copy of the parent cell and has the same gene structure, there is no change and the chromosome number remains the same. What then happens at fertilisation when two cells join together? In this instance, at some stage, special sex cells are formed, and this is done by what is called a reduction division, or *meiosis*. The chromosome pairs actually separate, so that each daughter cell, in this case a gamete, has only half the number of chromosomes, or $\frac{1}{2}$N. Consequently, when gametes from two separate parents, that is male and female, unite to start a new life, it is a matter of chance as to which of the chromosomes come together as pairs once more (Fig. 18).

The two complete sets of chromosomes thus obtained from its parent by a plant or animal may be compared to two sets of playing-cards distinguished by different coloured backs. In meiosis and gamete formation the double pack is separated into single packs but irrespective of their colours. Each separated pack will contain a complete set of cards, but they could be of either colour. The possible combinations of the cards when the packs rejoin, that is, when chromosomes unite into pairs once more, is enormous, and this is what gives rise to variation in the successive generations, on which evolution thrives.

Actually cell division and chromosome separation has been seen in operation under the microscope. It vindicates the Mendelian laws which at one time were rejected as inoperable. This was because any changes which we call mutations, and Darwin called sports, were found to be harmful in most cases, and also seemed to occur at such long intervals, in discontinuous steps. Mutations did not appear to fall in with Darwin's conception of a gradual evolutionary change for the better. Later work by geneticists, however, can now

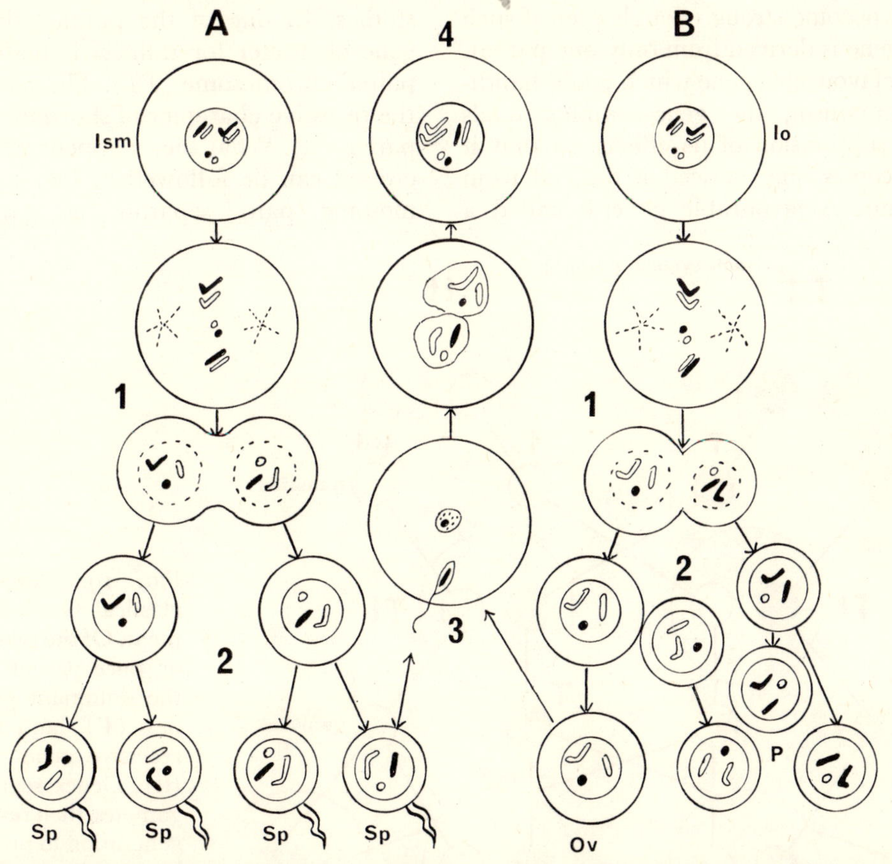

Fig. 18. Meiosis. The formation of male gametes (A) and female gametes (B) accompanied by the reduction division (meiosis); 1, reduction division of the chromosomes at the first maturation stage which is followed by 2—the second division. Note the four male gametes (SP) but only one functional ovum (OV). Three of the female gametes called polar bodies (P) remain unfunctional. 3, sperm and ovum unite in fertilisation with a return to the full number of chromosomes. 4, the zygote, commencement of a new individual. ISM, Immature sperm-mother cell and IO, Immature ovum.

demonstrate that these widely spaced mutations are really the extremes of a whole series of changes, minute and almost unnoticed as they may be, which result from the combined interaction of the other genes on the chromosomes. In other words, there is a gene complex which sorts itself out by the usual segregation and recombination of the chromosome halves at fertilisation.

The result is a slow and gradual change in character of the species, just what Darwin hoped for. A mutation which may benefit the organism will gradually assume this condition by the selection process of the whole gene-complex which will favour the mutant. Its effects will become strong enough even if such a mutant gene is derived from only one parent. With an unfavourable gene which could handicap the organism, the gene-complex tends towards a suppression of its effects, so that it usually becomes only noticed if derived from both parents. A favourable gene is called a dominant (e.g. Mendel's tallness in his peas), and an unfavourable one a recessive (the short peas).

Sometimes the dominant and recessive genes react to produce a mixture in the offspring, especially noticeable with colour characters, and what Darwin called blended inheritance. This, however, is only superficial, for the genes themselves are not altered, and will retain their identity until the next mutation occurs. It is merely the way in which the genes combine and react on one another, that either produces or represses the characters which they represent.

A case of dominance and recessiveness may be taken from Mendel's own garden pea studies. In one of the parents the dominant gene character for tallness is matched on the paired chromosome (TT). The same occurs for the recessive character of shortness in the other parent (tt). What then happens when these are crossed can be followed in Fig. 19. The chromosome pairs separate as gametes, then

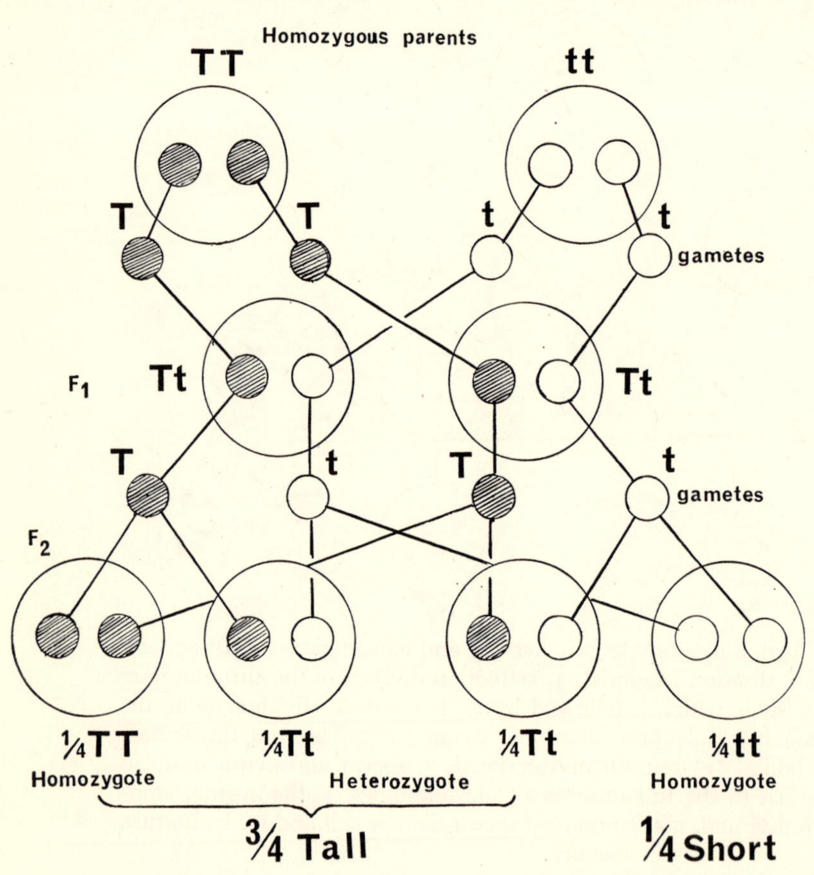

Fig. 19. Gregor Mendel's classical Garden Pea experiment. Of the two homozygous or pure parents, one carries the dominant genes for tallness (TT) and the other the recessive genes for shortness (tt). Genes segregate in the gametes, then reunite in the F_1 generation to produce tall but heterozygous offspring (Tt). Further segregation and recombination produces, in the F_2 generation, $\frac{1}{4}$ homozygous tall offspring, $\frac{1}{2}$ heterozygous tall and $\frac{1}{4}$ homozygous short. This gives a 3:1 tall/short ratio.

recombine in F_1 where T and t come together. Since T is dominant all these plants will be tall. In F_2 there is further recombination, in which a ratio of every three tall plants appears with one short plant. However, of all these plants only one-quarter has both of the tall genes (i.e. TT). The same applies to another quarter of short plants (i.e. tt). Both these will breed true tall, or true short, plants. They are called homozygous. The remaining half (i.e. Tt), however, are untrue, or heterozygous. They may look tall, but like their F_1 parents they can produce both tall and short offspring.

Lack of knowledge of this mechanism led earlier breeders and geneticists to reject the mutation theory. Character differences, we now know, are not sudden as a result of the abrupt appearance of a single mutation, but have been built up slowly from past selection of the gene-complex. It is estimated from a number of test cases, including those on man, that a given gene will mutate in about one in a half million individuals. This is sufficient to provide the gene-complex with enough material on which to work a gradual change. This is the essence of natural selection which governs evolution.

Acquired Characters

At one stage, before Darwin's idea of inheritance by means of an internal change was voiced (i.e. by a sport or mutation), an alternative theory was offered, in which evolutionary changes are brought about by the effects of the environment. Examples were given, such as the pigmentation of the skin when exposed to sunshine, the thickening of an animal's coat in a cold climate, and the increase in muscle development with exercise. By contrast a skin would pale in darkness, the coat would thin out in warm surroundings, and muscles would soften with lack of exercise. Ultimately, these might die out through lack of use. This became known as the theory of Acquired Characters, or Lamarckism, after the great French naturalist, Jean Baptiste Lamarck, who inspired it. The continued use and misuse of a character under the influence of the environment would cause it to improve with each generation, or die out. Since such characters are not passed on in this way, they do not affect the mechanism of evolution. A child, for example, does not inherit strong muscles and a powerful body from his father simply because the father during his lifetime built a strong physique during his work as a blacksmith. The son could just as well have been born to grow up a weakling, unless he followed the same trade. On the other hand, if the boy inherited a 'strong physique' gene from his parent, then, regardless of his trade, he would grow into a powerful man. The gene which determines this quality would be his birthright. Superficially, these two young men might look similar, but the first son would have built up his strong muscles through work, and the other boy would have been born with them.

Isolation

Once a change of character is under way within a species, then a new variety, and possibly later a new species, is in the making. As a species spreads out, the genes may become segregated so that they can no longer recombine. Part of the species population may become separated from the rest by some form of isolation. A breeder will prevent a recombination of characters by careful selection and separation of the stock. He breeds in certain qualities and breeds out others. In nature the barrier is often due to a spread of the population, in which groups may become split off by some barrier such as a mountain, the sea, climate, and so on. The original gene complex has been broken up, and new complexes are formed by further mutations. It has been estimated that it takes about a million years for a new species to evolve. In the interval many varieties may occur, and there are many cases where the present-day distribution of a species has been mapped out to show a whole series of variations over its range (called a cline).

In summing up, it can be stated that the evolution of plants and animals to produce new species is brought about by the occurrence of variation due to mutation of the Mendelian genes. This is acted upon by natural selection, and stabilised by isolation, so that the differences between population prevents interbreeding. They become new species.

Classification—the *Systema Naturae*

Some understanding of the way in which animals and plants are scientifically named and grouped is necessary in order to see where each species fits on to the 'tree of life'. This is the work of the taxonomist. The famous *Systema Naturae* devised by the Swedish botanist, Carl Linné (more usually known as Linnaeus), was the first serious attempt to classify and relate animals and plants on a comparative basis of their structure. With the flowering plants Linnaeus used a sexual system, and grouped them according to the numbers of stamens and pistils. Plants with one stamen were called Monandria, with two Diandria, with three Triandria, and so on. The basic unit in classification is called the Species. This may be described broadly as 'a group of animals or plants which resemble one another and can interbreed'. In most cases where two different species happen to cross, the offspring are infertile. There are exceptions to this, since a species is only an arbitrary unit made for convenience. Also, since life is in constant change, one species may eventually given rise to two new ones. This is made clear in Darwin's great work *The Origin of Species* published in 1859. In Linnaeus' time there was a more rigid attitude towards life's origin, in which each species remained immutable from the time it was created. In spite of these two opposing views on the formation of a species, it is remarkable that the *Systema* fits in so well with modern classification.

The science of classification, called Taxonomy, uses the species as a basis in the same way as Linnaeus did. It is called the binomial system, where two names are used. The first is written with a capital initial letter, and is the name of the genus to which the species belongs. The second name, called the descriptive or trivial epithet, written with a small letter, makes up the full specific name. The two names are always written in italics or underlined. A species name is then followed by the name of the author, i.e. the person who described it, followed by the date.

For example, Linnaeus named and described the common frog in his *Systema* as *Rana temporaria*. This should therefore be written as *Rana temporaria* Linnaeus, 1758 (i.e. the frog of temperate countries). Sometimes an author makes a mistake in his classification, and this has to be corrected. Linnaeus placed the common toad in the same genus as the frog, and called it *Rana bufo*. It was later renamed *Bufo bufo* (actually the toad belongs to quite a different family). In such cases the original author's name stands, but should then be put in brackets —*Bufo bufo* (Linnaeus) 1758.

Evolution is a slow and gradual process, so that a species within our lifetime remains fairly constant. However, localised geographical and other differences of shape, size, colour, and so on, can sometimes be noticed in certain areas of the species range. This is where evolution is at work slowly separating off a species into new ones as it adapts itself to different or changing environments. Where such sub-groups appear to be fairly constant the species may be split into sub-species or races. These are given a third name. The common frog, for instance, is properly called *Rana temporaria temporaria* in Britain and most of Europe. Another sub-species is known to occur in Spain. It has poorly webbed feet and is called *Rana temporaria parvipalmata*. Some species have been divided into many sub-species for different parts of the range, especially where they become separated by river, mountain, or sea barriers. Island races are a common occurrence, as happens in Britain.

The original specimen from which a new species is described is called the *type specimen*. If only one exists it is the *holotype*, if more, these are *syntypes*. These specimens, which become valuable additions to any museum or private collection, can be referred to when confirming identification of any further specimens.

A named species is placed with any other closely related species into a genus, genera into families, these into orders, then into classes, and these make up the major groups called phyla (animals) and divisions (plants). In some cases it may become necessary to make sub-genera, sub-orders, etc., where the groups are large or unwieldy. The larger groupings form a natural classification, since their origin and distinction is clearly shown from fossils. With the smaller groups such as species and genera, this is not always possible to determine, and is a more artificial way of classification; it can lead to much disagreement among systematists. Some appear to have conservative views over subdividing a species, and others will make this a rule. In scientific jargon this is known as 'lumping' or 'splitting'.

Since taxonomy is international an International Commission of Zoological Nomenclature, made up of prominent zoologists from different parts of the world has been formed. It meets from time to time to regulate the naming of species. This is published in the form of Codes. One strict ruling is the 'Law of Priority'. This states that the first name given to any species must stand in the classification table, provided that it is placed in the correct genus. This holds for names dating from the tenth edition of the *Systema*, published in 1758.

Occasionally further names are given to a species which has already been named by someone else. In this case the original name must stand. The other name or names are called *synonyms*. However, where a synonym becomes very well known, the Commission may agree to its use. A proper way to show this, is to add the later name in brackets, as in the case of the bluebell—*Hyacinthoides* (=*Scilla*) *non-scripta*. Scientific names are the tools of natural history classification, and should be used as much as possible, since they are international. Popular names, such as the English names, can sometimes cause confusion. For example, the above-named bluebell refers to the English bluebell. It is quite different from the Scottish bluebell which refers to a campanula, i.e. the harebell *Campanula rotundiflora*. The American 'robin' is really a species of thrush, and names like tortoise and turtle, mushroom and toadstool, crocodile and alligator, frog and toad, may sometimes lead to misunderstanding. Scientific names are in Latin or latinised Greek, the common language used by scholars in Linnaeus' time. This tradition has been retained in modern classification.

The two species mentioned above, the Common Frog and the English Bluebell, are placed here in the classification table:

Kingdom:	ANIMALIA
Phylum:	CHORDATA
Sub-phylum:	VERTEBRATA
Class:	AMPHIBIA
Order:	Anura
Family:	Ranidae
Genus:	*Rana*
Species:	*Rana temporaria*
Kingdom:	VEGETABILES
Division:	SPERMAPHYTA
Class:	ANGIOSPERMAE
Sub-class:	MONOCOTYLEDONES
Order:	Liliales
Family:	Liliaceae
Genus:	*Hyacinthoides* (=*Scilla*)
Species:	*Hyacinthoides* (=*Scilla*) *non-scripta*

Part 1: Section D

The environment

Organisms are slaves to their environment. It is the conditions imposed by their surroundings which control their lives. These are called ecological factors, and may be classed under three headings: physical, chemical and biotic.

Physical factors. These are of three kinds—climatic (light, temperature, rainfall, etc.), edaphic (soil) and topographic (type of landscape).

Climatic factors are overriding since they supply the main needs for living. On a world-wide scale they determine the major belts of vegetation. The amount of rainfall, range of temperature and degree of light combine to produce, for example, a tropical rain-forest, a desert, a temperate grassland, or an arctic tundra. In a comparatively small area like Britain the general climate is broadly the same, a temperate one within those latitudes which support a woodland belt of deciduous trees. Within these limits, however, such a climate can vary in relation to the position of the land and sea. Britain is surrounded by sea on the western edge of a large land-mass. It enjoys, or perhaps suffers from, an indifferent maritime climate of unsettled weather which has become a national topic for small talk.

By contrast her Continental neighbours live in a more clear-cut continental climate of better defined summers and winters, with greater degree of heat and cold. Edaphic and topographic factors are closely linked, and to some extent affected by the local climate. The geological nature of the bedrock, from which the minerals in the soil are derived, will vary from place to place. For instance, a hard, mountainous rock presents a rugged landscape exposed to the elements, and is thinly covered with a poor, often leached, soil which may support only a limited and specialised mountain flora. In contrast a soft rock such as chalk lying in a valley will be base saturated and rich in river deposit, and could support a heavy woodland flora.

Chemical factors. The materials of soils which originate from rock and humus are of different chemical composition, and may be acid, alkaline, or of neutral quality (see pH values in Appendix 15, p. 233). A concentration of the mineral salts can range in intensity from that of fresh rainwater entering a pond or river, to the saltiness of an estuary. Chemical differences will produce their own set of plants for each different habitat (e.g. acid heath, alkaline chalk-down and neutral oakwood).

Biotic factors. These are the conditions imposed by life itself, and serve as a reminder that no organism is entirely independent of its neighbours. The presence of one plant or animal species may be of vital importance to another, especially if it is a question of competition for food or living quarters. Any upset in the balance of a food-chain can have far-reaching consequences (see p. 63). Britain with its high human population must consider man as the most influential biotic factor. Almost every part of the countryside has felt the impact of his presence, usually to the disadvantage of the native animals and plants. This is expressed only in an ecological sense, since nobody would deny the value to man of such activities as agriculture, building, transport, forestry and mining. However, nature usually suffers from such invasions, since the natural habitat is disturbed. Sometimes there are interesting compromises reached, in which a man-made habitat is then exploited by wildlife (see the Bombed Sites, p. 123).

The Habitat and the Climax

Britain's position on the globe gives her a maritime temperate climate within certain ranges of temperature, rainfall and daylight. These produce a set of seasonal conditions that influence the lives of the animals and plants through the year. Thus we speak of springtime as a season of birth and re-awakening, summer as a growing season, autumn a time for fruiting and dying, and winter as a time for rest. This is the overall pattern. Locally the factors previously discussed will control the community gatherings of flora and fauna which are associated with each peculiar habitat, such as woodland, mountain, heath, pond, and so on. In each of these some dominating factor can usually be noted. Water dominates a pond, light controls a wood, topography a mountain, and so on. These are physical factors. On a heath a chemical factor, the acidity, has importance, and in a town man is the principle factor, a biotic one. Within each habitat there is a tendency for the plants and animals in the community to build up towards a peak population of numbers and species, in which one or two play the role of dominant. They are usually the largest or physically strongest members.

In Britain the natural and commonest habitat is the deciduous woodland, in which one of the tree-species is the dominant over the woodland community. Where this condition is found the habitat is said to have reached its climax. As long as the various factors at work in the woodland are in harmony, the woodland climax will continue to exist. However, with man's continual interference, this is seldom long-lasting. Tree-felling, a fire, or some introduced animal or disease may set back the woodland, or even destroy it altogether. If such a bared area should occur, a so-called biological vacuum is created. Immediately fresh plants and animals will begin to fill it, so that the community gradually builds up over the years to its former climax (see Project 5, p. 161).

The Biosere

The above recolonisation of the bared area, if unchecked, will pass through stages of growth and replacement, in what is called a biosere. Fresh organisms begin to cover the bared ground. Lichens (on bare rock) and algae (on exposed soil) are usually among the first pioneers to enter the 'vacuum'. This is the first seral stage towards the climax. Mosses, and perhaps liverworts, then take their place, adding humus to the ground as they die off. These may then be followed by grasses and various herbs (annuals followed by perennials). Grasses are usually common at this stage. These are then slowly replaced by a scrub of young bushes and trees, and finally the trees themselves take over, with one or two species as dominant. Oak grows well on clay, beech on chalk and pine on sand.

Keeping pace with the successive plant stages is a changing animal population usually in the order of the minute protozoan animals as first arrivals, then various invertebrates such as arthropods, molluscs and worms, followed by smaller, then larger, vertebrates. Eventually a badger colony may exist under the roots of an oak tree, where originally at the start of this biosere the algae and protozoa were in sole possession of the exposed ground.

A biosere which commences on dry land and ends as a woodland is called a xerosere. Such a wood could also originate from a water area, and is then the culmination of a hydrosere (Fig. 20). For instance, in a deserted and drowned gravel pit micro-organisms first appear in the water, followed by filamentous algae, then the higher flowering plants as the mineral content rises. Small invertebrates, especially the planktonic kind, then larger forms (insects, worms, snails, etc.) join the invasion, followed by fish, amphibians and water fowl. One of the larger fish such as a pike, or a species of water bird, becomes the dominant animal. A growing reed-bed overshadows the waterside, and by its slow increase the open water is invaded, so that the habitat changes from a marsh to a swamp, and a swamp to dry land. At this stage in the biosere the land species begin to take over, until finally a woodland community is reached as the climax growth (see Project 4, p. 159).

A biosere may be checked at any one of these stages, often by man, but also by some local and natural factor. For example, on mountain tops in the Highlands, a summit heath of grasses or heathers becomes the climax. Due to the exposure and short summer season trees cannot grow there, and the biosere is checked at the grass stage. On the southern downlands this grassland stage is also a common feature, but this is largely due to a biotic factor. Centuries of sheep rearing, and the presence of the rabbit, have held the climax beech woods in check. Today, with the loss of rabbits due to myxomatosis, and the removal of sheep from many areas, the chalk biosere has moved on to scrub, even beech woodland, in the last twenty or thirty years (see Chalkland and Mountains on pp. 82 and 104).

Tolerance

An inspection of any community within a habitat will probably reveal that some of the species are much more tolerant to their surroundings than others are. For example, the

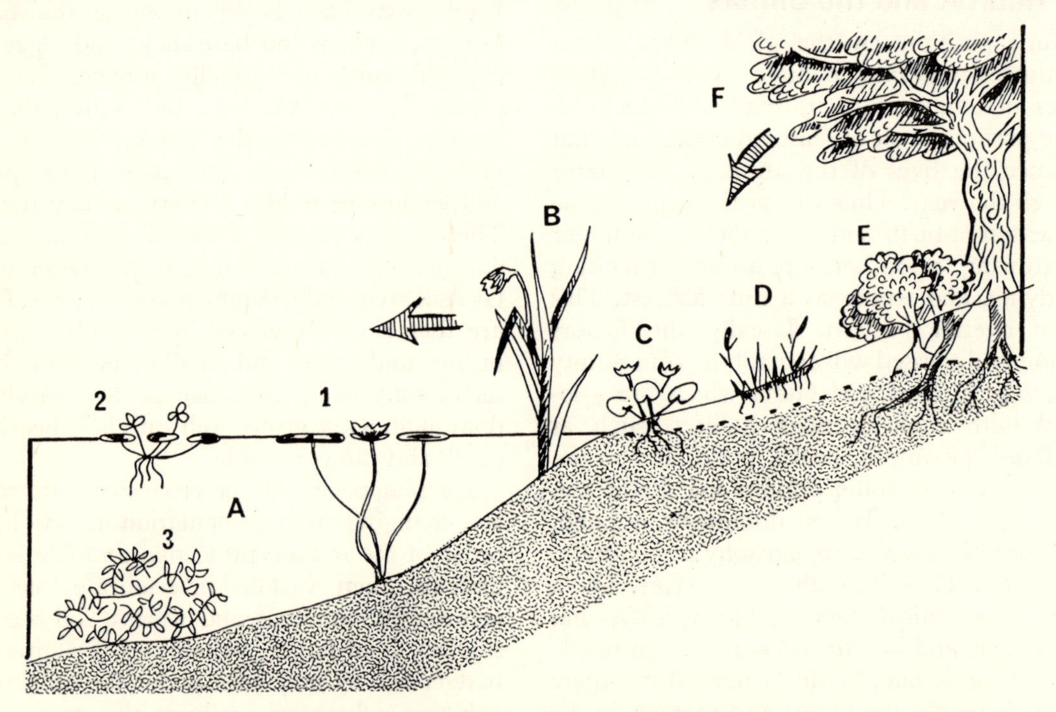

Fig. 20. A Biosere. Arrows denote the direction of plant invasion towards a climax woodland. The pond will ultimately become dry land supporting trees. The seral stages leading towards this are shown in zones, A, aquatic zones (1, rooting; 2, floating and 3, submerged aquatics); B, swamp zone; C, marsh zone; D, grass zone; E, scrub, and F, woodland; Waterlogged ground is shaded.

hawthorn tree (*Crataegus*) commonly seen as a hedgerow plant (see p. 128) can be found on chalk as well as acid soils, in sand-dunes, on waste ground, in hedgerows, woodland clearings, and on mountain sides. On the other hand the cowslip (*Primula elatior*) is seldom found off the chalk (a *calcicole*), and the heather (*Calluna vulgaris*) is restricted to acid soils (a *calcifuge*). Among mammals the rabbit, a somewhat questionable alien since Norman times, has made such a remarkable onslaught over most of Britain, that up to the time of its check by disease, it could be found firmly entrenched on acid heaths, chalk hillsides, cliff-tops, mountain slopes, sand-dunes, farmland, in fact, in any area with workable soil. In contrast, our rarest mammal, the pine-marten, is only found in places such as the remoter mountains, although perhaps not entirely from choice. Persistent persecution by man, unlike that of the rabbit, has driven it out of its former lowland woods.

Competition

An animal's tolerance towards its environment is termed its ecological amplitude. This is the sum of its adaptable qualities. The hawthorn already mentioned is apparently well endowed with this, whereas the cowslip's way of life is more rigid. Such potentialities for adaptation and survival are usually due to variations within a species which are bound up with its hereditary mechanism. However, since living things are in constant competition with their environment, and with each other, situations can arise in which an animal or plant is prevented from exploiting its adaptability. The pine-marten is now confined to the mountains because of the biotic factor which is man. But given full protection and no disturbance it might well spread out of the mountain fastness and return to its former haunts in lowland valleys close to its old enemy. The cowslip is reported to be spreading in many rabbit-free areas on the downs, but it is doubtful whether it will ever stray from its beloved chalky soil.

Plant Zones

In any habitat where the community has reached its climax, the final dominant species is often the largest (e.g. tree in a wood, fish in a pond, grass on a heath). This is termed a primary climax which can go no further. A secondary climax occurs where the biosere is checked at one of the stages by a natural or man-made factor (e.g. fire, animal grazing, ploughing, disease). In a full climax community, however, not all is lost among the subordinate species, and examples of each seral stage in the succession may be found in the shadow of the dominant. This is clearly seen in the typical British habitat, a woodland, where the various stages appear in growth levels (or as zones in a pond community). Since plants are static, these layers and zones are fairly well defined (see Woodlands and Pond, pp. 73 and 111, also Fig. 20).

Animal Niches

Animals also fit into the community life by taking over certain zones, from below ground to tree-top, or from pond-bottom to the surface, but these are less well defined since animals move about. A better term is a niche. This is a position within the community which is linked with a certain food, and is better applied to animals which must constantly search for it. Suitable divisions of animals on this basis are herbivores, carnivores, omnivores and scavengers. Parasites can be added as specialised feeders. A herbivore like a grey squirrel has a fairly wide range of plant diet, and this can partly explain its success, persecution notwithstanding. The native red squirrel, on the other hand, appears to be a much more restricted feeder belonging to pine-woods. With carnivores the badger is omnivorous and nocturnal, whereas its tiny cousin the stoat is a less catholic feeder, and probably more vulnerable to change in feeding habits, quite apart from its unpopularity with man. As a result the badger is still fairly well entrenched over much of Britain, whereas the stoat is becoming localised, even rare, in many places.

A niche also has a functional meaning, in which movement and home territory, often related to the food supply, may vary considerably.

A single pair of eagles will range over an

area of ten or more square miles in their search for food, especially during nesting. A robin pair may confine themselves to a single garden. Consequently the population figure for the robin niche is far higher than that for the eagle. Robins can find all the grubs they need in their garden, whereas eagles must range far and wide in search of hares, grouse and the occasional sick lamb or deer calf. In addition they have bigger mouths to fill.

The Food Chain, Food Web and Food Pyramid

Living things derive energy from the food which they eat. Plants take it from the soil, atmosphere, and the sun's rays. This is transferred to the animals which eat the plants, called the primary consumers. A second animal eats the first, becomes a secondary consumer, and so on. This forms a food chain (Fig. 21). It does not necessarily follow a straight line of consumers, but may involve a number of feeders in a complex pattern of cross-feeding. In addition to the herbivores, the carnivores, scavengers and parasites will all take part in the food sharing. This is called a food web (Figs. 21 and 50).

The transference of energy in a food chain may be likened to a pyramid (Fig. 21). So many plants measured by weight or numbers are needed to feed so many insects. These in turn keep a certain number of insectivorous birds supplied. The small birds become the prey of a larger hunting bird, say, a sparrow-hawk. At each higher level in this pyramid the numbers lessen within the food area. Consequently the plants occupy a crowded niche, the insects less so, the small birds even less, and

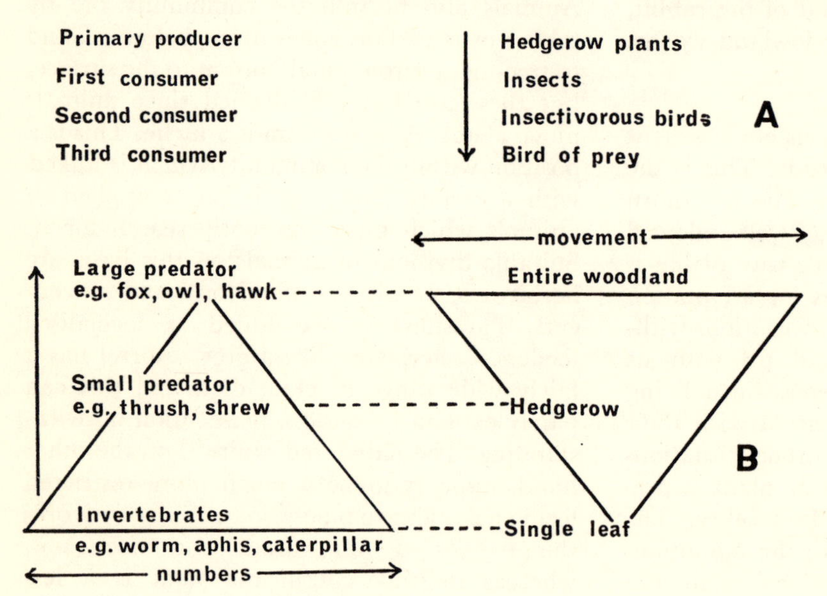

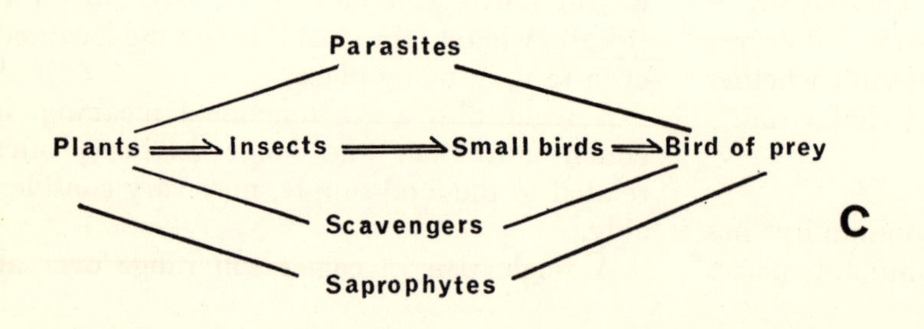

FIG. 21. A, A simple food chain mentioned in the text. B, The same chain illustrated as a food pyramid ranging from the 'key' to the 'apex' species and from a single leaf to an entire woodland. C, A food web based on the above chain. (See also Fig. 50.)

the hawk may have the whole woodland to itself.

There is here a subtle connection between each set of plants and animals, and their reproductive potential. In this example the plants produce large numbers of seeds which maintain their numbers against the attack by the insects. One could argue the other way around, by saying that because of the large numbers of plants the insects in turn can keep up their numbers. Even so, there must be less insects by weight than plants, otherwise the food will not go around. Similarly, there is an even smaller number of insectivorous birds. Even with two or three clutches of fairly large broods per season, this is nothing compared with the enormous progenies of the insects. In the case of the hawk there is usually only one family of a few youngsters reared each year. The bird-of-prey, man apart, has few natural enemies. Notice, too, how the death rate decreases towards the pyramid apex. Within a season countless numbers of plants, insects and small birds are sacrificed to sustain a single family of hawks.

'Key' and 'apex' species. The hawk which heads this food chain at the top of the pyramid is called the 'apex' species. Its existence is directly affected by the presence or absence of the animals or plants lower down the pyramid. At the base are the 'key' species, in this case the insects. This would apply to a woodland. Other important key animals are the plankton in the sea, the tadpoles in a pond, and the worms in the soil (including those along the sea-shore). Their presence and numbers may well decide the distribution of fish and whales in the oceans, newts and aquatic beetles in a pond, and waders along the shore.

A ratio can be worked out between an animal's size and food requirements, and its daily movement. An apex animal like the hawk will cover the entire woodland, whereas a key animal such as one of the insects may live its entire life on a single leaf. Within these limited food areas competition among the small animals can become intense, especially if the same food is shared between different species. In such cases the food niche becomes highly specialised, as illustrated in the project on the oak tree (p. 171).

Upsetting the food chain. In the constant struggle for food, each animal species within a habitat is striving to maintain its numbers, by holding on to a food niche against competition. An interesting discovery has been made, that if a gap appears in the food pyramid, then the fertility rate of the next consumer above or below may be affected. For example, if there were a shortage of small birds in the area, and consequently less food for the hawk, then this hunter might cease breeding for a year or two. Such is said to have happened to the buzzard in the absence of rabbits. On the other hand, if the hawk is missing, then the smaller birds might increase their numbers. Each link in the chain adjusts itself to the available food supply. Sometimes when things become acute, an enterprising animal may overstep its niche, with interesting if unpopular results. For instance, the fox in Britain, in the absence of rabbits, or because of increasing numbers, has become even bolder and more adaptable in its hunting, and is now openly entering farms and built-up areas much more readily, to the annoyance of poultry-keepers, sheep-farmers and cat-owners. It is fast becoming a suburban animal which lives in places safe from the hunt.

When man interferes by introducing a strange animal into the area, a food chain is sometimes upset, and severe competition follows. The introduced alien competes for food with the rightful resident. In Australia the kangaroo held the niche as a grass feeder. Then the sheep and rabbit were introduced. Were there no protection of sheep, war on rabbits and shooting of kangaroos, one wonders which of these three species, the indigenous kangaroo or the alien sheep or rabbit, would eventually monopolise the grass niche, Probably the last, for it has a high breeding potential and few natural enemies in Australia. As it so happens the problem has been partly solved by the introduction of myxomatosis.

This dangerous practice of introducing foreign animals or plants can have very serious consequences, even to the extent of ruining the plant life and baring the land of its soil by erosion. This is now happening along the wooded hill slopes bordering the fertile Canterbury Plain in New Zealand, where the British red deer and Australian opossum are playing havoc with the trees.

Food chains and pyramids can be so delicately balanced, that any outside interference, especially by man, can lead to near extinction of a member of the chain, especially the apex animal. An example of this which has received much publicity has been the use of certain insecticides whose toxic effects are persistent. The poison meant for the insects is carried right through the pyramid to reach the highly vulnerable species at the apex. The above mentioned sparrow hawk is an example. It is now an established fact that many birds-of-prey in Britain have decreased as breeding species, and that this 'pyramid of death' is a contributary factor.

The Micro-habitat

Within any community area can be found, but often overlooked, little pockets of animal and plant life which live their secret lives in hidden places such as under logs, holes in the ground and rocks, in hollow trees, rubbish dumps and similar undisturbed places. These are called micro-habitats. Because of the confined space and sheltered position they often show striking differences in the degree and range of light, temperature and humidity, even in their chemistry. The surrounding environment may be wet one day and dry the next, lit up and warmed by the daytime sun, cooled and darkened at night. Meanwhile the micro-climate under the stone or in the hollow tree is more equable. Temperature and humidity vary far less, and the tiny space is in constant darkness or gloom. Such a micro-habitat will attract a population of small animals which are adapted to dark and damp places (less so with plants which need light). The animals rarely emerge from their secret world, except after dark. Even their behaviour pattern may differ from the more free-living animals outside. They automatically seek the darkness. This can prove a life-saver, for dark places usually mean moisture and freedom from frost or drying winds. This powerful attraction towards darkness in the wood-louse, for example, may be connected with its need for moisture which is a legacy from its sea-ancestry.

Responses stimulated by light, moisture and other factors are well exemplified among dwellers of hill-streams and rock-pools (see pp. 119 and 93). There is a strong urge to cling to something, or to hide away. This is essential if the animals are to remain within the community, in a habitat which is constantly buffeted by moving water. Because of these critical factors which are so closely linked with behaviour patterns, the hidden world of the micro-habitat can sometimes be overlooked, yet it is everywhere around us—under a log, in a rubbish tip, a hollow tree, in the soil and leaf-litter, even in the cellar beneath our home (see Project 10, p. 173).

Life in the Soil—the Litter Flora and Fauna

In soil, especially when it is enriched with humus such as can result from the leaf-fall on a deciduous woodland floor, a concentration of living things may become as high as anywhere on earth, or even in the sea. In an acre of beechwood litter it has been estimated that there may be a population of some 175 million mites quite apart from all other species which inhabit such suroundings (Fig. 22). Varied as this may be, the animals show certain common characteristics. Most are small in size, and are reluctant to expose themselves, except perhaps after dark or rainfall. Their greatest danger lies in desiccation, and this can be far more serious in a small animal where surface area is large in relation to volume. As a life-preserver various mechanisms for conserving moisture are found in these litter animals. There may be a wax-like covering over the body which cuts down water loss. Since this is also impervious to gas exchange during respiration, special breathing organs are required, which at the same time can control any water loss. Spiracles in some, and lung-books in others, are common mechanisms, let into the body and controlled by muscles so that they function only when the

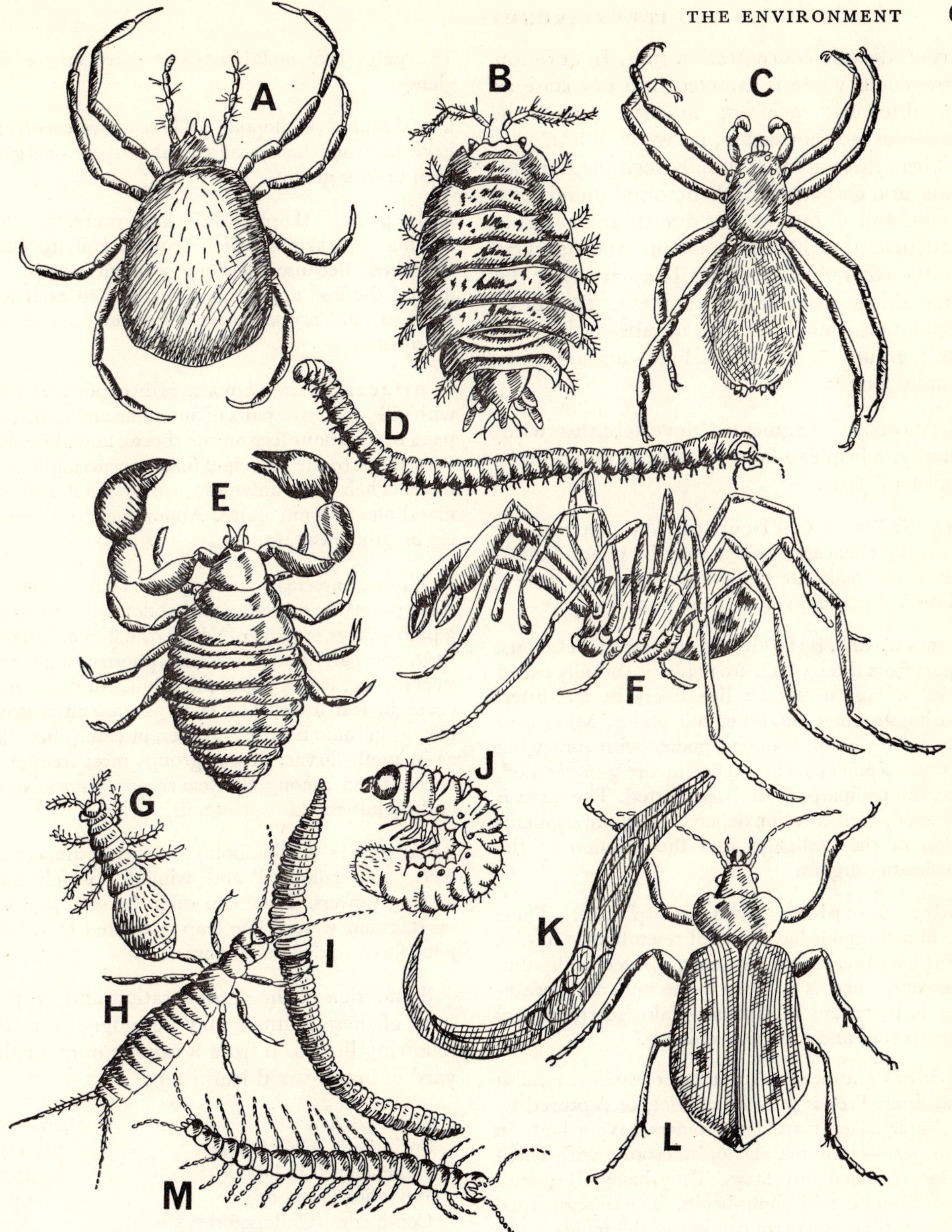

FIG. 22. The Litter Fauna. A selection of animals commonly found in soil and leafmould. A, mite (Acari); B, woodlouse (Isopoda); C, spider (Araneae); D, millepede (Diplopoda); E, false scorpion (Pseudoscorpionidae); F, harvestman (Opiliones); G, springtail (Collembola); H. bristletail (Diplura); I, earthworm (Annelida); J, beetle grub-cockchafer *Melolontha* (Coleoptera); K, nematode or eelworm (Nematoda); L, beetle *Calasoma* (Coleoptera); M, centipede (Chilopoda).

carbon dioxide concentration rises. In addition nitrogenous waste is excreted in a dry state as insoluble uric acid in insects (guanin in arachnids) so that, again, no water is lost.

These litter animals usually seek the damper spots, and gatherings of them found under logs, stones, and so on, are not due to any sociable tendencies, but because moisture and darkness are the common attraction. They are after the same thing, a stable environment. The litter population is made up of a number of invertebrate groups, of which the following are more usually found:

ARTHROPODA. Segmented invertebrates with, usually, a firm exoskeleton, many-jointed limbs and 'foot' jaws.

Arachnida or Chelicerata. The first segment bears the chelicerae (which may be poison fangs), the second pair the pedipalps (clawed in some), followed by, usually, four pairs of limbs.

Mites (Acari). By far the commonest soil residents. Apart from those which are parasitic (usually called ticks) a host of species live freely in leaf-litter, feeding on fungi and other soil plants. Mites usually have plump, rounded bodies with somewhat obscure segments. The chelicerae are 3-segmented, and the pedipalps 6- or 7-segmented. The larva is 6-legged. Identification depends upon the segmentation of the pedipalps and the position of the respiratory organs.

False Scorpions (Pseudo-scorpiones). These small arthropods have a slight resemblance to true scorpions because of the large, clawed pedipalps. However, there is no 'sting in the tail', and breathing is by means of trachae. False scorpions are carnivorous and hunt tiny animals.

Spiders (Araneae). A number of spiders inhabit leaf-litter. Prey is pursued on foot, or captured by a simple silken trap. True spiders have a body in two parts—prosoma and opisthosoma, with chelicerae modified into fangs. They have silk glands and breathe with lung-books. The 6-segmented pedipalps act as sperm carriers in the male.

Harvestmen (Opiliones). Resembling long-legged spiders, these harmless arthropods have the two body halves only faintly divided, and the segmentation partly hidden. Breathing is with tracheae. Of the 8 long legs, the second pair are longest. The palps are unclubbed and there are no silk glands.

Centipedes (Chilopoda). These active carnivores have flattened bodies and a single pair of legs to each body segment.

Millepedes (Diplopoda) are rounded, slow moving vegetarians which tend to coil up when disturbed. Because the body segments are in two halves, the legs appear as if doubled to each true segment, and are far more numerous than in the centipedes.

Crustacea. Mainly aquatic arthropods breathing with gills, with two pairs of antennae and numerous pairs of biramous legs on the thorax and abdomen. Woodlice are the principal forms occurring in leaf-litter. Their first antennae are vestigial and the mandibles without palps. Abdominal 'feet' act as air breathing organs.

Insects (Insecta). Arthropods with bodies in three parts (head, thorax and abdomen), antennae, 3 pairs of legs, 2 pairs of wings (the flies or Diptera have one pair) and a metamorphosis usually in 4 stages—egg, larva, pupa and adult. All stages may occur in leaf litter. The larva is sometimes grub-like, as in some beetles and flies, or caterpillar-like, as in moth larvae. Insect groups most frequently encountered among the leaves are beetles, moth and fly larvae, and springtails.

Springtails (Collembola) are tiny, primitive insects, pale coloured and wingless, which have leaping powers. They can often be seen jumping on stagnant water. The leap is assisted by a lever joint fitted under the abdomen.

Some idea of the concentration and proportions of these animals may be judged from the following list taken from a sample of one cubic yard of beech-wood leaf-litter:

Mites (Acari)	320,000
Springtails (Collembola)	51,000
Millepedes (Diplopoda)	170
Woodlice (Isopoda)	140
Centipedes (Chilopoda)	75
Spiders (Araneae)	25
False Scorpions (Pseudo-scorpiones)	8
Harvestmen	6
Other insects	500
	371,924

The litter flora is composed mainly of plants which can exist in darkness and damp surroundings, such as the bacteria and many ground fungi. These, together with their animal neighbours may be termed the soil-makers. Apart from breaking down the leaf litter, they can reduce between them a fallen tree to powder. Their value as decay and humus agencies cannot be over-emphasised. Such organisms are also a useful indication of soil conditions. Apart from direct chemical or physical tests on the soil, the presence or absence of some animal or plant species may give a clue. For example, the woodlice which stem from a sea ancestry require moisture to some degree, and can be used as indicators of soil humidity. The pill woodlouse (*Armadillium*) is fairly resistant to dry surroundings, in which the garden slater (*Oniscus*) would probably die. With chemical determination as to the acidic or alkaline nature of a soil, flowering plants such as the ling, *Calluna* (a calcifuge) and the cowslip (a calcicole) are a help. In some cases, such as with mosses, a whole series can be related to different soils, from a rich woodland to a poor heath soil (see Appendix 1, p. 212). In deciding where to plant his trees, or how to prepare the soil before planting (e.g. by drainage or using a fertiliser) a wise forester will first look at the mosses in the area before selecting a suitable species of tree as a growing crop.

THE WEATHER

The climate in Britain is such that it produces seasonal changes in the weather, and it is the temperature range over the year which has the most effect on life's activities. This quickens in summer and slows down in winter, in particular among the plants. Plants and animals can only survive within certain limits. Of these the higher temperatures are usually more dangerous to life, and death may result from heat stroke or desiccation. To avoid this, animals tend to hide away in shade or in enclosed places. Below vegetation, in cracks, holes and other covered spots the temperature is often lower, and the humidity higher, than it is in the open. Plants growing in dry places, however, cannot move away. Desiccation is avoided by having special adaptations for conserving water (see Xerophytes, p. 100). Where transpiration occurs normally the water which evaporates from the leaves helps to keep them cool. During winter the low temperatures, which only become lethal below zero, are avoided in a similar manner. Animals hide away in places free from frost, and tend to go to ground, especially those which hibernate. Water creatures are safe below the ice-covered pond at temperatures above zero (see p. 119).

Apart from a gradual rise and fall in the seasonal temperatures there are a number of local and temporary changes which occur at day and night, during the presence or absence of cloud cover, or because of the nature of the soil and its cover, also its height above sea-level. By day the sun's rays warm the surface on which they shine, and not the air through which the heat passes. The warmed soil in turn heats the air above it. At night the soil loses heat by radiation, and the air cools. This may be checked by a cloud cover, so that the radiated heat is reflected back. This is why clear nights feel colder than cloudy ones.

Air has a low specific heat content, and it takes a lot of heat to warm the cold soil. Specific heat of soil is high, so that it has a greater effect on temperature changes. Bared soil will heat and cool more rapidly, depending on its nature. A wet and compact soil such as clay is a better conductor than a loose and dry one such as chalk or sand. Covering to soil such as grass, trees, snow, or a garden mulch acts as an insulator to retain heat. Temperature will also vary with height and latitude (about 1°F for every 270 feet up or down, and the same for every 40 miles north or south). Hills and neighbouring valleys can show a marked difference in temperature. The sheltered valley is usually warmer than the exposed hilltop, especially in still air. With fog about, however, the hill is in warm sunshine and the valley in a wet and cold atmosphere. Valleys may also become frost hollows as the cold air streams down the hillside. It may pile up against an obstruction such as a wall, to form a frost trap. Windy nights can reverse the temperature.

Moving air over the hill can keep it low, whereas it remains more stable in the still valley air below.

The biggest temperature changes take place at ground level. In summer the day reading may go as high as 100°F, then drop to almost zero at night, giving a range of some 70°. As the heat penetrates the soil then slowly cools at night, there is a considerable drop in the temperature range the deeper one goes. At a depth of only two feet the fluctuation may be as little as 1°. Depending on the soil texture, rate of conductivity, and type of covering, also the time of year, the soil readings will vary.

Snow on the ground in winter can be a life-protector. The air temperature may be at sub-zero, the ground surface frozen hard, but somewhere below ground it is above freezing. Below the frost-line hibernating animals and dormant plants are safe. Even active animals such as voles, and mice shrews can use the snow as a retreat by tunnelling under its frozen crust.

As an example of how weather conditions can vary in different habitats so as to produce localised or micro-climates, a sand dune and a woodland may be compared (Fig. 23). In the dune the sun has direct contact with the loose and dry sand, so that it warms up by day. The warmed air above can reach a high temperature, especially in the sheltered dune slacks. At night, unless the sky is overcast, there is rapid heat loss from the ground. Condensation in the form of dew is a result, and ice may even form on winter mornings. The range between day and night temperatures is high, much more so than in a tree-covered woodland. Here the leaf canopy acts as a radiation barrier, and the temperature range is reduced. Air is warmer at night and during winter inside the wood than it is outside, and correspondingly cooler by day and in summer. The carpet of leaves acts as an insulation barrier between soil and air.

Standard weather maps for large areas show isotherms which are usually adjusted to sea-level temperature. This gives a kind of macro-climate for the whole country. Readings for a small area which is covered during a field survey need to be more exact, and instruments should always be used. Relying upon the 'feel' of the air on one's body can be deceptive. A sunny day in spring may appear to be warmer than a similar one in autumn, yet temperatures are the same. This is because we look forward to warm weather in spring, but have already had it by autumn. A windy day, even during sunshine, feels colder than an overcast day with still air, even at the same air temperature. Standing in a winter gale can feel bitterly cold, but sitting behind a sheltering wall can feel much warmer. Animals will use such sheltered spots, especially when the sun

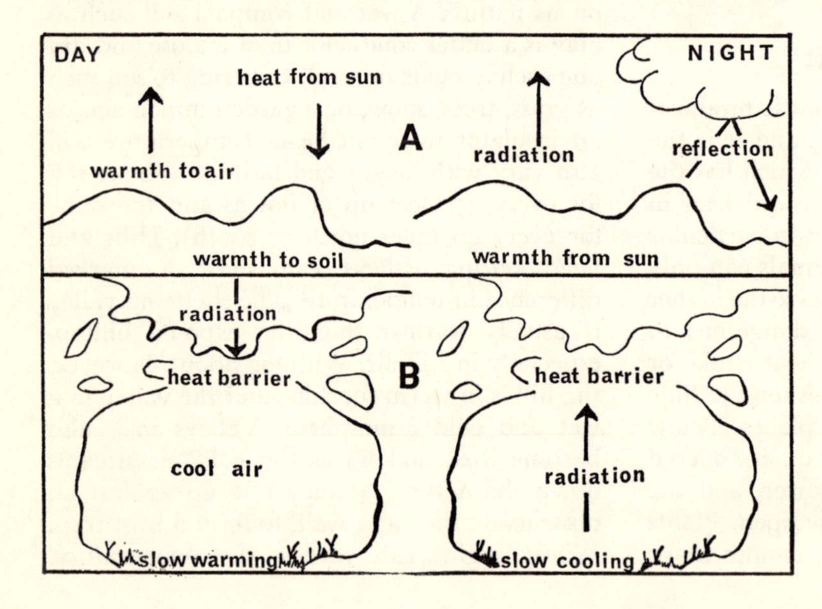

Fig. 23. Heat dispersal during day and night periods in A, an open sand-dune and B, a closed woodland. The former exhibits greater temperature extremes.

provides warmth, even in winter. When snow is lying about, a butterfly, bee or adder may be seen abroad. It is the wind that makes us feel cold, due to evaporation from the skin. Dry summer weather when the lawn turns brown makes us feel thirsty, yet no such feeling arises on a similar dry day in winter. Summer showers soon dry up but winter rain lies about, yet the amount of rainfall may be the same.

Temperature

To obtain reliable readings a reasonably accurate thermometer is required, also an effective screen. For field-work something portable is needed, and can easily be made.

For taking air temperatures it is important that the thermometer bulb should not come into contact with any object, since it may then give a wrong reading. If time allows, readings should be taken over one year so as to show seasonal fluctuations, also at different times of the day to show the diurnal rhythm. If three thermometers can be used at the same time, these can be fixed at different heights, say, at 6 feet, 3 feet, and at ground level. Below ground, readings should also be taken by lowering a thermometer into a sunken tube. The use of a maximum and minimum thermometer will help to cut down the work, as it will show any temperature fluctuation which has occurred since the last reading. Using a thermograph will give a continuous record. This apparatus consists of a pen attached to a lever which is operated by an expanding and contracting strip of metal. The pen moves over a rotating drum on which is fixed a paper scale. A continuous record of temperatures day by day for as long as a week is marked on the paper. The machine can be set against a standard thermometer, so as to give a direct reading. Results by either method can be entered on graph paper, and should make interesting comparisons for different localities. Exposed and sheltered places should be tried out, and differences will be noticed, however small. Places to try as examples are a sheltered ditch, a hollow tree, a rubbish heap, under a cow pat, the inside of a thick bush, a sunken pipe or tunnel, and so on. These are all exploited by animals, and the readings should be checked against those of the outside air.

Rainfall

For recording this a rain gauge is required. Daily readings taken over a period can be graphed, and should include records of snowfall, together with depth and duration. Snow can be measured as rain if melted inside the rain gauge. Snow-drifts make interesting records if measured by sounding with a measuring rod. These can then be drawn as cross-sectional diagrams. Place the rain gauge in an open position so that it receives the rain direct from the sky and not from a dripping tree, but at the same time avoid wind which will tend to blow the rain at a slant.

Humidity

Air humidity has an important effect on vegetation, also on many animals which require moisture. It can also control the rate of leaf transpiration. The relative humidity, that is the percentage saturation of the air, can be obtained by using a wet- and dry-bulb thermometer. The difference in the two readings can be used to obtain the relative humidity by referring to a prepared scale. Alternatively a pocket hygrometer can be used. It should be protected from rain and direct sun.

Evaporation

The water loss from the leaves and other surfaces is in direct relation to the water concentration in the atmosphere. To obtain a relative set of readings under different air conditions, an atmometer can be used.

Transpiration

The rate of water loss from leaves can be tested with cobalt chloride. This is deep blue when dry and turns pale pink if wet. Strips of filter paper are soaked in a solution, and can be kept in a tight bottle for field use. Water vapour given off by a leaf is enough to change the colour, and the time taken to do this is checked by reference to ordinary colour paper. Cut further strips of paper in three colours—pale blue, pale pink, and cobalt blue (i.e. the colour of the dry cobalt-blue paper). Stick one strip of each colour paper, and also a strip of cobalt blue paper, onto a microscope slide. Place this on the underside of the leaf to be tested, so that the cobalt paper is in contact with the leaf pores; place another, similar slide on the other side of the leaf, and hold together with a paper clamp. The change, if any, of the colour of the cobalt paper can then be watched until it reaches the colour pale blue or pale pink, and the time taken noted. Select only dry leaves, in the open as well as in the shade. It will be found that the transpiration rate varies according to the plant species and the leaf position. Within a tree or bush the evaporation rate, and consequently the transpiration rate, is slower than on the outside. Figures obtained by this method are only relative.

Light

Light is an obviously important factor to plants, especially in shady places such as woods. To record absolute measurements in terms of candle units is not always possible in the field, and relative figures may have to be used. This can be done with a photographer's light-meter. It should be held in the same position for each reading, pointing half upwards towards a white cardboard sheet held just above it. This will give a comparable and diffused light for each reading. The meter is adjusted so that the needle covers the whole range of the scale, and the figures obtained are converted into fractions of the maximum light intensity, i.e. in one or other open spot within the area. Choose a day when the sky is clear or overcast, not cloudy, so as to obtain comparable readings.

Exposure

According to where readings are taken, the temperature, humidity, light and so on will vary with the locality. This can be noted from visible sightings. Trees will have more algal growth on one side than the other. Sloping ground facing south will tend to be warmer and may show richer vegetation and earlier growth. Open ground may dry out in summer, and wet patches persist under cover. Cold air will settle in hollows and may delay the vegetation. An interesting feature in built-up areas is the earlier spring growth of some vegetation, such as among bulb plants in sheltered spots, or leaves on trees standing close to artificial lighting in the streets.

Part Two

Major Habitats in Britain

Part 2: Section A

Woodlands

Under natural conditions much of Britain would be covered by deciduous woodland, since it lies within the latitudes and climate which can support this kind of vegetation. Such well-known trees as the oak, beech, ash, hazel and birch are common to the landscape. In the Scottish Highlands, however, these deciduous trees give way to needle trees, or conifers. Here the altitude governs a climate more in keeping with that of Northern Europe, the home of conifers which are adapted to withstand more severe winters, as well as a physiological drought when water is frozen.

Natural stands of Scots fir (*Pinus sylvestris*) still exist in places as remnants of the old Caledonian Forest. Elsewhere conifers occur as escapes, or have been planted (see p. 76). To the south, in Mediterranean lands, occur the evergreen plants, many of which are grown as ornamental shrubs, but do not concern us here.

Today much of Britain's woodlands have disappeared. Tree felling began with the Neolithic farmers who settled in the south and cleared spaces among the woodlands in order to make room for their dwellings, crops and farm animals (see Chalkland, p. 82). At various stages through history trees have also been cut down over much of the country for such needs as timber for fuel, houses, furniture, ships, vehicles and tools, and to make space for roads and spreading towns. This has severely curtailed the woodland habitat, to the detriment of many plant and animal species which normally live among trees. The hunting tradition in Britain has partly helped in this extermination, but paradoxically has also helped to save certain species. Were it not for the strict protection laws of the continental invaders, such as the Normans, who brought with them a love for sport or venerie, much of what is left of our woodlands today might also have disappeared.

Before Duke William arrived, Britain was divided into small kingdoms occupied by various tribes who lived off the land, and treated it as a communal property on which to hunt for food and to grow crops. The Normans set aside large tracts of this 'waste' land in order to provide sanctuaries for the animals which they wished to preserve for hunting. Such Crown property, known as a hunting forest, was subject to a special Carta de Foresta, or Forest Charter, in order to protect the trees and the game. The Normans were well aware that in order to hunt it is necessary to preserve. Today many pieces of Crown land still remain as unenclosed common land, or as remnants of the old hunting forests and chases, such as Enfield and Thetford

Chase, and Epping (see Part 3), Sherwood and the New Forest. Elsewhere private estates and woods, parks belonging to the crown or a municipality (e.g. town parks), and plantations of the Forestry Commission, make up a scattered patchwork of trees over the countryside. Much of this is preserved for amenity and to safeguard the country's natural beauty.

In a woodland it is the trees which dominate and influence the community, since they take the lion's share of daylight. The degree of light which filters through the tree canopy of leaves is a main controlling factor on woodland plants. For this reason a large number of flowering herbs bloom and set seed in the spring when light is available (e.g. primrose, lesser celandine, wood anemone and bluebell). During the summer shade they make their food with their leaves, and store this in underground organs (Fig. 24). The later summer flowers are usually found along the borders or in clearings where more light can penetrate. Woodland plants flourish in great profusion because of the shelter given by the trees against drying winds, excessive sunshine and extremes of temperature. Tree roots bind the soil to prevent erosion, and a rich leaf litter accumulates, to decay and provide plant nourishment. Trees also hold in moisture transpired by their leaves and retained in the still air. In effect, all these beneficial factors help to maintain a rich flora, not only among flowering plants but also among the lower forms such as mosses, ferns and fungi which can tolerate shade, and in the last group live as saprophytes.

A woodland in Britain may contain one of the richest of floral communities. In turn this supports a concentrated and varied animal population (see p. 212). Although the commonest species may vary from wood to wood,

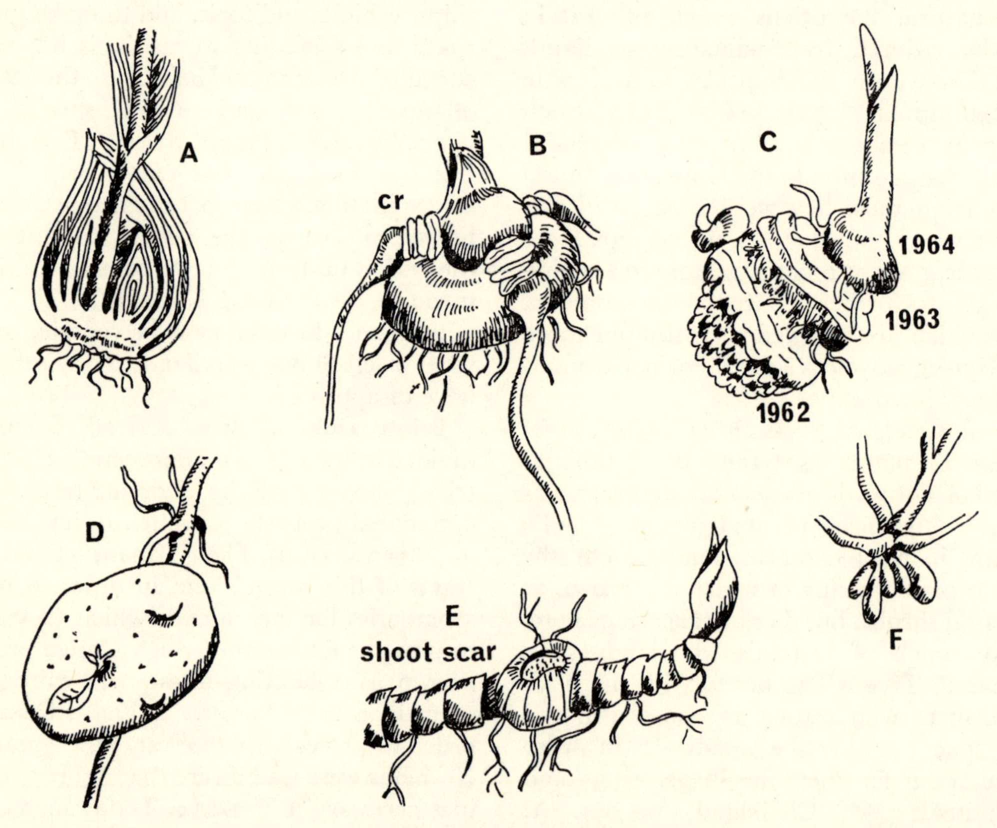

FIG. 24. Underground storage organs. A, tulip (a bulb); B, crocus (a corm); C, arum lily (a corm); D, potato (a tuber); E, lily-of-the-valley (a rhyzome); F, lesser celandine (a root tuber); cr, contractile root.

depending to some extent on the dominant tree species and type of soil, a similar pattern of layering may be found (Fig. 25). At the top is the tree canopy of leaves and branches forming a roof over the woodland. Below this comes a mid or shrub layer of subordinate trees and bushes. At ground level is an association of woodland herbs including grasses and ferns. A fourth layer of mosses, lichens and fungi may also occur on the ground itself, the so-called Bryophyte layer. Here the fungi help in the important role of decay agents.

In a woodland climax, that is, when full woodland maturity is reached and the canopy is a closed one, one or other tree species becomes the dominant plant. Oak and birch usually occur on non-calcareous soils, whereas beech prefers chalk or limestone, as also does ash. Sometimes a wood contains a mixture of co-dominants, such as in an oak-ash wood found on the marls of mountain slopes.

Certain plants and animals may turn up with regular monotony in a variety of woodlands because of their tolerance to a range of different factors. Other species are more restricted. For instance, the foxglove (*Digitalis*) may be found under oak, pine or birch, whereas the conspicuous fly agaric toadstool (*Amanita muscaria*) is rarely far from birch. The blue-tit occurs in most woodlands, whereas the crossbill is seldom far away from conifers.

Although there is considerable mixing of trees in Britain, due to introductions and escapes, a number of woodlands with characteristic individuality can be recognised.

Pedunculate Oakwood (*Quercus pedunculata*). Of all British types of woodland this supports the richest and most varied community of plant

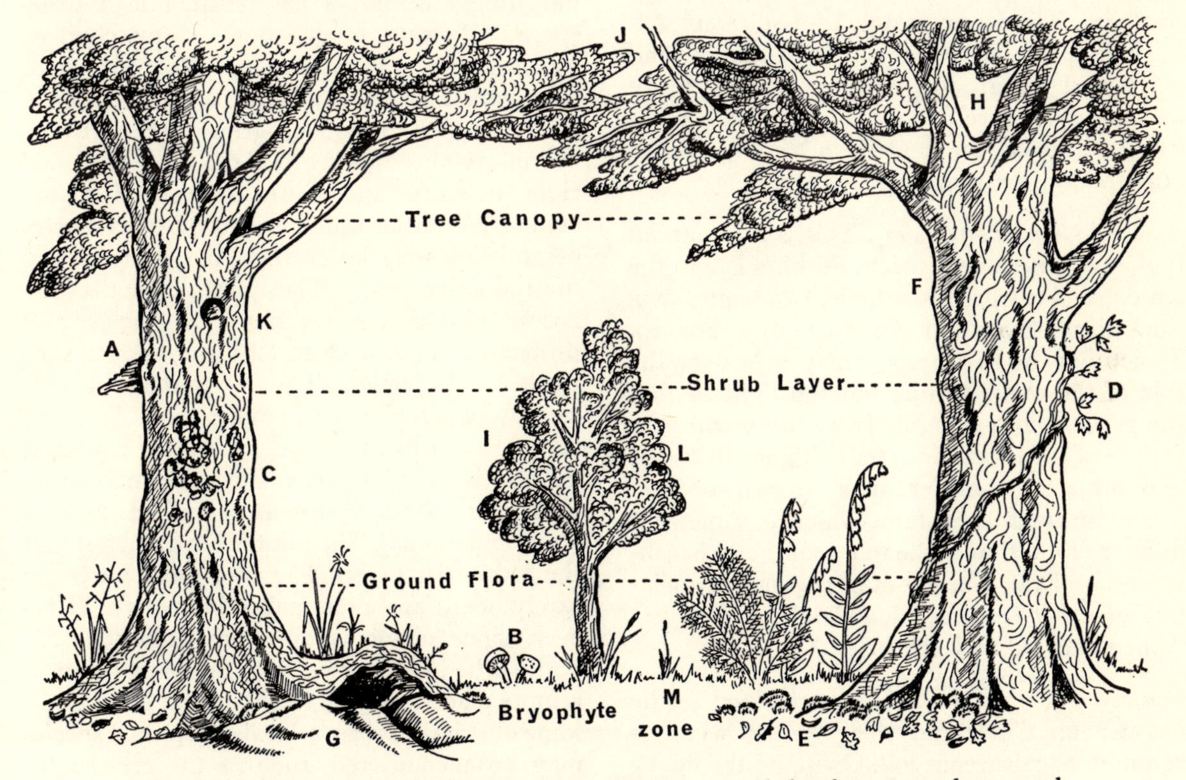

Fig. 25. The Woodland Habitat, showing various growth levels among plants, and a number of homes occupied by animals. Plants. A, a parasite (bracket fungus); B, a saprophyte (toadstool); C, an epiphyte (lichen); D, a climber (ivy). Animals. E, in leaf-mould (litter fauna); F, under bark (beetle grub); G, below ground (badger); H, in tree (squirrel); I, in bush (dormouse); J, on branches (woodpigeon); K, in tree-trunk (woodpecker); L, in bush (song thrush); M, on ground (woodcock).

and animal. Situated mainly on heavy clay soil, in sheltered lowland areas, and with just the right combination of light, moisture, humus and soil, there is hardly a square foot of ground which is not covered with plants. The community swarms with animal life, from invertebrates such as insects, to birds and mammals. Every niche in the food chain is occupied, such as the leaf-litter, herb, bush and tree, and every plant product, from root to fruit and leaf, has its particular animal diner. A single oak provides a home for an astonishing wealth of animals, even after it falls down and decays (see Project 9, p. 171).

Durmast Oakwood (*Quercus sessiliflora*). Growing usually at higher altitudes, and on more basic soils, this kind of woodland is common on mountain slopes, such as the slates in Wales, sandstones in Yorkshire, and granites in Scotland. The loamy, sandy and siliceous soils in these places are less fertile than the lowland clays, and the ground flora less varied. Ferns and mosses figure highly in the numbers of species, and lichens flourish in the unpolluted atmosphere (see Mountains, p. 104).

Ashwood (*Fraxinus excelsior*). This is a woodland typical of limestone districts, flushing late in the season, and casting little shade. Consequently a thick shrub layer and ground flora builds up. Sometimes the shrub layer may take over the role of the trees, cutting out the light so that the ground flora dies off. Ferns are a common feature of ashwoods. Situated at higher altitudes, and subject to heavier rains, an ashwood on limestone supports a flora which is somewhat different from that of the more southerly beechwood on chalk. Each contains its own association of calcicoles (see Chalklands, p. 82, and Appendices 2C and 2D, p. 213).

Beechwood (*Fagus sylvatica*). As explained in the chapter on Chalkland this type of wood is common to calcareous soil. Owing to the heavy shade caused by the mosaic positioning of beech leaves, only a limited amount of flowering plants can grow. A number of these, as calcicoles, are peculiar to beech woods. The heavy shade (as much as 80 per cent. of available light is cut off) enables shade lovers to take over. Shade also inhibits bacterial activity in the decay process, so that the fallen beech leaves lie around mummified and of little value as fertiliser. The rich autumn tints, the rustle of dry leaves beneath one's feet, and the long vistas through the stately trees, are some of the attractions of a beech wood. Much of the animal life is hidden below the thick leaf carpet which attracts a high population of litter fauna composed of mites, spiders, centipedes, etc. (see p. 64) and produces a heavy crop of fruiting fungi during the autumn months.

Pinewood (*Pinus sylvestris*). Apart from the remnants of the Caledonian Forest in the Highlands, all our pine woods have been planted. Since the First World War an intensive Government campaign to make up the loss of our timber resources has resulted in a widespread planting programme. Many conifers, including the native pine, will thrive on poor soils which are sandy or acid, or lacking in base elements. They will grow successfully on our mountains and heathlands. From Scotland right down to the south coast of England, conifer plantations maintained by the Forestry commission may be encountered on the more unproductive soils. The public usually has access, although some areas are enclosed as protection against deer and the commoner's cattle and ponies. The permanent needle canopy of a regimented conifer wood lends an air of gloom to the scene. The thick carpet of needles is slow to decay and acid in quality. Humus is poor and supports a limited range of plant and animal. The impression gained is that the habitat is deserted of life, since birds are rarely heard and mammals seldom seen. However, since conifer plantations are often left undisturbed for long periods, and may occur in remote places, e.g. the Highlands, it is here that some of Britain's rare or localised animal species may be encountered, such as the crested tit, capercailzie, red squirrel and pine-marten (see appendix 2K, p. 217).

A feature of many heathlands in the south is the occurrence of scattered pines which are probably escapes from a neighbouring plantation. If deliberately planted they would grow in ranks.

Birchwood. Natural stands of the common or downy birch (*Betula pubescens*) occur above the durmast oak woods on our mountain slopes, where they can withstand a wetter climate (see Mountains, p. 104). In the more sheltered lowlands the silver birch (*Betula verrucosa*) is commonly found. It is a rapid coloniser of open spaces caused by fires and clearance, and competes readily on the poorer sandy and gravelly soils. Like the pine in the south, it is a feature of heaths and commons. This tree even succeeded in growing on the bombed sites of our towns after the Second World War (see p. 123).

A year's study of a woodland community, from spring to autumn, will reveal a constant succession of growth, and it may be asked how the soil can accommodate and nourish so much plant life. This is largely possible because of the architectural growth of the plants both above and below ground level. Above, as already noted, the plants grow in tiers, from lowly moss to tallest tree. Similarly, below ground, there is a layering of roots, in which the trees and shrubs go deepest, with herbs and grasses near the surface. In between are the plants with storage organs, and which have perennial growth (Fig. 26). In a sense each plant finds its own level, and need only compete with its own neighbouring kind, or with another species invading its territory at a similar growth level. Plants will also extend growth of shoot and root to fill up a gap in the habitat, and this explains the variety of growth shapes within a species. A tree may be spear shaped in a crowded community, but more bushy shaped in an open clearing. Plants with underground

Fig. 26. Layering of underground organs among woodland plants of different height, ranging from fungus mycelia to tree roots. a, hairy woodrush; b, toadstool; c, wild arum; d, moss; e, wood melick grass; f, wild hyacinth; g, hazel and h, tree.

storage organs will each find a particular depth of growth. As a seed the plant commences near the surface, but then it gradually sinks. Attached to some storage organs are special contractile roots. The lower end is firmly anchored in the soil, then becomes wrinkled as it contracts, and so pulls the storage organ down after it (Fig. 24).

Further characters found among woodland plants show their adaptation to a darkened environment. Leaves usually grow larger in shady places, even on the same tree. They also position themselves in relation to the light source, broadside on in shade and at an acute angle in direct sunlight. Mutual overshadowing in woodlands is to some extent lessened by etiolation, that is, the lengthening of the internodes. Summer flowers are mostly inconspicuous, since attractive colours and mechanisms would be largely wasted in the undergrowth. Instead, many flowers are scented to attract the insects (e.g. honeysuckle, wood sage, woodruff and wild arum) (Fig. 27).

The foregoing assessment of our woodlands is given in general terms. Each community will reveal its individuality, depending on the the influence of each factor at work such as the amount of light, soil condition, humus and water supply. In addition there are the biotic factors to consider, such as felling, replanting, fires, grazing, and so on. Coppicing or pollarding of trees can change the whole character of a woodland for many years to come. Human intrusion caused by shooting and trapping, or by over-eager naturalists (!), may result in a woodland community markedly different from that of a nature reserve in which conservation is practised and life left undisturbed. Interesting ecological studies and surveys may be carried out at all times of the year in our woodlands, one of the commonest and most easily reached habitats in Britain (see Part 3). There is always

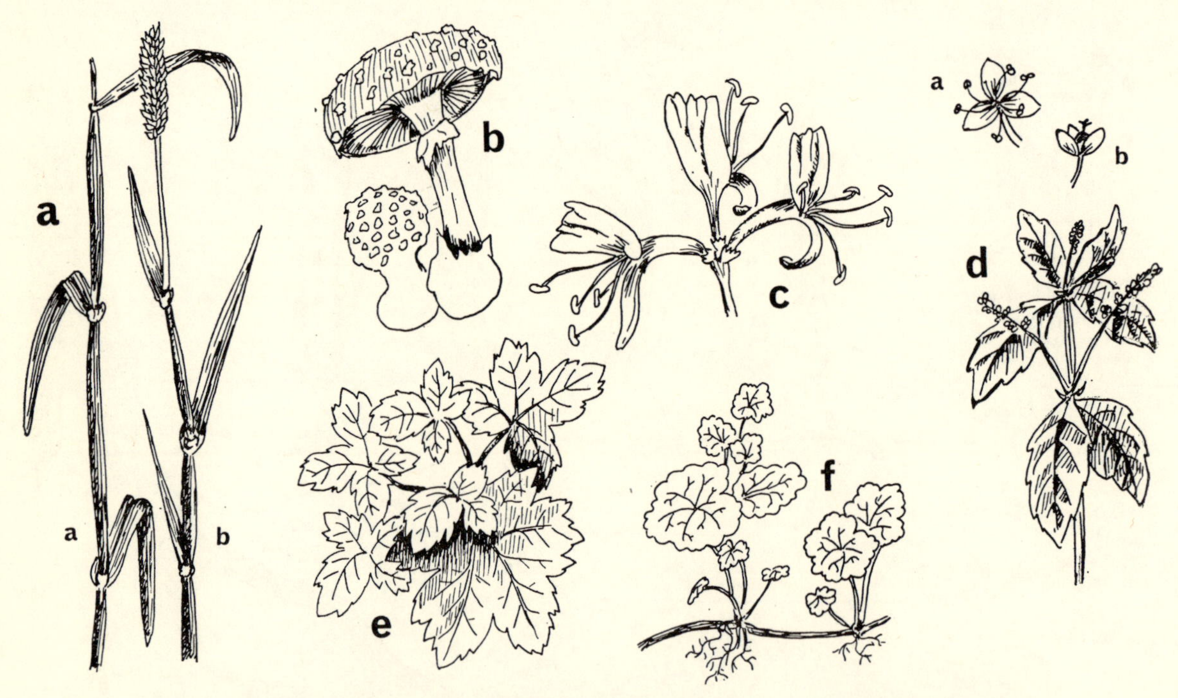

Fig. 27. Plant adaptations to shady places (woodland); A, wood soft grass in a, shade (note long internodes and reduced flower head) and b, in sun; B, a saprophyte (toadstool) lacking chlorophyll; C, honeysuckle with scented flowers and deep corolla tubes adapted to pollination by hawk moths; D, dog's mercury with inconspicuous scented flowers in shade a, male and b, female flower; E, leaf mosaic in sycamore; F, ground ivy, a creeping ground plant.

something doing. Where an area has been checked by felling, a fire, or by overgrazing, a project in woodland succession may be followed through from the initial bryophyte association to the final climax of trees.

A list of typical woodland plants in Britain is given in Appendix 2A (p. 212).

Woodland Animals

The shelter and food which is available in woodlands attracts a wide variety of animal life, perhaps more so than anywhere else in Britain. Woodlands are the country's natural climax and animals from greatest to smallest form part of the community. The list given in Appendix 11B on page 220 was selected from the Epping Forest project mentioned on page 169, and illustrates the kind of food sharing which occurs on and below the trees. The vertebrates are chosen for this project, whereas the more numerous and less easily noticed invertebrates are discussed in the section on the litter fauna on page 64, and in the oak tree Project 9 on page 171.

Forestry

Although Britain's climate and soil can support a plant climax in which deciduous trees are the dominants, little today is left of these natural woodlands. Because of the widespread destruction of native trees, also the establishment of many aliens in forestry, it is often difficult to work out the normal woodland climax of any given area. Even soils have changed in places where a higher fertility once prevailed under the cover of trees, areas which have for many years been exploited in the grazing and cropping activities on cleared land.

Much of the land today is in an artificial state of growth, particularly in the south and south-east, and this will tend to increase as the demand from a growing population rises. Because of this demand it has become a practice to grow timber on those unproductive soils which are unsuitable for agriculture. These are the areas called 'waste' lands in the economic sense, and usually occur on rocky slopes, on outcropping hard rock, or on infertile sands and gravels. They are covered by a turf of heath or chalk plants devoid of a permanent crop of trees. This may be because of earlier felling without any subsequent regeneration, or because the trees are kept in check by animals (e.g. sheep and rabbit) by firing (e.g. grouse moor) or by movement of people (e.g. a common).

The tree species which should ultimately grow best in the replanting programme is determined largely by two factors—the soil moisture content, and its fertility.

Forestry's aim is to produce timber for the nation, and over the years this has had a strong influence on the countryside and its wildlife. Natural historywise, this may be deplored, in that a former habitat is completely altered by the introduction of some alien tree. Take as an example the former Breckland of the Norfolk and Suffolk border. For centuries this was a barren waste of heathland on a dry and unproductive soil with the lowest annual rainfall for the whole country. Between the two world wars it was transformed into a large forestry plantation, mainly of Corsican pine, called Thetford Chase. Today relicts of the former breckland community can be found in the occasional plant, such as the Breckland Catchfly (*Silene otites*) or a bird like the Norfolk 'thicknee', a migrant plover seen nowhere else in the British Isles. On the other hand the increase in the red squirrel and roe deer, and frequent occurrence of the winter crossbill can be put down to the presence today of large stretches of conifers.

Such conditions where the original flora and fauna still survive in company with new arrivals are of interest both to forester and naturalist. A study of such areas can sometimes reveal the kind of community which once existed, and which is 'natural' to the soil. Dr Mark Anderson of the Imperial Forestry Institute at Oxford has prepared a classification list of seventeen wasteland communities for Britain, naming the climax tree in each case. The following is a brief summary (see also Bibliography):

Dry grass herb community. Dry areas rich in lime and soluble matter, e.g. downland and limestone hills, also East Anglian brecks. A chalk sward containing calcicole herbs. Cocksfoot grass (*Dactylis glomerata*) and sheep's fescue (*Festuca ovina*) with beech as dominant. Can support Austrian and Scots pines.

Moist grass herb community. In valley bottoms on chalk or fertile porous sediments, e.g. Old red sandstones. A regular water supply. Ash, sycamore and elm.

Grass-rush community. Heavy soils with a little peat or humus, rich in minerals. Water-impervious over coal measures, heavy boulder-till soils and Silurian shales. *Deschampsia* grass and *Juncus* rush in herb layer with alder and ash.

Rush community. Wet areas favouring the accumulation of peat. *Juncus* common and nearly pure in local hollows. Willows, and when drained, Sitka spruce.

Dry-grass community. On more basic well drained rocks and slopes, also on dry limestone. A few calcicoles but *Brachypodium sylvaticum* (a moss) predominates. Ash and beech, with introduced Japanese larch.

Fern community. Productive, moist soils of fertile boulder tills along river valleys. Mainly oak and Douglas fir, with larch and pine on steeper slopes, especially in north. Herb layer of ferns a feature, e.g. *Dryopteris dilatata*, *D. felix-mas*, *Athyrium felix-foemina*. The ubiquitous bracken often masks such communities, and should be ignored.

Sedge community. Localised on fertile clay patches in hilly places associated with streams and springs. Alder and willows, with spruce after drainage.

Molinia community. Peaty areas on hilly districts over Silurian schists and Carboniferous limestones, mainly in hollows or on lower slopes receiving water. Almost pure moor grass (*Molinia*) well suited to turf planting, with Norway and Sitka spruce after drainage.

Grass-heath community. Local on loamy, dry soils on slopes devoid of drift material. Mainly on silurian hills and Highland slates supporting a *Calluna-heath community*, with *Deschampsia* and *Festuca*, also the mosses *Hylocumium squarrosum* and *Polytrichum commune*.

Nardus-molinia community. Non-fibrous, peaty soils less than 6 inches deep on silurian, schistose and limestone slopes, well supplied with water. Berries like *Vaccinium myrtillus* may occur. Good for Sitka and Norway spruce when drained.

Cotton-grass community. On upper slopes and plateaux above the last community with cotton grass dominant. Less fertile soil with some *Molinia* and *Juncus squarrosus*. Best for Sitka spruce.

Calluna-heath community. Dry, infertile soils, typical of grouse moors. *Calluna* with a bottom layer of *Hylocomium*, although the mosses are mainly burned off. Ideal for Scots pine with birch intermixed. This is forest land rather than pastoral or agricultural.

Vaccinium community. Associated with the natural Scots pine woods of the West Central Highlands, on coarse moraine. Will degenerate into a *Calluna* moor if cleared of trees.

Myrica community. On fertile hillsides with high rainfall. Contains fibrous peat and a sparse *Calluna* cover. Useful for Sitka spruce after drainage.

Lichen community. A dense bottom layer of grey lichens with sparse *Calluna* cover, locally on granites of North-east Scotland, and leached sandstone of North-east England. May be the result of excessive moor fires. Unplantable and infertile, except for pines.

Erica-tetralix community. The Cross-leaved heath mixed with *Calluna* but little else on bare, exposed moorland on heavy soils. Impervious and moister soil than last community which favours heaths. Lack of minerals prevents much else. A difficult soil only suited to pines with some birch or alder.

Calluna or *sphagnum-moor community*. This is not the same as the *Calluna* heath above. *Calluna* dominates with an underlayer of red *Sphagnum acutifolium*. Bog Asphodel (*Narthecium*), *Erica tetralix*, *Molinia*, *Scirpus caespitosa* and Cotton grass may also occur. After burning the *Scirpus* and *Molinia* recover first, then the *Calluna* (see Epping Forest Project 5, p. 161). Fibrous peat occurs to a depth of 15-30 feet on a so-called High moor. After burning may resemble a *Molinia* moor, but is unplantable even with drainage or turf-planting. Lodgepole pine does best, with Scots pine a poor second.

NOTE. In making out this list the author warns against too much emphasis being placed on certain plant indicators which can mislead as they tolerate a wide range of site conditions. Examples are bracken (*Pteris*), ling (*Calluna*) and purple moor grass (*Molinia*). These can sometimes dominate a community. In places one meets examples of vegetation which do not readily fit into any of the above categories. It may then help to search for the more exacting, and perhaps fewer, specimens of plants which could be the relict species of the former community before it was interfered with by man. To improve conditions foresters will often carry out drainage, apart from adding fertiliser to the soil.

Part 2: Section B

Chalklands

Much of England south-east of a line from Dorset to Yorkshire is covered by a soft white rock known as chalk. This was originally the bed of a shallow sea laid down during the late Cretaceous Period (Chalk Age) between 70 and 90 million years ago. It now forms the familiar landscape of curved hills and valleys of the Downs and Wolds, and is covered by a short, springy turf grazed by sheep, or arable farmland under the plough. The ridges which criss-cross the downlands are usually topped with strips of woodland, mainly of beech trees, called hangers. Where a ridge meets the coast, tall white cliffs are exposed, as at Dover, Beachy Head and Flamborough Head.

Originally a sea-bed of soft ooze, the exposed chalk was later subjected to folding by severe earth movements in southern Europe. These produced the Alpine Chain of mountains, from which shock waves can be traced across Europe into Britain, in the form of more gently folded rocks. Further sea deposits then filled in the 'troughs' of chalk so that only the much eroded 'crests' of each wave remain exposed today as escarpments.

In Britain the Chalk has its main centre on Salisbury Plain in Wiltshire (Fig. 28). From this radiate five main chalk ranges as shown on the map. This open, pleasant and undulating countryside is popular with holiday-makers and ramblers, and is of special interest to botanists since our richest flora occurs on chalk soils, including some of the rarest of wild orchids.

The Downs were first occupied by the earliest invaders of the New Stone Age. Traces of their burial mounds and encampments are visible along the hilltops. To make rooms for their dwellings, animals and crops, these Neolithic farmers cut into the woodlands. It is this activity of forest clearance which has given the present atmosphere to the Downlands, an open, largely treeless landscape in which the former woodlands have been held in check by later farmers and their animals, especially the grazing sheep. Since Norman times the rabbit has also played its part in arresting the return of the trees, and has rightly been called a landscape gardener.

A major factor which determines the presence of chalk plants is the Chalk itself. A large number of British plants have a preference for soils containing a percentage of lime, and are known as calcicoles or chalk-lovers. Some species occur nowhere else, and can be used as indicators of chalky soil. The principle ingredient is calcium carbonate ($CaCo_3$) which may be as high as 90 per cent. in the purer chalk. In some places foreign matter becomes mixed with chalk, or the chalk content is weakened by leaching. It becomes washed out by rainwater, especially on sloping ground. In other places the Chalk is overlaid by a later deposit, as in the case of the Boulder Clay in East Anglia (Fig. 28). This was the result of glacial action during the Ice Age. Stones of flint are a common feature of chalk, and are one reason why Stone Age Man settled there. Flint is easily worked into tools, and these implements as well as flint fossils of sea-urchins from the Cretaceous Sea are commonly found on the Downs.

Chalk is a soft and porous rock which becomes weathered into rounded contours. Where it is exposed it readily absorbs rain water. It is well aerated and soon warms up in the sun. Standing water is usually scarce, unless collected in special man-made dew-ponds. Springs and rivers arise where the Chalk meets an underlying bed of harder rock, such as clay or sandstone.

The biological influence of man and grazing animals have an important effect on chalk ecology, in preserving the grassland stage, and preventing a return to former woodland. If left undisturbed a patch of bared chalk will slowly return by stages to a climax woodland—bare chalk with early colonisers (mainly mosses and annuals)—grass turf with perennial herbs (the chalk flowers)—chalk scrub (bushes and climbers)—chalk woodland. All stages in this biosere can be found during a walk over the Downs. This natural succession, and the fact that the Chalk once supported a wooded climax, is shown by present events. Since the loss of rabbits due to the virus disease of myxomatosis, as well as the removal of sheep from many areas,

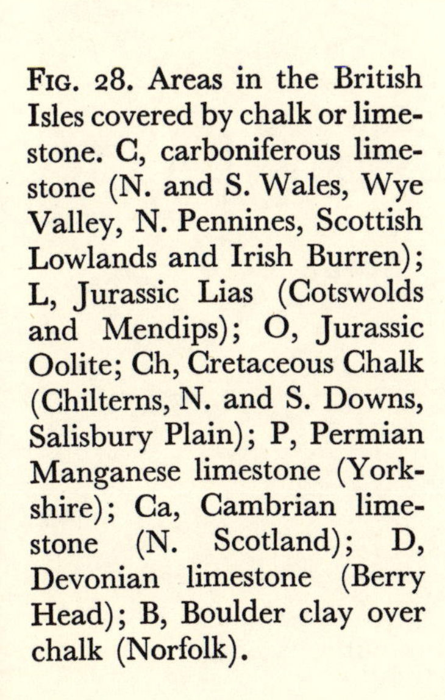
FIG. 28. Areas in the British Isles covered by chalk or limestone. C, carboniferous limestone (N. and S. Wales, Wye Valley, N. Pennines, Scottish Lowlands and Irish Burren); L, Jurassic Lias (Cotswolds and Mendips); O, Jurassic Oolite; Ch, Cretaceous Chalk (Chilterns, N. and S. Downs, Salisbury Plain); P, Permian Manganese limestone (Yorkshire); Ca, Cambrian limestone (N. Scotland); D, Devonian limestone (Berry Head); B, Boulder clay over chalk (Norfolk).

the grassland stage is slowly reverting to scrub, even with young trees appearing here and there. This is a rebirth of chalk woodland. On the Downs and Wolds a number of habitats may be encountered, each one a part of the above biosere, and each supporting its own peculiar flora (Fig. 29).

Bare chalk. This is found on arable land where the plough maintains a permanent first stage of annuals which set their seeds in the previous season, or are carried in by wind and birds. Such pioneer colonisers are also found on bared chalk resulting from fires, turf cutting, or in fresh diggings in chalk pits. If left alone this primary stage will slowly change to grassland.

Chalk grassland. It is in this second stage that the greatest variety and numbers of chalk plants are found, growing as perennial herbs in a thick turf mat of grasses. Due to the exposed conditions in such surroundings the chalk plants tend to grow in rosette form, or as creepers, hugging the ground where the air is kept moist by evaporating dew. Some species check their transpiration rate by having modified leaves which are reduced in surface area, have thick cuticles, or are hairy (see Xerophytes, p. 87). Although the porous chalk beneath the turf layer quickly absorbs rainwater, this is as readily drawn back to the surface by capillary action. Some of it is retained by the spongy turf mat, and available to the surface annuals. Many perennials, however, grow deep roots to an average depth of eighteen inches in order to tap the lower water-level. Incidentally, it is the thick root-mat of chalk turf which gives its springy character (Fig. 30).

Chalk scrub. Where grassland is left undisturbed by the plough and grazing animals, a third stage of dense scrub of bushes and climbers will grow up. This is often a feature of the steeper slopes of an escarpment where the plough cannot operate, as also on the sides of old earthworks. A cover is given to tree seedlings and a woodland may eventually result.

Chalk woodland. Apart from woods which may be privately owned, the semi-natural beech

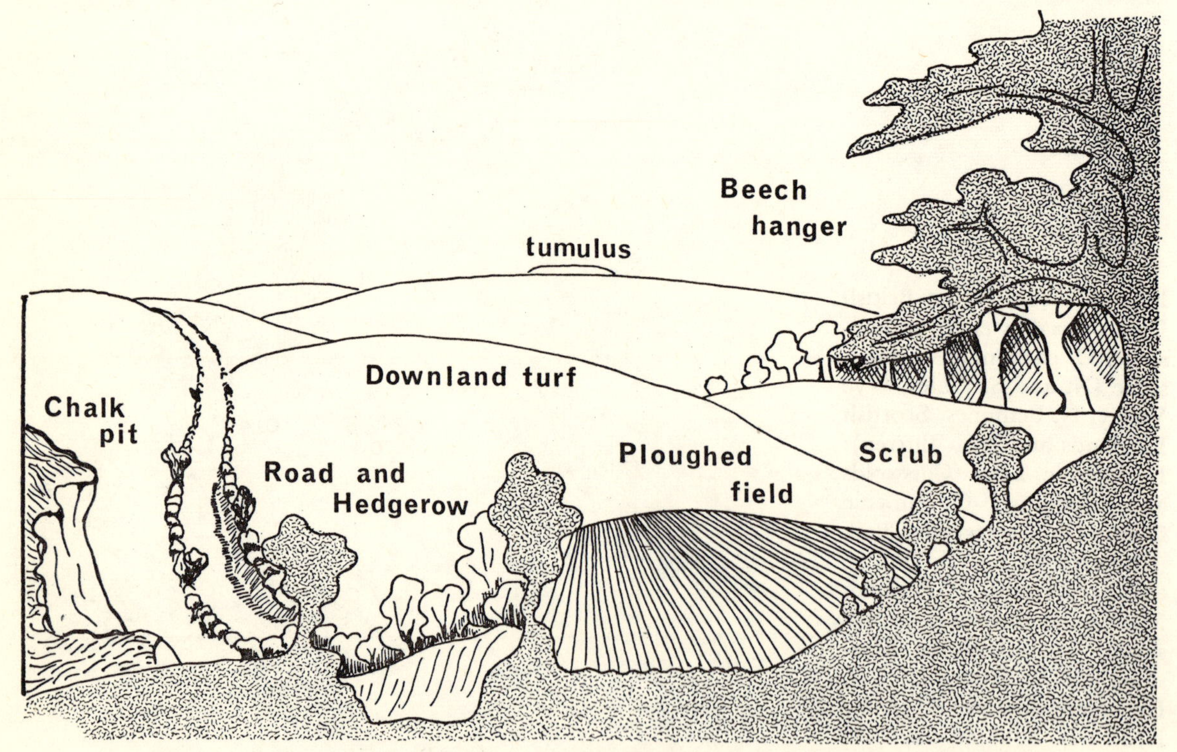

FIG. 29. Chalk downland habitat showing different plant communities (see text).

hangers along the escarpment tops are a common feature of the Chalk. Where the branches join above to form a closed canopy in the beechwood, the final stage is reached in the Chalk biosere. In the heavy shade of a beechwood the ground flora is limited. Accumulating leaves tend to make the ground acid, and together with the lack of light, these factors can mask the presence of the underlying chalk. In fact, where humus collects in hollows, acid-loving plants such as foxglove (*Digitalis*) and heather (*Calluna*) may actually be found in association with nearby calcicoles such as the white helleborine orchid (*Cephalanthera*) and clematis. In modified form a hedgerow bordering a field or road which crosses the Chalk,

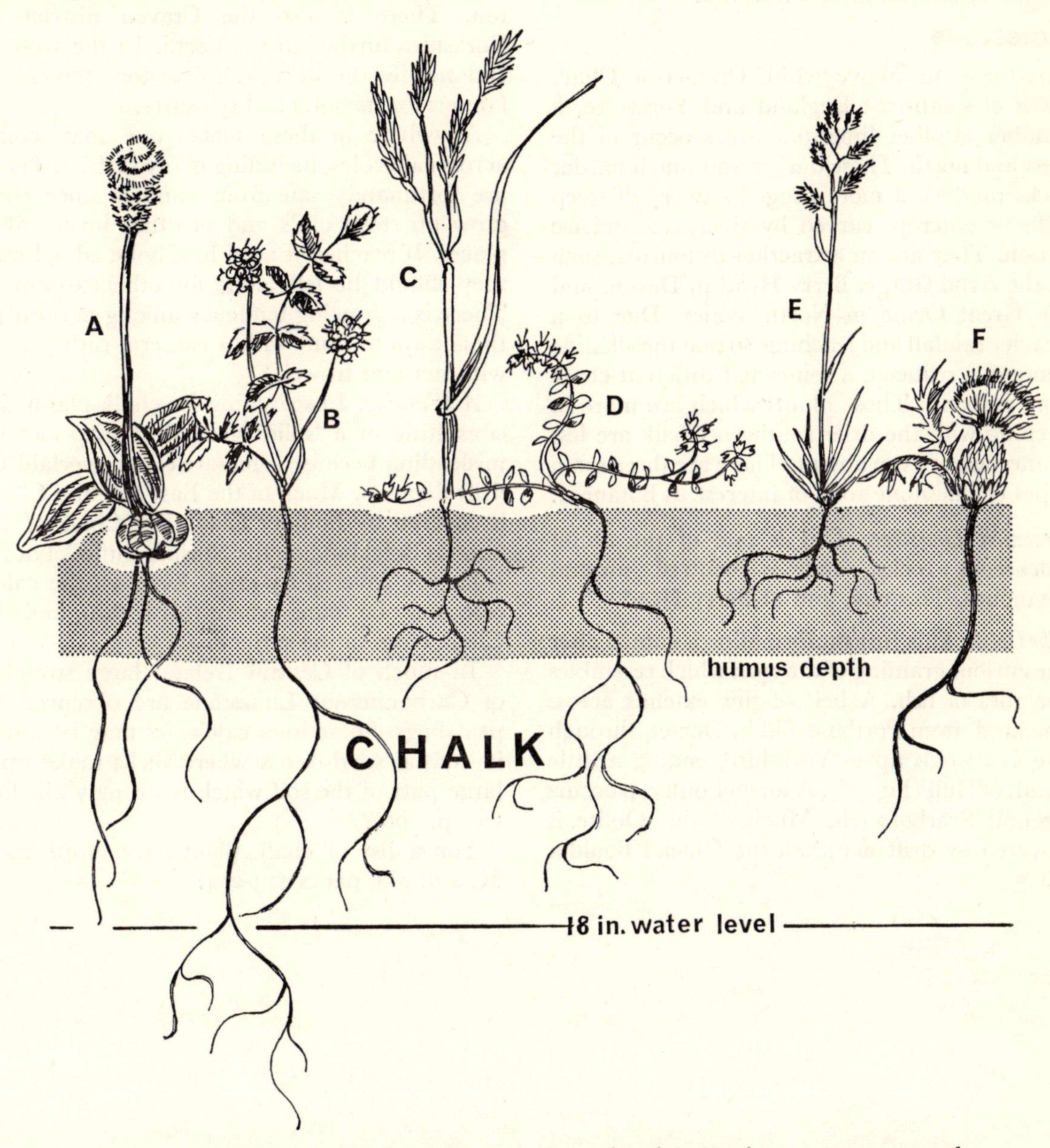

Fig. 30. Layering of roots in chalk turf. Note also the creeping or rosette growth above ground. A, hoary plantain; B, salad burnet; C, upright brome grass; D, wild thyme; E, sheep's fescue grass; F, stemless thistle.

especially if a ditch is present, can harbour a rich chalk woodland flora.

Chalk pits. A derelict chalk pit which remains undisturbed may prove to be a treasure trove for the seeker after rarities, since it may conceal orchids and other vulnerable plants. Such a place is usually kept secret by conscientious botanists, and carefully preserved.

Limestone

Apart from the above 'white' Cretaceous Chalk areas of south-east England and Yorkshire, a number of older limestone areas occur in the west and north. These darker and much harder rocks produce a more rugged scenery of steep cliffs or outcrops carved by river, sea and ice action. They are an attraction to tourists, such as the Avon Gorge, Berry Head in Devon, and the Great Orme in North Wales. Due to a heavier rainfall and leaching, so that the alkaline content is reduced, a somewhat different chalk flora is found. Those plants which are more in keeping with the drier southern Chalk are less numerous, or even absent. There are three main types of limestone rock of interest to botanists:

Devonian Limestone. This is one of the oldest calcareous rocks, and is centred around Plymouth, Torquay, and at Berry Head.

The Oolite. This Jurassic formation is named after the curious graining of the rock which resembles the roes of fish. A belt of this extends across England, from Portland Bill in Dorset, through the Cotswolds up to Yorkshire, ending a little south of Hull (Fig. 28). A further outcrop occurs around Scarborough. Much of this Oolite is covered by drift material, the Glacial Boulder Clay.

Carboniferous Limestone. This splendid scenic countryside contains a number of beauty spots of steep grandeur, such as the Cheddar Gorge in the Mendips, the Avon Gorge near Bristol, and the Great Orme in North Wales. Other areas include the Gower coast and Tenby Cliffs in South Wales, the Wye Valley, and the Derbyshire Dales between Matlock and Castleton. There is also the Craven district of Yorkshire further to the north. In the west of Ireland lies the Burren, a limestone pavement famous for its botanical treasures.

Anywhere in these places one may come across calcicoles, including many rarities. These are fortunately safe from vandals since they grow on steep cliffs and in other inaccessible places. If reached it need hardly be added that they should be left alone for others to enjoy. There is a growing tendency among naturalists these days to hunt with a camera, rather than with net and trowel.

A Warning. In searching for chalk plants the consulting of a 'solid' geological map can be misleading because some areas are overlaid by drift deposits. Much of the East Anglian Chalk is covered by Boulder Clay. The Chalk of the famous Breckland on the Norfolk-Suffolk border is largely covered by sand. Even so, the calcicoles may be found where this is thin, and the Chalk close to the surface.

In much of Central Ireland large stretches of Carboniferous Limestone are obscured by peat bogs. Sometimes calcicoles may be found in coastal sand-dunes where shells make up a large part of the soil which is strongly alkaline (see p. 100).

For a list of chalk plants, see Appendices 2C and 2D, pages 213-214.

Part 2: Section C

Heath and moorland

By definition a heath is land dominated by members of the Heather family. In Britain these plants usually grow on infertile soils, acid in quality, which occur in the more open places, and at different altitudes. On soils of a heavy and non-calcareous nature formerly occupied by oak woodland, a scrub of grasses and low-growing herbaceous perennials, mixed with prickly shrubs such as gorse, hawthorn, bramble and rose may be found. Such communities occur on the open spaces of heath-like aspect known as commons. These have been preserved as recreational areas, or as grazing areas for the commoners' cattle and ponies. Although the prickly plants are fairly safe from these animals, the grass is steadily grazed, and this helps to check the succession which might lead to woodland. Periodic firing, as practised on the grouse moors in mountains, also helps to maintain the open nature of a common.

On the more sandy and gravelly soils which are poor in mineral content, due to the underlying base-deficient rock, leaching and rapid drainage (e.g. on some mountain summits), a true heath will grow. The ling or heather (*Calluna*) is a dominant plant, and may form a continuous carpet on old heaths. The deep shade cast by the branching, woody ling keeps out most competitors, except in the open gaps and boggy places. Co-dominants are the acid-tolerant and low-growing shrubs, such as dwarf furze (*Ulex nana*) and whortlebery (*Vaccinium*). Some heathland species prefer the drier, raised areas, such as the bell heather (*Erica cinerea*), whereas the cross-leaved heath (*E. tetralix*) keeps more to the wetter places (Fig. 31). Bog moss (*Sphagnum*) is a sure indicator of places to avoid. Here, a community of bog plants may be found rather similar to those in a dune-slack (see p. 101). A distinct reddish colouring seen from a distance may reveal a whole carpet of insect-catching sundew (*Drosera*).

Due to the shallowness of the peat layer, a heath absorbs large quantities of rainwater, but loses this as readily by evaporation in dry spells and from wind. Below, the sand or gravel remains dry. In consequence most heath plants show xerophytic adaptations against transpiration (Fig. 32). Some have their stomatal pores lying in grooves (heathers), others have reduced leaf surfaces (spines in gorse and reduced leaves in broom), or thick cuticles (whortleberry). Many species remain evergreen, and have a tough branching system to withstand wind and desiccation. Typical heathlands belong to the south, and are scattered throughout parts of Surrey, Hampshire and Dorset on the Bagshot sands of the Hampshire

Basin. On many of these the Scots fir and birch have become established, either as escapes or in plantations. The more natural habitat of such trees is in the mountains to the north (see Mountains, p. 76).

In areas of high rainfall a heathland may eventually turn into a moorland (e.g. in Yorkshire, Devon, Wales and Scotland). Such upland moors have arisen on basic, mineral-deficient soils once dominated by woodland, and traces of the former trees, especially birch, can be seen embedded in the peat. This may be of considerable thickness, giving the soil a greater capacity for holding water. The rapid growth and decay of moorland plants, such as bog moss and cotton grass (*Eryophorum*) adds continually to the peat layer which may be up to 15 feet or more deep on some moors.

The high acidity of moorland water, often stained brown with peat, limits the number of species to a monotonous repetition of heathers, heaths, berry plants, bog-moss, cotton-grass and lichens. Xerophytic foliage is a common feature, a little unusual perhaps for such wet places.

FIG. 31. A Surrey heath. A, Drier slopes containing a, heather (*Calluna vulgaris*) and b, bell heather (*Erica cinerea*). B, Wetter hollows with c, cross-leaved heath (*E. tetralix*); d, bog asphodel (*Narthecium*) and e, cotton grass (*Eriophorum*). C, Heath bog with f, sundew (*Drosera*); g, bog myrtle (*Myrica*) and h, marsh pennywort (*Hydrocotyle*). D, Heath pool containing i, bladderwort (*Utricularia*). E, Introduced pine (*Pinus sylvestris*) along ridge.

There is a possibility that the cold and poorly aerated soil, together with drying winds, lowers the rate of root absorption, so that a relatively low amount of transpiration becomes important. An interesting character of some moorland plants, especially the heathers, is the absence of root hairs. Their function is taken over by fungal threads, called mycorrhyza. These also occur in the Scots fir (*Pinus*) which can exist on poorer soils, and often forms part of a moorland landscape.

A glance at moorland vegetation will give a clue to the type of soil. Dry, shallow soils to a depth of about 4 feet will support a heather moor. Although acid the soil is fairly high in mineral content. In deeper but less rich peat whortleberry may be dominant. In very deep peat which is almost pure humus and lacking in minerals, and often very wet, the cotton grass takes over. The xerophytic nature of this sedge, its narrow leaves with thick cuticle and numerous air canals, enables it to grow in waterlogged peat poor in minerals.

Sometimes it may come as a surprise, when testing a peat soil, to find it is alkaline instead of acid. This may be due to its origin. An acid moor develops from a former woodland over a mineral-deficient rock, whereas an alkaline moor originates from a fen situated over or near to a calcareous rock. Such fens occur in East Anglia and the New Forest.

In Ireland there exist many ancient moorlands, some on calcareous bedrock, on which are found raised bogs appearing as low and rounded hillocks. In Wales, at Cors Tregaron Nature Reserve near Aberystwyth, a raised bog exists on the bed of an old glacial lake. A river still passes through it. Gradual filling of this lake with peat has resulted in a low hill about 25 feet high at the centre.

Such a raised bog builds up in the following manner. In succession a thick bed of *Sphagnum* moss grows up, which then breaks down into a humus to provide anchorage for a heather bed. In turn this dies down to form more peat which holds moisture like a sponge. The moss returns and the whole process is repeated (Fig. 33). A study of pollen grains found at different levels in such a bog gives some idea of former climates. Pollen grains are well preserved in peat, and many have a characteristic appearance which makes identification of the species possible. The presence of this or that species in each peat sample at different levels will give some indication of past climates. Thus, a dominance of grass pollen grains would suggest a cold tundra type period, alder a wet period, birch sub-arctic, and so on. In some places, notably Denmark, traces of cereal crops have given information about farming practised by Neolithic man (Fig. 33).

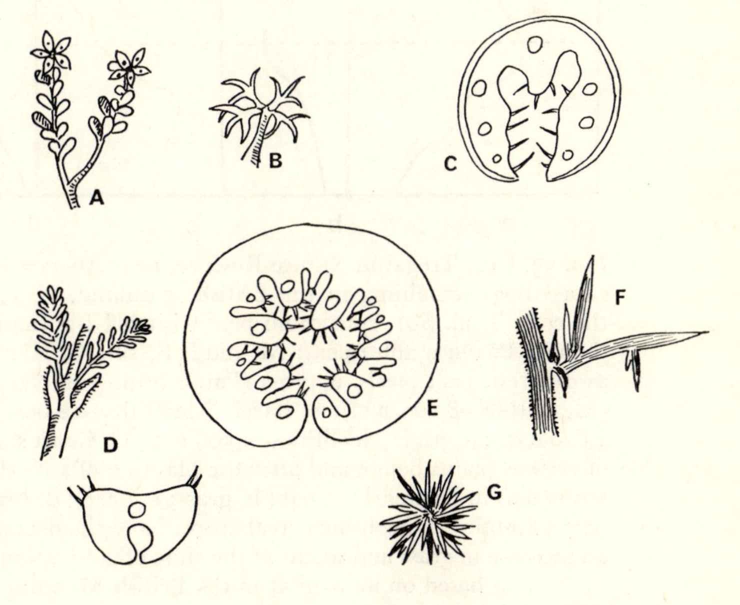

FIG. 32. Xerophytic plants showing adaptations to dry surroundings by means of cutting down transpiration. A, stonecrop (*Sedum*) with succulent leaves (in sand dunes); B, branched hair of lavender (sea coast); C, cross-section of rolled leaf of sheep's fescue grass (*Festuca*) on chalk downland; D, Heather (*Calluna*) rolled and reduced leaves with cross-section; E, cross-section of rolled leaf of marram grass, *Ammophila* (sand dunes); F, reduced leaves of gorse, *Ulex* (heath); G, peltate hair of sea buckthorn, *Hippophae* (sand dune).

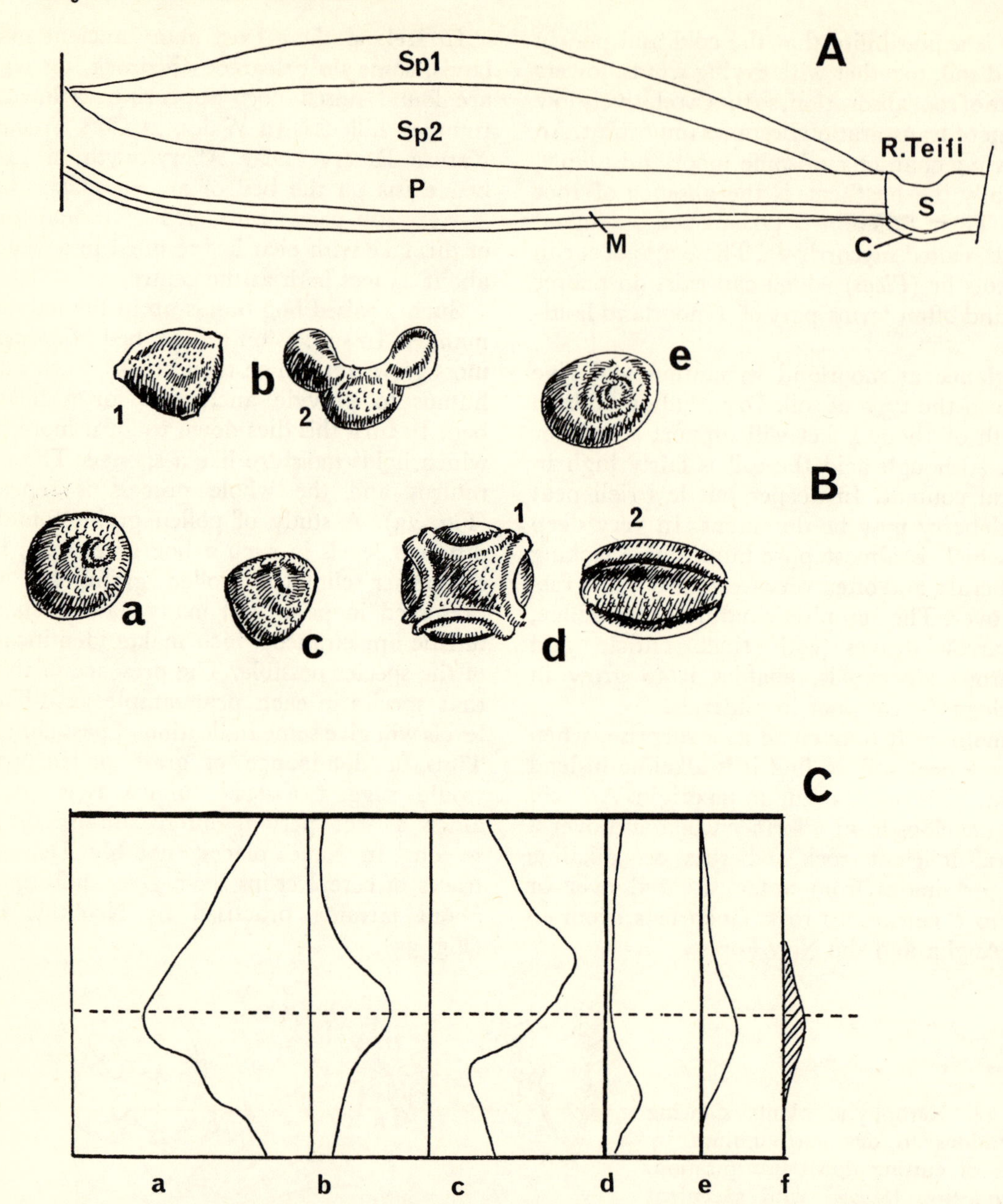

FIG. 33. Cors Tregaron Nature Reserve, near Aberystwyth, N. Wales, one of the few raised bogs remaining in the British mainland. A, cross section of bog including the river Teifi. Sp1, sphagnum peat with pH less than 5; Sp2, peat with pH more than 5; P, phragmites peat; M, mud; S, silt; C, lake clay; B, pollen grains found at different peat levels. a, grass (Palaeolithic—10,000 B.C.); b1, birch and b2. pine (Mesolithic—8,000 B.C.); c, hazel (Mesolithic—6,000 B.C.); d, alder (Mesolithic—4,000 B.C.); e, grass (Neolithic—2,500 B.C.) C, Chart showing the relative abundance of various plants before and after the planting of a Neolithic cereal crop by a shifting settlement in Denmark. a, oak; b, grass; c, hazel; d, bracken; e, weeds of cultivation (e.g. plantain); f, neolithic cereal crops. Note the decrease in trees and bracken with an increase in grass and weeds at the time of cultivation (along dotted line). Sketches based on an exhibit in the British Museum (Natural History).

Moorland Animals

Animal life of moorlands and heaths consists of creatures which either avoid man or seek a home in open places. Ground-nesting birds or those which nest in low bushes (e.g. grouse, curlew and whinchat) find their food on the heath plants or in the soil. Others hunt from the air, and require space in which to fly (e.g. short-eared owl and merlin). Small mammals may be abundant in such places (e.g. field-vole) or use them as retreats from enemies such as man (red deer). Reptiles, the few that we possess, are found mostly in this kind of habitat, and include the rare smooth snake (*Coronella austriaca*) known only on the Bagshot heaths in southern England. The viviparous lizard (*Lacerta vivipara*), the only reptile native to Ireland, is mainly a heathland animal. The abundance of invertebrates, such as spiders, form its prey, and the lizard in turn is hunted by the smooth snake or adder. A typical insect of our heaths and moors is the handsome emperor moth whose caterpillar feeds on the heather. A list of heath and moorland animals and plants is given in Appendices 2G and 2H, p. 215.

Part 2 : Section D

Sea coast

Britain's coastline presents a number of habitats whose features are sufficiently distinct to support individual and separate plant and animal communities. In some places these areas are quite isolated, but in others they may merge one into the other. For instance, in one area a sand-dune grades into a heath. In another, a shingle bank will separate the open sea from a salt marsh, and this in turn grades into a salt pasture with grazing sheep on the landward side. A good area for investigating the different coastal habitats is in the region of Poole harbour and Swanage on the Dorset coast (Fig. 34).

Such plant succession is particularly noticeable on the flatter coastlines where land is gradually encroaching on the sea, as along the Norfolk coast. In contrast the Atlantic side is largely composed of tall, solid cliffs which fall directly into the sea.

Britain's coastline is made up of three main materials, each supporting its individual community life—rock, sand and mud.

Rocky Coast

The more spectacular sea-cliffs of Britain occur along the coast of Scotland (where the Highlands reach the sea), in the West Country (Devon and Cornwall), Wales and much of Ireland. Here many of the rocks are of a hard and rugged quality, such as granites, slates, basalts and limestone. The coast ends abruptly in high cliffs with small, often inaccessible coves in between. Other tall cliffs found more on the east and south coasts are made of softer rock where an escarpment of chalk reaches the coast (e.g. Beachy Head). Here the tall headlands jut into the sea, and there are wide bays in between frequented by holiday-makers.

On Britain's coastal cliffs grow plants which can withstand exposed conditions and salt spray, finding what shelter they can between cracks or on narrow ledges which they share with the cliff-nesting sea birds. Long roots are sent down after water and anchorage (see also Alpines, p. 107).

In some places, particularly on calcareous cliffs of limestone (e.g. the Gower peninsula and Great Orme Head, both in Wales, and the Burren in West Ireland) a veritable rock-garden may blossom in early summer. Here the sea pink (*Armeria maritima*) vies with the sea campion (*Lychnis maritima*), rock-rose (*Helianthemum nummularium*) and bloody cranesbill (*Geranium sanguisorbum*) for colour. Brightly coloured lichens add to the scene. In more inaccessible places rarities may lurk, such as the maidenhair fern (*Adiantum*) or the semi-parasitic broomrape (*Orobrache*).

The sea-cliffs are the temporary summer home of thousands of nesting shore birds which use the ledges and cliff tops, often concentrating in large colonies (Fig. 35). Far below, on the stony beaches and in the caves, the Atlantic seal suckles her pup in late summer. Many of these die from drowning during autumn gales. Waders and gulls forage along the shore-line, joined by the scavenging fox, rat, crow, jackdaw and wandering otter.

It is between the tidal reaches that, paradoxically, the greatest assemblage of animals may be found. Periodically this area is inundated, then exposed with the changing tides. Each rock pool harbours a natural aquarium of marine life which somehow survives the two great hazards of tidal movement—the beating waves, and exposure at low tide which may lead to desiccation. Many cling tightly to the rock face in this uneasy world. Familiar examples are the limpet, barnacle, mussel, various sea-snails and sea-anemones (Fig. 36). The risk of drying out is minimised by retiring under some kind of shell which either closes up or clamps to the rock. Alternatively, there are damp hiding places for shelter and safety among the sea-weeds, or between cracks in the rock. Such a corner may harbour a crab, an octopus or blenny. Others seek hiding by burrowing into the exposed mud or sand, especially along the flatter shore-lines where this soft material builds up. Various marine worms make tubes, and betray their presence at low tide by the worm casts. Certain crabs, molluscs (e.g. cockle, gaper and razor-shell), starfishes and sea-urchins, burrow out of sight as the tide recedes. All this is very necessary at such a time when shore birds are on the lookout for a meal.

The flora of the tidal area is composed almost entirely of algae (sea-weeds) whose colour to some extent determines their position on the beach (Fig. 37). Depending upon the length

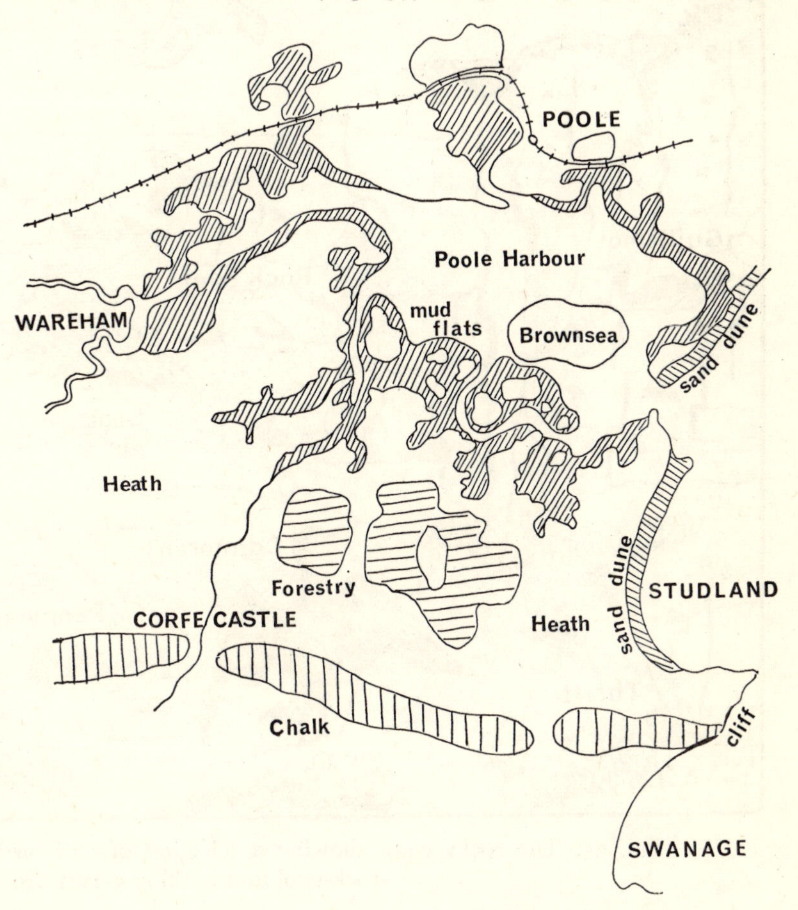

FIG. 34. Swanage and Poole Harbour, Dorset. A good botanical centre within easy reach of a number of plant habitats such as woodland, both conifer and deciduous, heath, saltmarsh, sand dune, chalk, river and cliff. The shingle habitat of Chesil Bank is not very far away. (See Fig. 39).

of the colour-waves and their power of penetrating water, the brown sea-weeds go deepest, next come the 'reds', and the green sea-weeds occur at the top of the beach. Some, especially the wracks (*Fucus*), can be used as indicators to the various zones. The slimy nature of these sea algae is due to an abundance of mucilage which helps to retain moisture, and also protects the plant during exposures which can last for many hours. To maintain their position on the shore sea-weeds cling to the rocks and stones by means of a hold-fast which takes the place of the root in a land plant.

As with the plants, there are zones of animals which are associated with them. The various species of periwinkle (*Littorina*) make useful indicators when zoning a shore-line. With animals, however, it should be noted that the occupants of a rock pool do not necessarily fit into this zonation. Such a pool is, in effect, a small portion of the sea itself which is held back at low tide. Life goes on normally here all the year round in such a natural aquarium.

Flat Coast

A study of zonation of coastal fauna is more clearly expressed along a gradually sloping shore-line which is covered by stones, sand and mud, rather than by rock (Fig. 38). Here the zonal communities are more clear-cut. Above

Fig. 35. The rocky coast showing a selection of cliff birds and plants. Note the rock-pool and basking grey seals.

FIG. 36. A selection of sea-shore animals and their habitats. A, dog whelk (*Nucella*), rocks and crevices; B, starfish (*Asterias*), rocks and sand; C, heart urchin (*Spatangus*) sand burrower; D, sea squirt (*Asidiella*), rocks and harbour works; E, mussel (*Mytilus*) rocks and piers; F, piddock (*Pholas*) in rocks and submerged wood; G, dahlia anemone (*Tealia*) rock crevices and seaweed; H, barnacle (*Balanus*) rocks and breakwaters; I, limpet (*Patella*) rocks; J, butterfish (*Centronotus*) under rocks and seaweed; K, top shell (*Calliostoma*) rocks. L, hermit crab (*Eupagurus*) crevices; M, shore crab (*Carcinus*) crevices and seaweed; N, shrimp (*Crangon*) sand; O, razor shell (*Ensis*) in sand; P, blenny (*Blennius*) crevices and seaweed; Q, flounder (*Pleuronectes*) mud or sand in bays or estuaries; R, gaper (*Mya*) in sand; S, ragworm (*Nereis*) in sand and mud; T, cockle (*Cardium*) sand and mud; U, periwinkle (*Littorina*) rocks and seaweed.

the high-tide mark, in loose shingle reached only by salt spray, lives a moving community of scavengers, such as the sea-slaters and sand-hoppers. These hide among the wet stones, avoiding the waves or dry places, and are common in the shore drift and debris cast up by the sea. Below this is a more stable shingle, mixed with mud or sand, and covered at high tide. Unlike the upper beach this supports a large population of animals of three kinds—stone dwellers (mussels and barnacles), browsers (winkles) and burrowers (bristle-worms). Still further down the beach, on what are called the sandy or mud flats, is an unstable zone of mixed sand and mud in which are found mainly burrowers such as bivalves (cockle and piddock), also burrowing worms. Exposed at low tide this part of the shore-line may be found covered with worm casts. Finally, at the lowest tide level the bedrock of that particular flat coast becomes exposed. This is usually one of the softer rocks such as clay or chalk commonly found along the south and east coasts of England. It is sufficiently stiffened at low tide to walk on. London Clay is a good example. Here, a number of surface dwellers can exist such as mussels, crabs and oysters, which are, in turn preyed upon by the wandering dog-whelk, starfish and piddock. The starfish is an all too familiar menace in mussel and

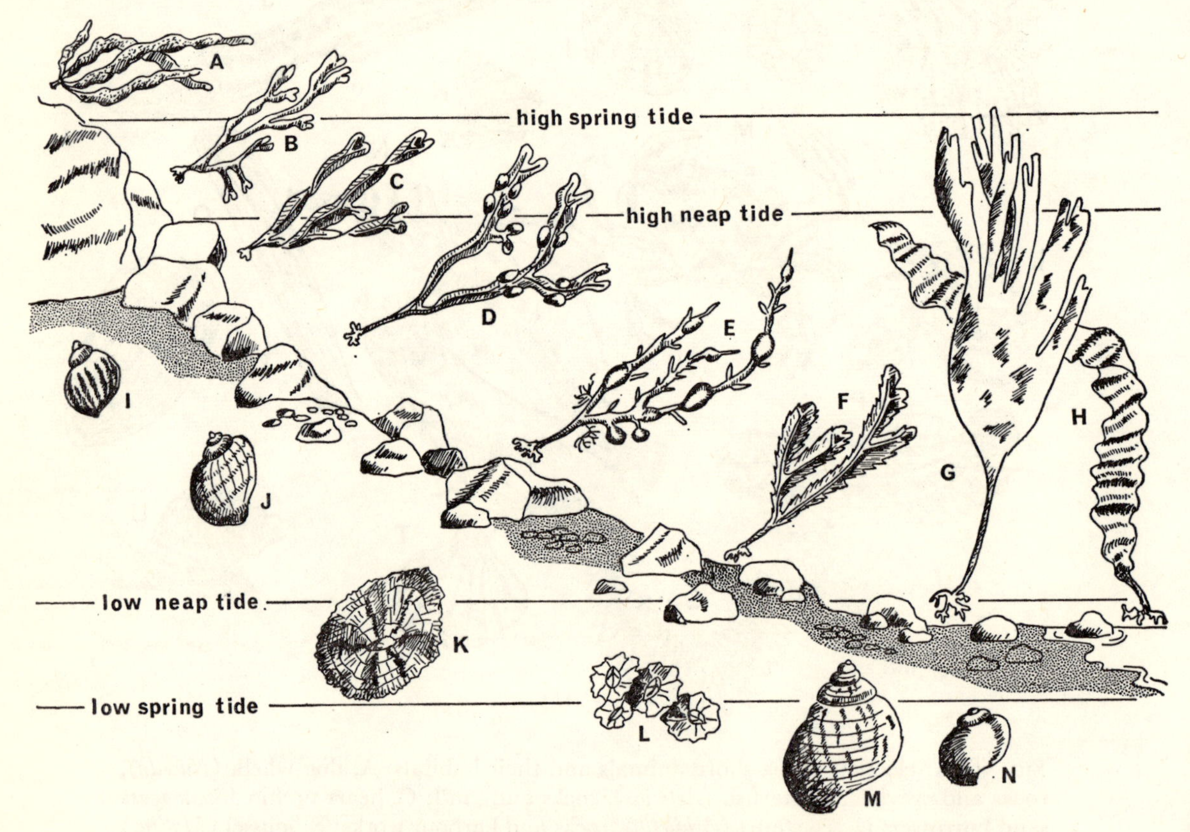

Fig. 37. Intertidal zonation of seaweeds, molluscs and barnacles on the rocky shore. A, *Enteromorpha intestinalis*; B, *Pelvetia canaliculata* (Channel Wrack); C, *Fucus spiralis* (Flat Wrack); D, *Fucus vesiculosus* (Bladder Wrack); E, *Ascophyllum nodosum* (Knotted Wrack); F, *Fucus serratus* (Serrated Wrack); G, *Laminaria saccharina* (Ribbon Oarweed); H, *Laminaria digitata* (Flat Oarweed); I, *Littorina neritoides* (Small Periwinkle); J, *Littorina rudis* (Rough Periwinkle); K, *Patella vulgaris* (Limpet); L, *Balanus balanoides* (Barnacle); M, *Littorina littorea* (Common Periwinkle); N, *Littorina littoralis* (Flat Periwinkle).

oyster beds. Sea-weeds will occur on these flat and 'soft' shores only where there happens to be anchorage provided by stones.

Shingle

Above the tidal zone on many parts of our coast the beach is composed of water-worn pebbles which have been thrown up to form a shingle beach. Originating as fragments which have fallen from eroded cliffs and rocks these have been built up as beaches, or as bars running parallel with the shore. Over the years the tides pile up more stones, and if the prevailing wind or sea current should carry the stones along the coastline, then a shingle spit will form (e.g. Chesil Bank, Dorset, Fig. 39). The stones lie in a graded series, deposited along the bank according to their size and the carrying power of the tides.

Barren and inhospitable as these shingle banks may at first appear, a special plant community found in no other area can still build up among the pebbles. The controlling factors are the extreme mobility of the loose stones, the shortage of fresh water and humus, and the exposure to hot sun and winter gales.

At first the pile of stones deposited along the shore by sea currents is buffeted by each tide

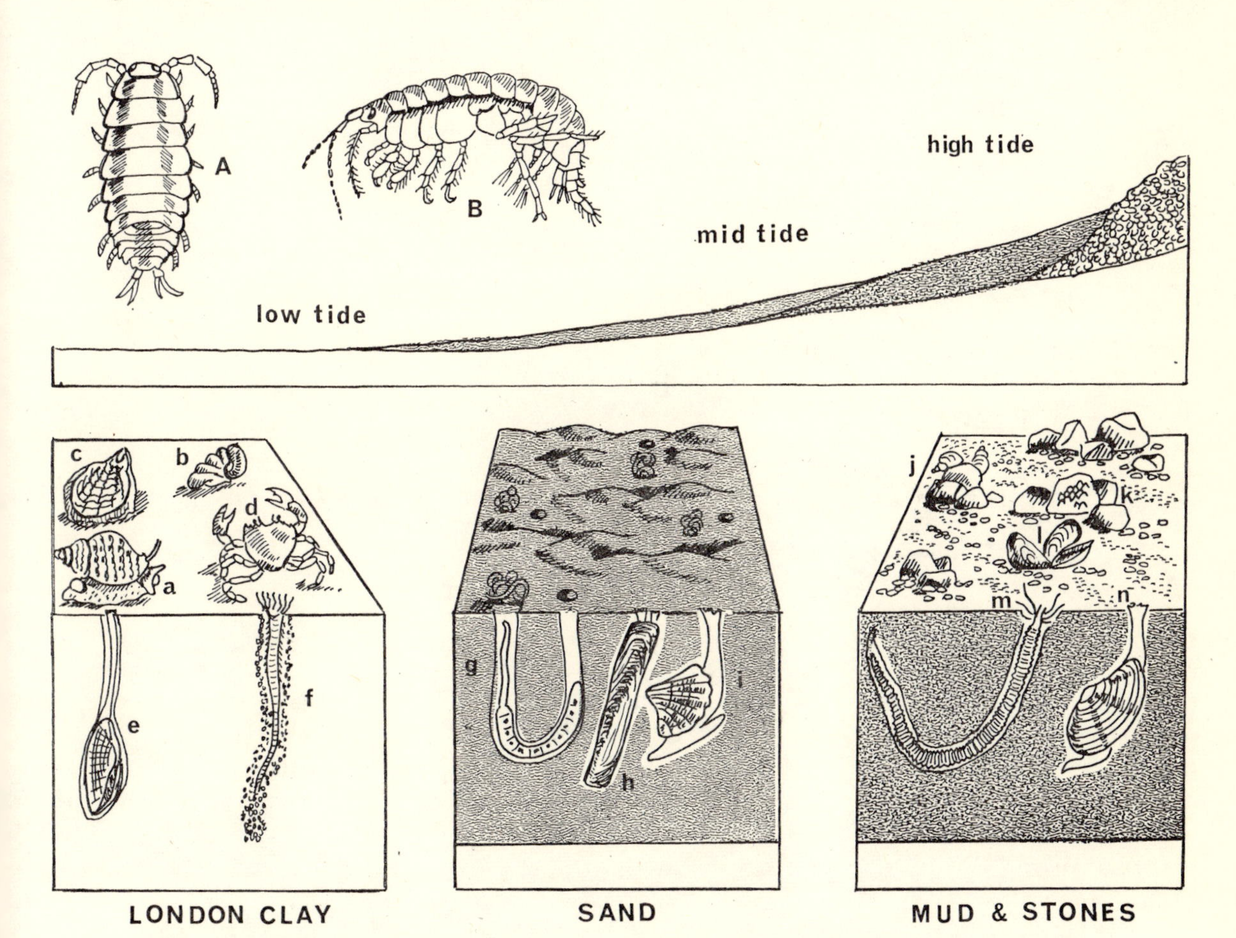

FIG. 38. The flat seashore on London Clay (Essex coast) showing low, mid and high-tide zones. A, sea slater (*Ligia*) and B, sand-hopper (*Gammarus*), two animals of the splash zone. a, whelk (*Buccinium*); b, slipper limpet (*Crepidula*); c, oyster (*Ostrea*); d, crab (*Cancer*); e, piddock (*Pholas*); f, sand mason (*Lanice*); g, lug-worm (*Arenicola*); h, razor shell (*Ensis*); i, cockle (*Cardium*); j, periwinkle (*Littorina*); k, barnacles (*Balanus*); l, mussels (*Mytilus*); m, ragworm (*Nereis*); n, gaper (*Mya*).

and gale. As the bank builds up, becoming more stabilised and growing beyond the reach of the waves, rainwater settles between the stones. Sand is also washed in, as well as drift such as plant remains cast up by the sea or blown across from the landward side. What little rain, humus and sand is able to settle will make possible the beginnings of a plant community on the shingle. The specialised plants which eventually grow up are characterised by their tough and penetrating roots, often many feet long, their low rosette growth above the surface, and a remarkable power for penetrating above the stones and sand which is thrown up with each winter storm. Leaves are of a fleshy and succulent nature, and the plant species mainly perennials which die down after each flowering season. The seaward side of the shingle bank or beach is usually bare of vegetation due to disturbance by the waves, whereas the shoreward side may be well covered.

Sand-dune

In course of time a shingle bank may become entirely covered by sand. The sand-dunes which result can be found in such areas as the Scolt Head region of the Norfolk coast. Old curved shingle spits entirely covered by sand can be

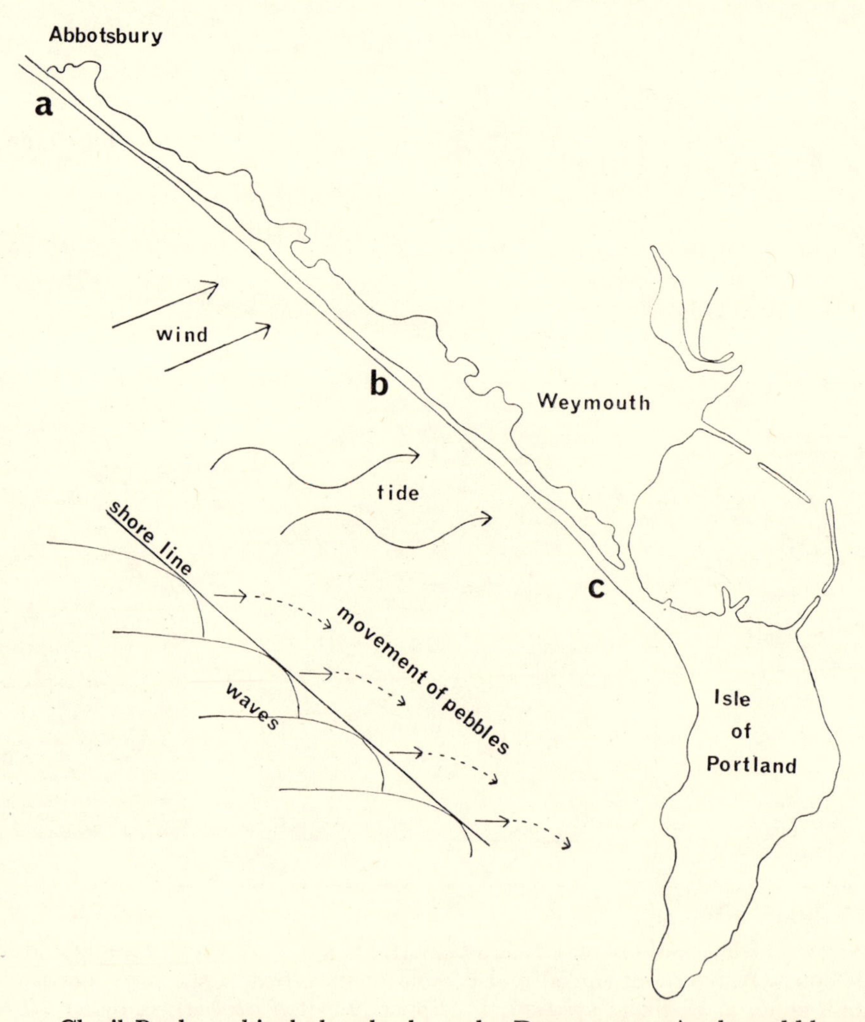

FIG. 39. Chesil Bank, a shingle beach along the Dorset coast. As the pebbles are moved along the beach by the tidal waves they are deposited in graded sizes of approximately $\frac{1}{4}''$ diameter at a; $\frac{1}{2}$-$\frac{3}{4}''$ at b; and 2-3″ at c.

picked out by their contours and vegetation (Fig. 40). In other areas the erosion along coasts in the shelter of headlands has also resulted in extensive sand-dunes (e.g. the Poole Harbour district in Dorset, Braunton Burrows in North Devon, and the dunes near Morecambe in Lancashire).

A sand-dune has much in common with a shingle, but being composed of much finer particles can retain more water. In its early stages a dune is very unstable, and this shifting character of the sand largely determines the type of vegetation. The importance of the shifting sand becomes obvious when crossing the sand-hills from the shore-line inwards. At first the hills appear bare of plants except for just a few, and have a yellowish look. Further inland the hills increase in size and merge together. Plant life increases, and the surface takes on a more greyish colour, due to a higher content of humus (Fig. 41).

The young or 'yellow' dunes near the shore are constantly moved by the wind, and the few plants species which can obtain a hold are perennials with a special method of overcoming burial. The well-known marram grass (*Ammophila*) becomes frequently buried during sandstorms but continues growth by means of a creeping underground rhyzome. In a series of steps this growth, often yards in length, keeps pace with the rising sand-hill so that fresh shoots can sprout at each new level. During a windy spell each tuft of leaves acts as a windbreak so that airborne particles of sand are deposited on

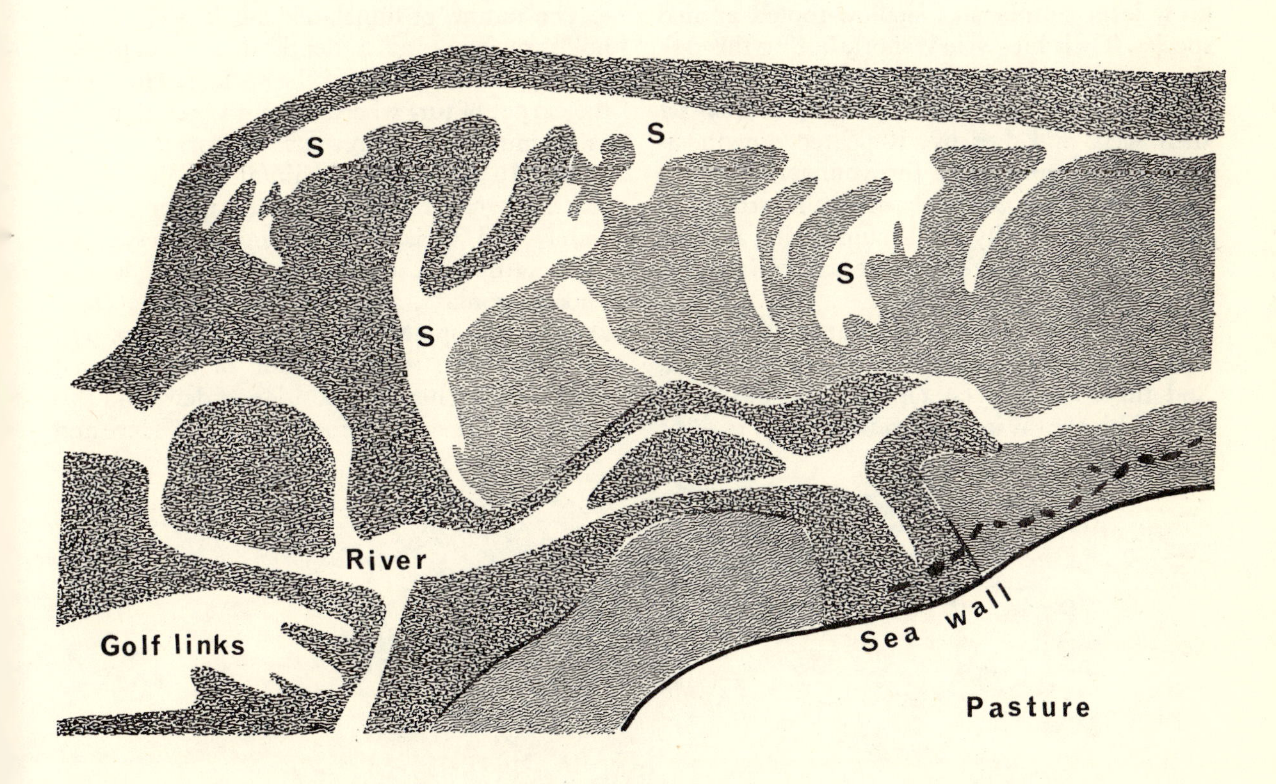

FIG. 40. Map of the Norfolk coast at Scolt Head. The outflow of the river (Norton Creek) is being slowly deflected westward by the accumulation of mud. Scolt Head 'Island' consists of a large ridge of shingle which has extended west forming a series of lateral hooks visible from the air. These curved ridges become covered with sand to form dunes(S), the spaces between silting up to form tidal marshes.

the leeward side of each obstacle. This may eventually smother a grass tuft, but as it continues to force its way out of the sand the process continues indefinitely. In this way a sand-dune builds up and becomes stabilised by the internal 'skeleton' of plant rhyzomes and roots which act as binders. Should the wind move to another direction the sand may become undermined, and a hollow or 'blow-out' results. If this goes deep enough it will expose the long dead rhyzomes of former years, and even the old shingle bed which is the foundation of the sand-hills. In the hollows where the humus collects occur a number of species such as the sea rocket (*Cakile maritima*) which are sheltered from the wind.

Behind the shifting sand are found the older, stabilised 'grey' dunes. In the numerous hollows grow large numbers of shallow-rooted annual species. Their life-cycle is short, lasting through the winter and spring when moisture is available. By summer they have mostly set their seeds and died off. Repeated growth, as with the perennials on the younger dunes, is less essential for survival, since the ground is kept more stable by the accumulating humus. Also, this retains more moisture, and can support a greater cover of vegetation.

In the oldest parts of the dunes at the rear occur those species less tolerant of shifting sand, and these are thickest on the lee side of the hills. This area is in an advanced stage of dune fixation, and the continuous sward of plants, including grasses and lichens, gives the surface its typical greyish colouring. This is where coastal golf-links may be found. Finally, at the fringe of the dune area, a scrub may develop, and trees appear.

Relative to this progress towards a fixed dune is a gradual alteration in soil chemistry. Samples tested at regular intervals from the shore side inwards show a change from alkaline to acid. In some young dunes a number of calcicoles may actually be found, where the chalky nature of the sand is due to the accumulation of fragmented shells of innumerable snails. With age and leaching, also accumulation of humus in the older grey dunes, the acid content rises, and in some areas the dune grades into a heath (e.g. Dorset coast).

The nature of duneland soils is such that it holds moisture at a depth due to capillary action, but dries out on the surface. However, the carpet of turf where there is vegetation acts as a mulch and prevents evaporation from below. In this kind of habitat the perennials reach down with deep roots to tap the hidden water, and their transpiration is arrested by such xerophytic characters as rolled leaves, sunken stomata, fleshy or hairy leaves, thick cuticles and spines (see also Heathland, p. 87). At night the strong radiation of the warmed sand causes a heavy deposition of dew, and this is used by the shallow-rooted annuals. Here and

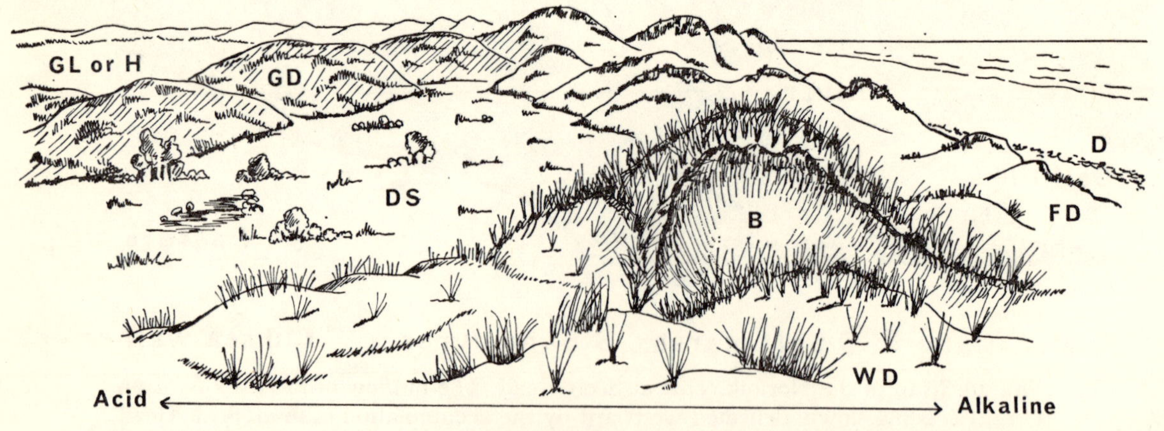

FIG. 41. The Sand Dune showing zoning from fore-shore to heath (H) or golf links (GL). D, drift line; FD, fore-dune; WD, white dune; GD, grey dune; DS, dune-slack; B, blow-out.

there in the hollows where fresh water reaches the surface, and may even rise and fall on the supporting tidal sea water below, a dune-slack may be found. In the damp and acid soil a bog community not unlike that of a heath bog will grow, and may contain some rarities in undisturbed dunes. A typical species is the marsh helleborine (*Epipactis palustris*).

Salt Marsh

Along the flat coastline of the south and east, especially in river estuaries like the mouth of the Thames, also inlets such as the Wash and Poole Harbour, areas regularly washed by the tides produce a salt marsh. Flat beds of mud or silt are laid down, and constant flushing by the tides produces a flat surface of soil which will eventually support a community of plants tolerant of sea-water. Similar conditions may arise in the shelter of a shingle spit especially if there is a river mouth near by into which a constant supply of mud is deposited, as at Scolt Head. As the mud flat is gradually built up, and the sea slowly pushed back, a succession of plant zones is formed until the original sea-bed is changed into dry land (Fig. 42).

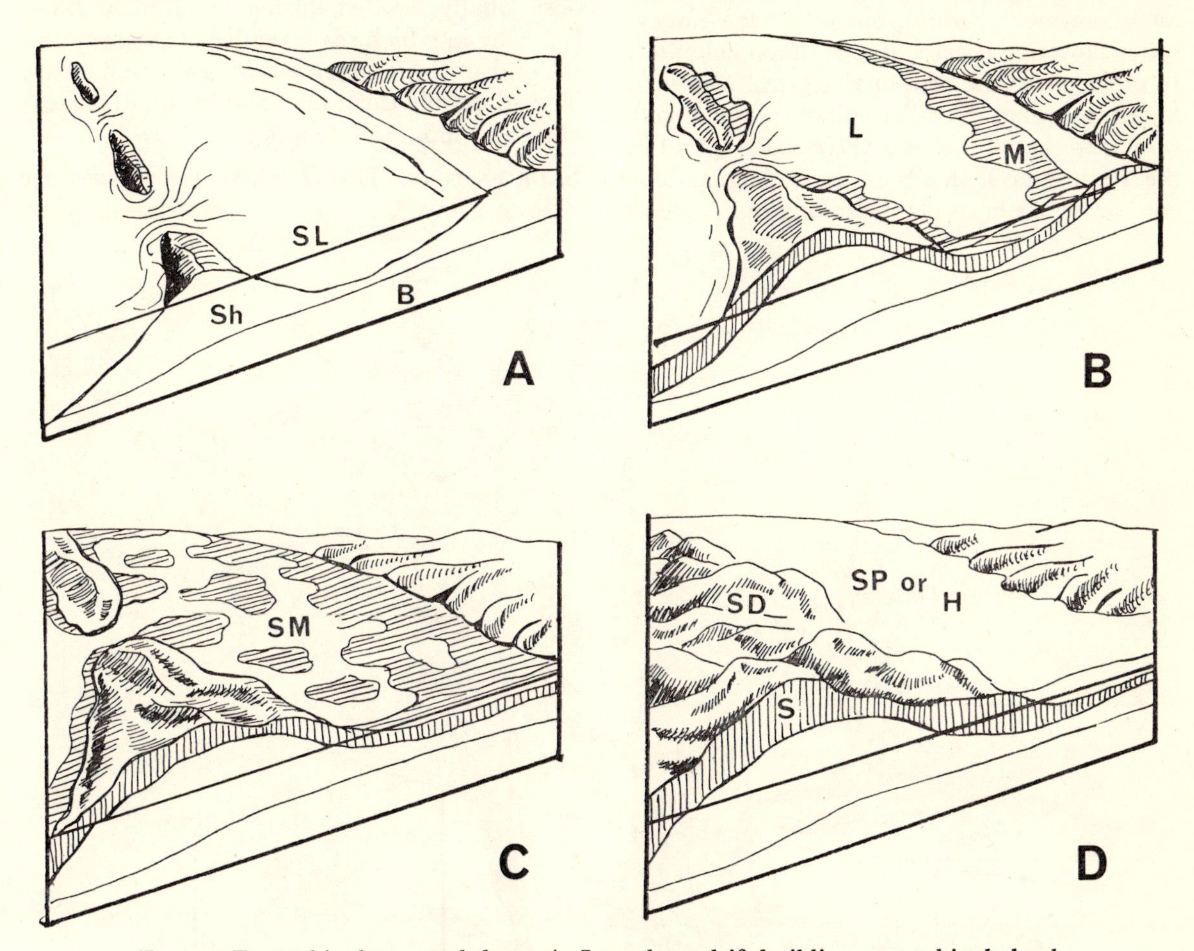

FIG. 42. From shingle to sand dune. A, Longshore drift building up to shingle bank on the sea bed. B, This gradually cuts off part of the sea to form a quiet lagoon which then silts up with land-eroded material. C, Meanwhile sand from the seaward side covers the shingle as the lagoon fills to become a salt marsh. D, A final sand dune covers the shingle bank and the marsh turns into a salt pasture or heath. B, bedrock; SH, shingle; S, sand; L, lagoon; SM, salt marsh; SD, sand dune; SP or H, salt pasture or heath.

BARE MARSH. The youngest part of the marsh on the seaward side is regularly flooded by the tides. At low water the sea drains off along water channels, and on the exposed mud can be found the earliest salt-marsh colonisers. These are small ribbon and thread-like seaweeds which help to bind the mud (e.g. *Enteromorpha*). (Fig. 43).

LOW MARSH OR SALICORNIA STAGE. This flowering plant, the glasswort or marsh samphire, continues the binding process. It can withstand periodic disturbance and is well adapted to sea-water. Along the south coast of England the grasswrack (*Zostera*), one of the few underwater flowering species, is a common coloniser. In other places the cord or rice grass (*Spartina*), a tall growing, reed-like plant plays a role similar to that of the reed (*Phragmites*) found in the freshwater fenlands of East Anglia. The cord grass has been planted along our coast in many places as a mud binder. In 1870 a hybrid, called Townsend's rice grass (*Spartina* × *Townsendii*), was noted in Southampton Water. This is a cross between the native *S. maritima* and an introduced American species *S. alterniflora*. It has proved a rapid salt-marsh coloniser.

MAIN MARSH. With further deposits of mud, and less flooding by the sea, yet still within the tidal reaches, more plants appear to form a fairly continuous cover. Sea blight (*Sueda*) is fairly common to this zone.

HIGH MARSH. Away from the sea, and only occasionally flooded during exceptional tides, is a zone in which sea lavender (*Limonium*) and sea pink or thrift (*Armeria maritima*) can flourish. The adaptable thrift may also be found on sea-cliffs and mountain ledges (see p. 92).

SALT PASTURE. This is where the grasses are able to grow. Cut off by sea-walls and dykes,

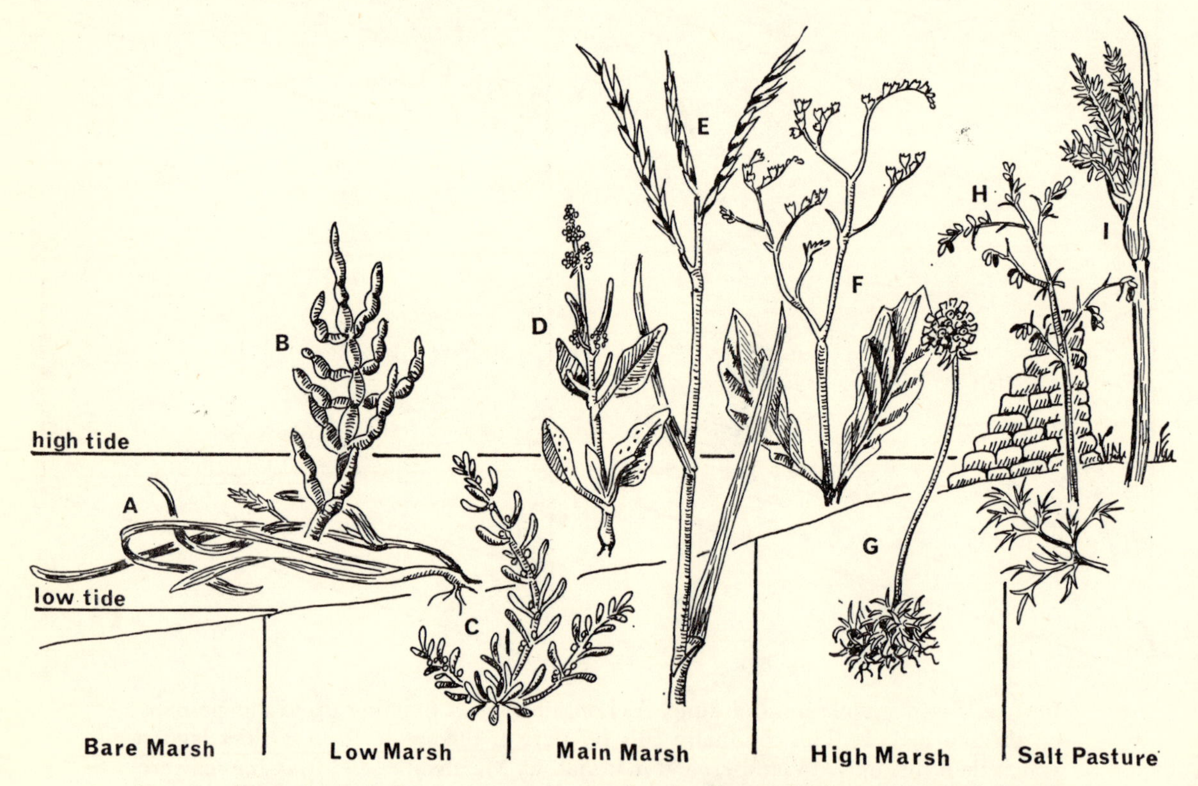

FIG. 43. Zonation of plants on a salt marsh. A, *Zostera* (Eel grass); B, *Salicornia* (Glasswort), C, *Sueda* (Sea blight); D, *Halimone* (Sea purslane); E, *Spartina* (Cord grass); F, *Limonium* (Sea Lavender); G, *Armeria* (Sea pink); H, *Artemesia* (Sea mugwort); I, *Juncus* (Sea rush).

and suitably drained, the marsh comes into cultivation as pasture for sheep, or arable land for root-crops such as beet. The lighter, sandier soils are used for raising bulb crops and for market gardening.

In examining a salt marsh community over its whole area certain species should be looked for. These help as indicators, marking out the different zones. Silver goosegrass (*Galium aparine*) and sea purslane (*Halimione portulacoides*) are found along the water channels and creeks. The tall sea rush (*Juncus maritimus*) and the sea mugwort (*Artemisia maritima*) with its silver-grey and downy leaves are conspicuous along the upper borders of the marsh, and along the sea walls.

A list of sea-shore plants and animals is given in Appendix 2I (page 215).

Part 2: Section E

Mountains

As one ascends the slopes of the higher mountains in places like Wales, the Highlands, Lake District and Pennines, a succession of plant zones will be encountered. The degree of exposure, chemical nature of the rocks, amount of rainfall and snow, and the average yearly temperature, will all determine to some degree their position and relative altitude above sea-level.

Typically there is an order of zoning of plants as one climbs upwards. Passing through a wooded zone, a scrub emerges which then gives way to a grass community on some of the slopes, or a moorland on others. Each of these may reach a summit. Here and there a so-called mountain 'flush' is discovered, a rich pocket of plant life where water and mineral salts have gathered, and this is in contrast to the bare rocks and jagged peaks where little seems to grow.

These mountain zones can be divided into two major regions, the Sub-montane up to about 2,000 feet, and the more exposed Montane beyond this (Fig. 44).

Sub-montane (Up to 2,000 feet—3 zones).

a. *Woodland.* The principal mountain trees in Britain are the durmast oak, the common birch the Scots fir and the ash. These four are upland species more in keeping with mountain slopes, but may also occur at lower levels as escapes or colonisers, or in plantations (see Heathland, p. 88).

The durmast or sessile oak (*Quercus petraea*) is still fairly extensive, whereas the Scots fir or pine (*Pinus sylvestris*), a true native only in Scotland, has suffered much from disafforestation, and to some degree from a changing climate. It has been replaced largely by heather moorland, but some remnants of the old Caledonian Forest still exist, and are now being preserved in places like the Beinn Eighe Nature Reserve. The common birch (*Betula pubescens*) is a ready coloniser, and will take over felled or burnt areas until replaced by trees with denser shade. Natural birch woods usually grow just above the sessile oaks, on the poorer soils. Similarly, ash (*Fraxinus excelsior*) will take over the more calcareous soils such as limestone.

The rugged and twisted nature of these mountain trees is due to winds and exposure, also to snow pressure. They are frequently scarred from wounds caused by falling rock. For trees to grow in this country it is estimated that at least two months in the year should produce a temperature above 50°F. Formerly many of the summits were extensively covered with trees, and their absence today is due to felling. The name of Scafell, now a treeless grassland peak in the Lake District, comes from Skogafel, meaning 'a wooded hill'. Whereas the oak woods support a fairly rich carpet of ground flora, especially on the wetter slopes, the pine woods are almost bare except for mosses and fungi.

b. *Lower grassland.* This is sheep country, especially in Wales and the Highlands, and occurs on the slopes above the woodlands. Sheep walks, as they are termed, are maintained on the better drained slopes of richer, browner earth in which the sheep's fescue grass (*Festuca*) and bent grass (*Agrostis*) grow. Below this, on wetter soil, there may exist a belt of mat grass (*Nardus*). Then comes an area of bracken (*Pteris*), which mixes with the birch or pine at the upper limits of the woodland zone.

Bracken appears to be spreading in our mountains, since this is avoided by sheep which have largely replaced the mountain cattle. Once established it is difficult to control, and almost impossible to eradicate, as it can resist the normal methods of soil clearance by burning (see Project 5, p. 161). At the lowest level of the grassland zone, in the alluvial soils of mountain valleys, are found the cultivated pastures of grasses like rye (*Lolium*) and meadow grass (*Poa*).

c. *Moorland.* Moors occur on acid soils, but are not easy to describe because local conditions result in different plant species becoming established as dominants (see Heathland, p. 87). The 'bonnie Blue Hills' for instance are obviously so named because of the dominant ling or heather (*Calluna*). To encourage its growth in order to provide food for grouse, periodic firing is practised. Such uncultivated moorlands are used in Scotland mainly for grouse shooting and deer stalking (i.e. on the grouse moor and deer forest). In such moorlands, especially on the wetter slopes, the deer sedge (*Scirpus*) and cotton grass (*Eryophorum*) are dominant plants. In the wet hollows are found communities of bog plants, many of which can also be found on lowland moors and heaths to the south.

Montane (above 2,000 feet—4 zones) (Fig. 44)

a. *Summit heath.* By continual leaching of rain-washed soils on the mountain summits, salts are lost, and the area becomes acid or base deficient. It is also exposed to winds, frost and heavy mists. Such a place provides a habitat

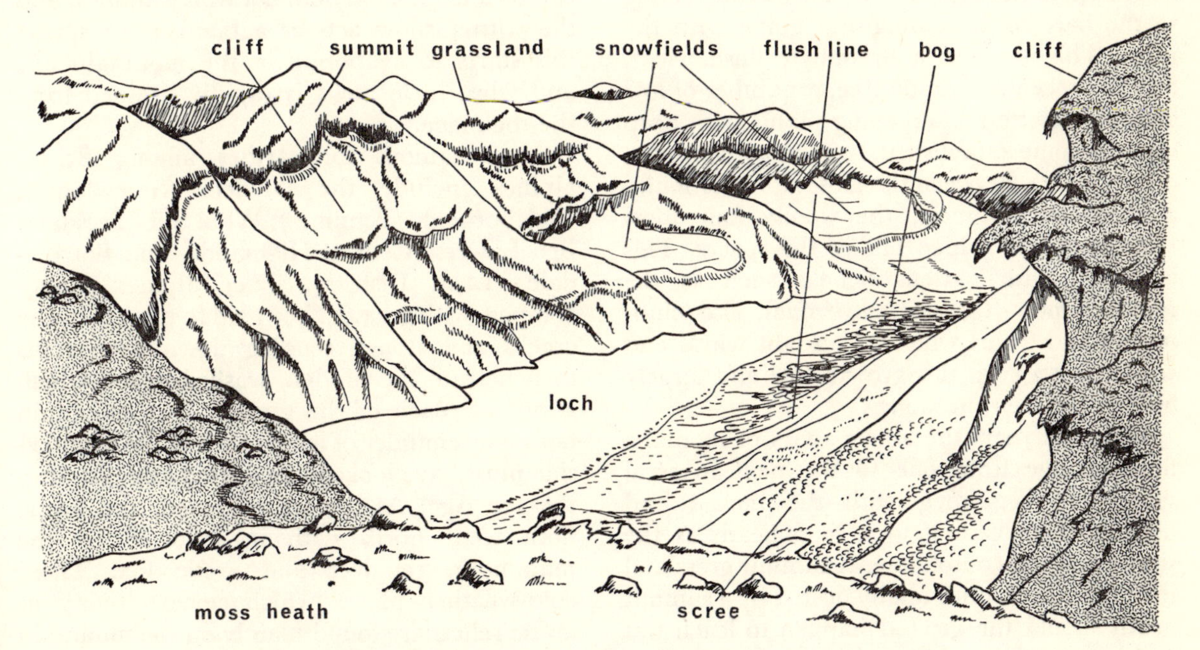

FIG. 44. The Montane habitat. The Highlands, over 2,000 feet.

for only the toughest of species, mainly shrubs which are low growing and show xerophytic characters. Ling is mixed with various mountain berry plants and lichens, and these occur with regular monotony. The principal grass is the mat grass (*Nardus*), a sure indicator of poor soil.

b. *Summit grassland.* This develops on the somewhat insecure and often exposed soils which are mixed with broken material, and are subject to 'frost heaving'. Whereas the summit moorland usually exists on the harder rocks such as granite, the grassland grows up on a richer base, although this may be less secure due to its looseness. On it develops a moss-heath. Dominant grasses are the mountain hair grass (*Deschampsia*) and sheep's fescue (*Festuca*). Only a few species of flowering plants can survive, and are to some extent kept in check by the grazing sheep. This is the main habitat for the various clubmosses (*Lycopodium*).

c. *Colonising zone.* The summit vegetation on the two nature reserves, Ben Lawers and the Cairngorms, both in the Central Highlands, show well the contrast between grassland and moorland (see p. 107). In other places, ranging from bare rock to broken scree and shifting gravels, a succession in colonising can be seen. First appear the mosses and lichens which cling to the bare rock, coming and going with the rains. This is called a Bryophyte flush. Next, on the broken rock and scree, a number of low growing rosette and creeping plants with deep roots become established (e.g. moss campion, *Silene acaulis*). Between the loose stones is found a characteristic moss, the woolly hair moss (*Rhachometrium*) which fills in the spaces. This is mixed with various lichens such as the reindeer moss (*Cladonia rangiferina*), also clubmosses and dwarf willow. Constant winds and winter frosts keep this growth low and largely hidden between the stones.

Over the years the above zones of montane flora change from one to the next in seral succession, beginning with the mosses and lichens, then leading to a moss-heath as the stones and soil consolidate. A summit grassland then takes over, until finally ousted by a summit heath should the ground happen to leach out and become acid.

d. *Mountain flush.* Here and there, in places where there is a steady downward movement of soil and drainage water, a rich flora called a flush will grow. This may be around a spring, in a gully, on a sloping rock face, or on a broad ledge. Leaching of the upper soils produces a rise in mineral content below, and this becomes washed into the rock fissures and between the stones. The accumulation of scree and broken rock provides anchorage and shelter. This is the home of the alpines, of which some 60 species occur on the mountains of Britain, some now very rare. Since these plants usually prefer the alkaline soils and drier atmosphere, their distribution is somewhat localised. It is constant dampness perhaps more than anything which excludes these plants from otherwise suitable localities. This can also explain the disappointment to many gardeners of failures to keep alpines, especially in southern and lowland districts. The wet winter soil, snow-slush and air pollution often encountered in gardens near towns is hardly an ideal habitat for alpines. The unheated and well-ventilated greenhouse is often the only solution to successful alpine culture. Alternately, polythene sheeting as winter protection against excessive dampness can be a help. Cold does not worry alpines, and the winter snow acts as a blanket. In spring this starts to evaporate in the mountain air, and what little melts is rapidly drained from the rock face.

Areas famous for rarities among British alpines include the Glyders (Snowdonia), Ingleborough (Pennines), Whin Sill (Teesdale) Ben Lawers (Central Highlands) and Borrowdale (Lake District). Since alpines cannot compete with grasses, they tend to grow between cracks and stones, sending down deep roots to penetrate the hidden, washed-in soil which is out of reach of the grasses. Our mountain flora is a reminder of what much of the countryside must have looked like during and just after the Ice Age. As the ice retreated, the arctic flora went north with it, followed by the mammoth, reindeer and arctic fox. Those cold-weather plants which survive today as arctic relicts are found mainly on the mountain tops, or sometimes at sea-level to the north where the mountains reach the sea. About 7 species are true alpines, only found elsewhere in the mountains of Central Europe. Arctics, found also in the far north, number about 20 species. The rest are called arctic-alpines, and occur both in the Alps and within the Arctic Circle.

Alpines tend to grow in tight rosettes (e.g. moss campion, *Silene acaulis*) or have a creeping habit which spreads over the rocks (e.g. mountain avens, *Dryas octopetala*). Successive decay and rain-wash provides a constant supply of humus within the rock crevices and stone cavities (Fig. 45).

The above is an overall description of Britain's mountain flora. Local factors will affect their position in altitude, and their distribution. The nature of the base rock and the amount of soil which it produces, also the rainfall, will have its effect. For example, on a similar latitude in the Central Highlands lie the two nature reserves with different vegetation—Ben Lawers to the west and the Cairngorms to the east (Fig. 46). Ben Lawers is composed of calcareous mica-schists which fracture easily and produce sharp and jagged peaks. A resultant release of mineral salts from weathering, plus a high average yearly rainfall of some 100 inches, provides a soil which supports a fairly rich flora. Alpines grow on

Fig. 45. Some British montane plants: a, Mountain Avens (*Dryas*) a creeping plant; b, Starry Saxifrage (*Saxifraga stellaris*) a rosette plant; c, Moss Campion (*Silene acaulis*) a cushion plant; d, Alpine Bistort (*Polygonum viviparum*) and e, Viviparous Fescue (*Festuca vivipara*) the last two with bulbils.

the peaks and in the flushes, and alpine pasture grows in the valleys. In contrast, the Cairngorms are of coarse, hardened granite which weathers more slowly in rounded contours. Rainfall per year is far less, from 40 to 100 inches. This results in poor, free-drained acid soil which can only support a moorland community of heathers, lichens and berries capable of surviving long periods of snowfall. In some places the snow may lie in pockets well into midsummer. Although no permanent snowfields exist in Britain, this could easily happen with only a slight average yearly drop in temperature. As it is our mountain plants are able to survive the winters. It is estimated that plant life in north-west Europe can only survive so long as the mean yearly temperature remains above 42°F. Temperature drops 1°F for every 270 feet in altitude in Britain.

It follows from all this, and coupled with the effect of high winds, heavy rains and long snowfall, also a short growing season, that only the hardiest of plants can survive in our mountains. Fertilisation of flowers and dispersal of seeds and fruits is made difficult, so that alternative methods of regeneration have been evolved. One of these, called vivipary, occurs where the fruits germinate on the parent plant, becoming temporary parasites (e.g. mountain sheep's fescue, *Festuca vivipara*). In other cases

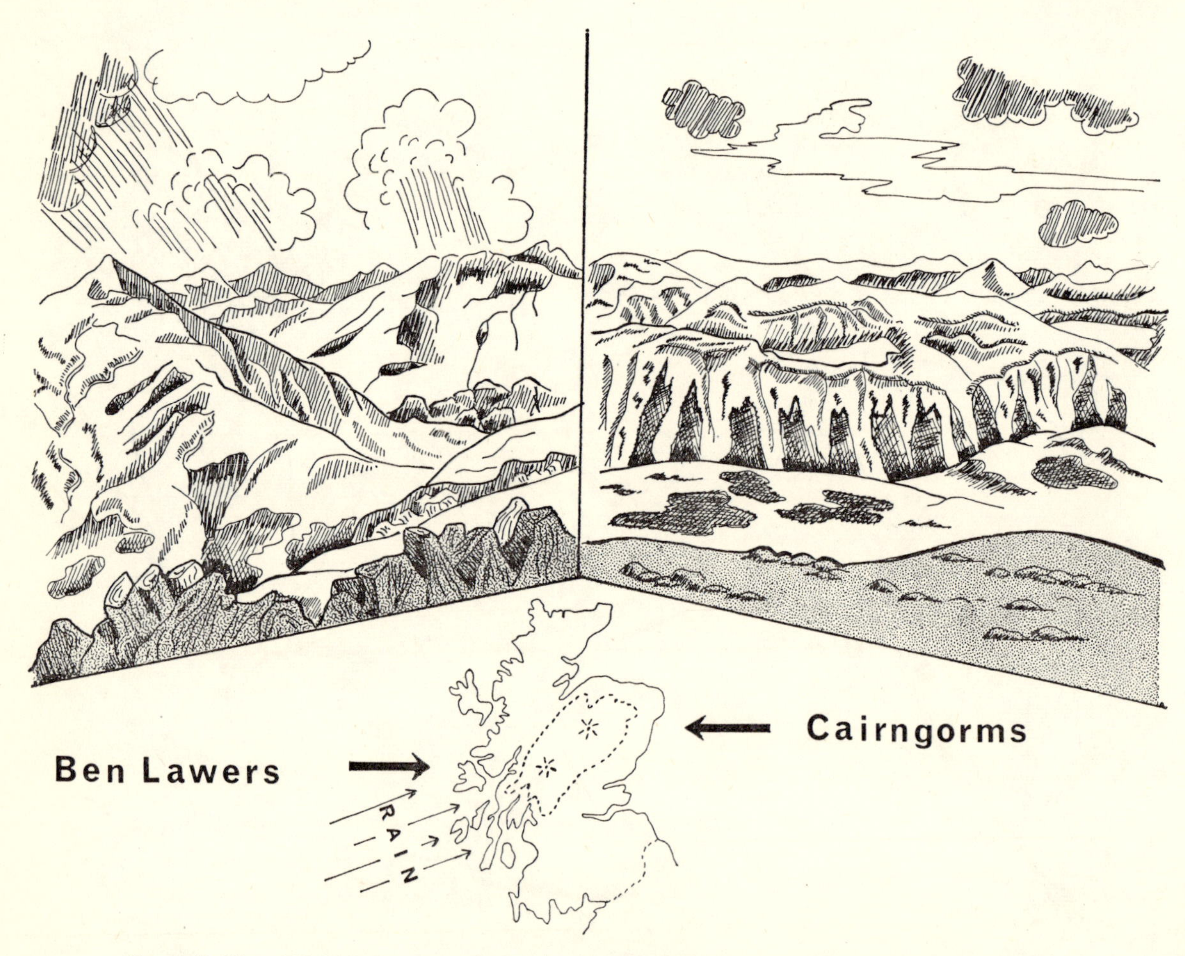

FIG. 46. Two Highland communities of mountain flora. Ben Lawers which is composed of a calcareous schist and supports grasses and alpines. The Cairngorms with more rounded contours of granite rock supporting mosses and lichens (based on a diorama in the Natural History Museum, London).

small and detachable buds, called bulbils, grow on the parent (e.g. alpine bistort, *Polygonum viviparum*) (Fig. 45).

A selected list of plants associated with mountain zones in Britain is given in Appendix 2J, page 26.

Mountain Animal Life

Whereas most mountain plants are established on high and often exposed ground which is their natural habitat, the same cannot be said for Britain's mountain animals. They are there, and sometimes nowhere else, due to man's persecution, and are forced to seek refuge in high and remote places. The red deer, for example, our largest and noblest land animal, is by nature a woodland dweller, and is often found at sea-level in forest country on the Continent. Yet the deer forests in which it must live in Britain are almost devoid of trees. The same applies to the open moors in the south west where it also lives. Centuries of hunting and extermination, especially of game animals and carnivores, have resulted in making our mountains a home for rarities. Today, an enlightened view, with some improvements in legislation, is helping to save many from extinction. World wars which caused an absence of game-keepers and a drop in field sports have given some reprieve to the mountain animals, and the present-day practice of promoting nature reserves, also conservation, is helping to build up dying populations.

A search for our elusive and often rare mountain animals will show that they tend to live within certain habitats linked with the plant zones already mentioned. Some are found on the summits or among the crags, others on the moorlands, some in the woodlands and some by the seashore. (A list is given in the Appendix 2K, page 217).

Mountain Lakes

A special feature and beauty of our mountains is the view of a clear mountain lake high in the hills. Many of these are of glacial origin where the moraines of retreating glaciers have dammed a former glacial valley, or where water has settled within a rift zone or catchment area. Many of the lochs and meres are long and narrow, and steep-sided in places, containing deep and clear water. Apart from birds of open waters, some of them rare, also certain fish which are considered to be arctic relicts, most of the plants and animals are concentrated along the shore-line. An interesting phenomenon of the open waters is the abundance and seasonal fluctuation of the plankton life. In the case of certain algae their abrupt appearance due to changes in temperature, length of daylight and increase in mineral foods can cause a remarkable colour change in the water almost overnight. This is known in the Lake District as the 'breaking of the meres'. For a list of mountain lake flora and fauna, see Appendix 2K, (p. 217).

There are two main types of lake which may be found in Britain, the 'evolved' lake and the 'primitive' lake. The former occurs on the softer rocks, has sloping shores and plentiful vegetation. Drainage is rich in dissolved plant foods, and products of decay in the water settle on the bottom to form mud. At times there is a large plankton population in the upper waters.

A primitive lake occurs in mountainous country, and has steep, rocky shores with little vegetation. The water is soft and clear but poor in plant food. Such a lake may remain unchanged for long periods. In the deeper lakes an important 'layering' of water occurs during summer periods (Fig. 47). The upper layer warms up through the sun's rays and movement caused by winds, whereas the bottom layer remains cold, dark and still. Light and heat cannot penetrate. In between is an intermediate layer where a sudden change in temperature is found, called the thermocline, at an average depth of 25 to 30 feet. The upper layer is called the epilimnon, and corresponds to the water of a shallow pond. Below is the hypolimnon.

Towards winter with steady cooling the body of the lake returns to the same density throughout, so that mixing of the waters is now possible as the thermocline breaks down. Substances which had fallen to the lake bottom, or are washed in by winter gales, can now circulate freely. Decomposition is hastened with an increase in oxygen supply. This all adds to the nutrient reserves of the lake, so that by spring there is abundant mineral food for the renewed burst of plankton, mainly of algae, which takes place. There is a peak in about April-May, followed by another in late summer.

Lakes which have a rich plankton population, as in the evolved kind of waters, are called *eutrophic* lakes (Greek *eu*—well or good, and *trophe*—nourishment). Primitive lakes low in foods are called *oligotrophic* (Greek *oligos*—little). Occasionally a third type of lake is mentioned by limnologists, called a *dystrophic* lake (Greek *dys*—bad) where the lake bottom is peaty, and the soft, acid water lacking in minerals and tending to inhibit decay.

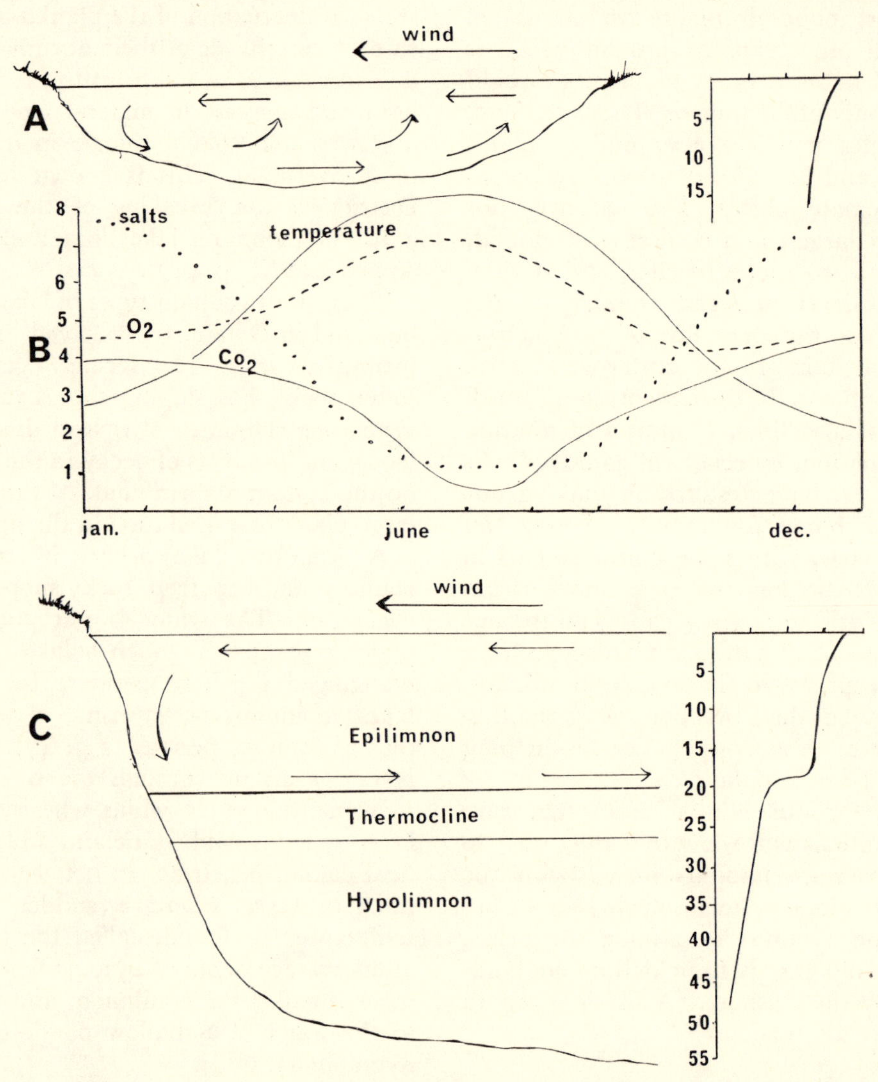

Fig. 47. Comparison between a pond and a lake. A, a shallow water area (pond) in which light can penetrate and winds distribute the water to give a fairly even temperature. B, chart showing the fluctuation in physical and chemical properties of a pond over the year. The vertical scale is arbitrary. C, a lake showing the epilimnon which resembles a pond, and the deep and static hypolimnon with a thermocline in between (see text).

Part 2: Section F

Pond, river and marsh

Beneath the land surface there is a water-level which lies at varying depths according to the nature of the soil and underlying rock, the topography at the surface, and the available water supply. Where the water reaches the surface on sloping ground, e.g. at a spring, a river is born. Water which settles in a hollow, or is arrested by some artificial or natural barrier (e.g. a dam or a moraine), will form into a marsh, or if deep enough, a lake. Such features are usually permanent, so long as there is a water supply. It is in the smaller pools or ponds of shallow depth that the habitat becomes unstable, especially in the case of man-made hollows which fill up with rainwater. Examples of these are the farmer's cattle pond, an ornamental pool, a hammer pond, dew pond, or some drowned digging for gravel, chalk, clay, stone, etc. During the last world war many ponds were created where bombs fell on open ground (see Project 4, p. 159).

If such a water-filled space is left alone by man, nature takes over and a process of colonisation begins. At the peak of its development the pond will exhibit a number of plant zones (Fig. 20).

a. *Marsh zone*. This is the area adjacent to the pond side or in the backwaters of rivers, where the ground surface is above water-level, but waterlogged below (see Glei Soils, p. 25). In this the marsh plants live with their roots and underground organs bathed in water. The species which grow there will to some extent be controlled by the chemical nature of the soil, and may differ widely in such contrasting places as a highly acid moorland bog and a marshy fen on chalk.

b. *Swamp zone*. This next zone occurs where there is standing water of shallow depth along the pond or lake side, or in the river backwater. This swamp community is a conspicuous feature, consisting of tall-growing plants such as reeds and rushes. Such plants grow with their lower parts and underground organs in water. In a well-marked swamp zone three belts of vegetation may be noticed, their presence and width depending on the slope of ground. On the landward side, barely covered by water, is a belt characterised by the sedges (*Carex*), followed by a strip of reeds (*Phragmites*), then a stretch of bulrushes (*Scirpus*) bordering the open water. Associated with these three common swamp plants are a number of species which are adapted to standing in water. Most have an upright growth with narrow and pointed leaves. This is probably a precaution against injury which could easily arise in a crowded community during a heavy rainstorm or with sudden gusts of wind. As in the case of land plants such as grasses, especially the cultivated cereals, each plant bends over to the wind, then rights itself, in a succession of movements which produce an attractive wave motion across a corn-field or reed-bed. The swamp zone is usually the richest and most crowded area in the whole pond community. It is here that the process of decay is most active, so that a constant build-up of rich mud in shallow water is going on.

c. *Aquatic zone*. On entering open water the true aquatic plants begin to show. Some may actually be found among the standing swamp plants, as in the case of pondweeds (*Potomageton*) and water lily (*Nymphaea*) seen between the reed stems. Aquatics also grow in zones according to the depth of water, and are of three types—rooting, submerged and floating. Rooting aquatics, such as the above pondweed and lily send up their floating or aerial leaves and flowers on long stalks. This growth is usually limited to a depth of 3-4 feet, and is seen just beyond the swamp zone. However, in a shallow pond or lake, the whole surface may be covered with the leaves and flowers of these aquatics. In extreme cases where there is deep water, as in some lakes, there are certain pondweeds which can reach the surface from a depth of 10 feet or more.

In between the rooting aquatics, and beyond, grow the floating aquatics their roots dangling in the water and acting as balancers. Here again, a small pond may be entirely covered, as in the case of the tiniest of flowering plants, the duckweeds (*Lemna*). Below the surface are found the submerged aquatics, such as the widespread and familiar Canadian pondweed (*Elodea canadensis*) Special adaptations for flowering are found among these submerged plants (see p. 114).

At the greatest depths where the flowering plants cannot penetrate, the stoneworts (*Chara*) take over, together with a number of filamentous algae as well as planktonic forms near the surface. Open waters, especially in deep lakes and reservoirs, are the home of vast numbers of microscopic plants and animals which need no anchorage (see p. 109).

In freshwater, in contrast to sea-water, most of the plant life is composed of flowering plants. Since these are of terrestrial origin a number of modifications have evolved for successful living in water (Fig. 48). Being of greater density than air, water gives support to an aquatic plant so that it does not require the same rigid skeleton found in land plants. The tissues are designed to take a pulling strain against moving water, so that the mechanical cells are arranged as a central core after the fashion of a root skeleton (see p. 32). Also, to lessen injury due to the pressure or friction of water, the submerged leaves in aquatics are either long and narrow, or broken up into fine threads so as to minimise resistance.

Gaseous exchange also presents problems. Bulk for bulk, water contains less oxygen than air does. Carbon dioxide, however, is readily diffused, so that an aquatic has plenty at its disposal. On the other hand, its submerged parts are without breathing pores (stomata), so that gases must pass through the surface by slow diffusion. The slow breathing rate of water plants is to some extent compensated by the thin foliage and cuticle, also much division of the leaves to increase the surface area. Consequently, with a good supply of carbon dioxide, a steady stream of oxygen is produced which can be seen rising as bubbles on sunny days. Much of this is stored within the plant, in special air reservoirs, and this helps to buoy up the plant and support the aerial flowers. In the case of floating leaves stomata occur on the upper side. This is covered with a waxy cuticle to keep the surface dry, and to help in buoyancy.

Generally speaking, submerged leaves are thin and divided, whereas surface leaves are flat and rounded. They tend to have centralised leaf-stalks which are flexible (e.g. lily-pads) so that the leaf-blade is not pulled under during disturbance or changes in water-level. Some plants have both floating and submerged leaves (e.g. water buttercup, *Ranunculus aquatilis*). The arrowhead (*Sagittaria*) may even grow three types—submerged, floating and aerial. Such variation in leaf structure is more usual in still water. In streams and rivers the leaves are mainly of the submerged kind. Some species like the arrowhead can adapt themselves to both running and still water. Others may even survive out of water, on wet mud, during periods of drought, and turn into land forms which grow erect and are self-supporting. The water milfoil (*Myriophyllum*) will sometimes cover a whole mud-bank in this way during a dry summer.

A system of aerating canals also exists in the underground organs of some swamp and marsh plants. Others develop a kind of aerating tissue consisting of swellings in the submerged stems and roots. Since much of the light which penetrates water is lost by reflection, and what enters by absorption, aquatics tend to send their leaves to the surface. The need for this varies with the depth and clarity of the water. An abundance of water, plus a good mineral supply, stimulates rapid growth among water plants. This must be obvious to anyone who

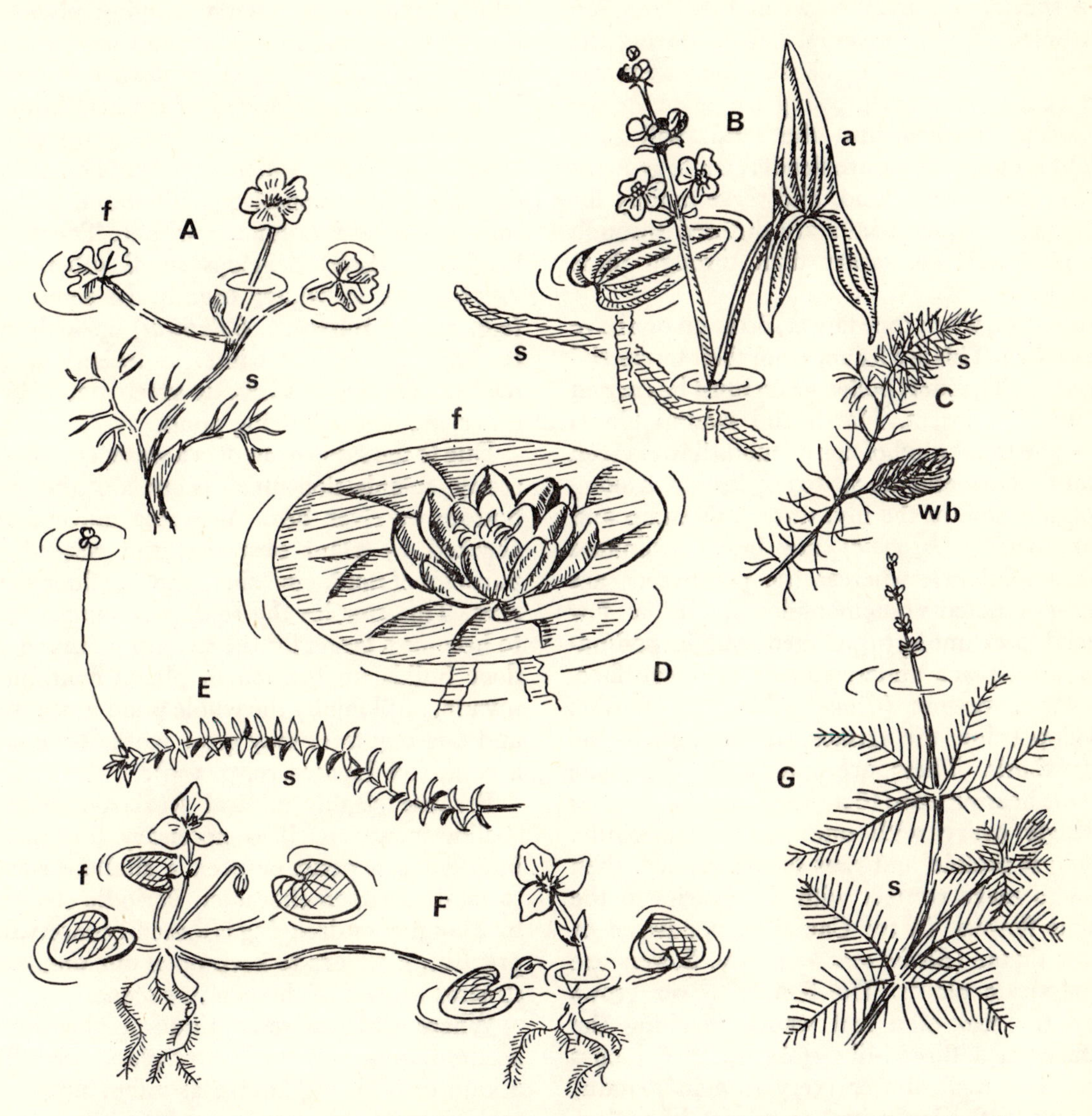

FIG. 48. Some plant adaptations to freshwater. A, Water Crowfoot; B, Arrowhead; C, Hornwort; D, Water Lily; E, Canadian Pondweed; F, Frogbit; G, Water Milfoil; a, aerial leaves; f, floating leaves; s, submerged leaves; wb, winter bud or turion.

keeps a well-maintained aquarium in which it may become necessary to prune the luxuriant growth. Over the British countryside the richness of a woodland flora (see p. 74) vies with that seen on the mud at the waterside of our rivers and ponds.

Vegetative reproduction is common in water plants, and usually by fragmentation of a growing branch. Some species rarely set seed (e.g. *Myriophyllum*). Prolific growth probably helps in colonising and dispersal, as happened in a spectacular manner when the Canadian pondweed (*Elodea*) invaded Britain during the last century and rapidly choked the waterways and canals. A small growing fragment in the aquarium, pushed into the sand, will soon sprout rootlets. These are slender, white growths of the adventitious kind, and serve mainly for anchorage. Mineral salts are diffused through the plant surface, rather than through these rootlets.

Aquatics, when they flower, do so on or above the surface. This is a legacy from their terrestrial ancestry. The inert pollen grain must be carried by wind or insect. Marsh and swamp plants grow the usual kind of conspicuous flowers seen in land plants and are visited by insects. Among aquatics those of the lily, water buttercup and water violet (*Hottonia*) are insect-pollinated (entomophilous), whereas the pondweeds are wind-pollinated (anemophilous). In a few special cases among submerged aquatics pollination takes place underwater or at the surface. In the hornwort (*Ceratophyllum*) the flowers develop below the surface, in the axils of the leaf whorls. These are unisexual, the male flowers bearing up to 16 stamens, and the females a single ovary with one carpel and ovule. Pollen is prolific and rises to the surface, then sinks to make contact with the ovaries in the female flowers. In the Canadian pondweed or water thyme (*Elodea canadensis*) the flowers are on separate plants. The female flower grows from the axil of a leaf whorl, reaching the surface on a thread-like stalk up to 6 inches long. The male flower (very rare in Britain) remains in the leaf whorl to release its pollen. This rises to the surface and floats towards the female flower (Fig. 48).

In the case of some floating flowers which may become submerged by a rise in water-level before they ripen, self-pollination may occur (cleistogamy). As with the pollen the fruits of many aquatics float readily and are dispersed by water currents. During winter the aquatics and waterside plants die down to their roots or storage organs. Floating and submerged aquatics which have no permanent attachment form so-called 'winter buds'. These are the tips of growing shoots with tightly packed leaves which break away from the dying plant and sink to the bottom. Next spring they grow into new plants. In the frogbit (*Hydrocharis*) summer buds also form at the end of runners from the parent plant. In strawberry fashion the growth spreads over the water surface. The minute duckweeds fill themselves with starch and also sink to the bottom for the winter. The arrowhead (*Saggitaria*) develops small tubers from its rhyzome as winter organs. In some cases (e.g. water starwort, *Callitriche*) growth may continue throughout winter, especially in mild seasons, making a pleasant green patch in an otherwise lifeless looking pond.

Due to the rapid growth in a water community ready colonisation may occur on the pond bottom or river bed. As water movement is checked by plant obstacles, the water-borne particles of mud, etc., sink, and the floor slowly rises. This enables the shallower aquatic zone to become a target for the swamp plants. As the floor builds up the marsh plants continue to invade, until finally the whole pond is occupied, and one day trees may grow where once stood a pond (see Hydrosere, p. 159).

Because of this natural seral succession of shallower waters, it is necessary for man to interfere if a pond or even a shallow river is to retain its open character. Periodic clearance by dragging or dredging is a useful conservation practice, if we are to keep open our ponds and streams as part of the country scene.

Where a hydrosere continues unchecked the accumulating debris of water plants may form a kind of peat bog known as a fen, supporting such plants as the ragged robin (*Lychnis flos-cuculi*) and meadow sweet (*Ulmaria*). This is alkaline in nature and high in mineral content, especially in the neighbourhood of chalk strata. Where trees such as the alder, buckthorn and willows take over a fen-wood or carr results. Such habitats may be seen in parts of East Anglia. Similarly an acid peat bog will occupy a hollow, or man-made digging on a moorland soil.

Animals in Freshwater

Freshwater animal life occurs in almost bewildering variety and numbers, and one is impressed with the continual struggle for existence which goes on in a pond, and the many ways there are of ensuring Darwin's 'survival of the fittest'. One thing becomes very apparent—there is hardly a corner in the pond or river which is not in some way occupied by some animal or plant species. A similar intensity of life and competition may also be found in a rock-pool by the coast (see p. 93).

Freshwater animals may be studied in their respective groups or families when surveying a pond. This can be of great value in the understanding of species distribution and population, but since a water community is to a large extent a 'closed' one, it is of equal interest to treat the animals in relation to their food niches. This requires an understanding of their respective ways of life, and how they set about seeking food—which is the aggressive carnivore, the harmless vegetarian, the scavenger or parasite, and so on.

As with plant zoning, similar divisions among the animal community may be noted (Fig. 49).

Neuston. This is the surface population first encountered when approaching a pond. Small, light-bodies animals, often with flattened shapes, are capable of riding on the water, or can hang from the surface film by molecular attraction. Many possess air reservoirs to increase their buoyancy. Mosquito larvae, pondskaters, springtails and whirligig beetles are some examples. Such life can only exist in the still waters of ponds, slow running streams and river backwaters.

Plankton (meaning wandering). Small animals with a high surface area to volume and a low density appear to drift aimlessly in open water, rising and falling with the weather conditions, time of day, and fluctuating with the available food supply. This also applies to the minute free-swimming algae which on occasions may turn the water green with their concentration. At times, too, the water swarms with water fleas and their allies, to give a reddish quality to the water. These are the permanent members of the plankton life, which is joined by others during their temporary larval stage (e.g. certain insect and mollusc larvae). To increase their lightness these tiny organisms are often equipped with numerous projections from the body and limbs. Plankton can exist only in still or very slow-moving water, and is a special feature of deep lakes and reservoirs (see p. 109).

Benthos. These are the bottom dwellers which occupy the mud and debris of dead plants in ponds, or remain hidden beneath the stones of streams. Mostly they are adapted for clinging or digging or they can burrow (e.g. caddis larva, planarian, snail, mussel and worm.) Others hide between the plants and are provided with attachment organs such as hooks and suckers (e.g. dragonfly larva and leech). Such hidden places are as much in use by prey as by predator, in a constant but deadly game of hide-and-seek (e.g. harmless tadpole and lurking pike).

Nekton. Free-swimming and usually powerful and large animals belong to this group. They swim about actively and boldly, and show modifications for free movement, such as streamlining, paddle shaped limbs, fins or flattened tails (e.g. fish, amphibian, beetle). This group also includes a number of visitors such as the otter, water-shrew and fish-hunting birds like the grebe. One or other of these may dominate the pond or river and form the apex to a food pyramid (see p. 63).

Since a water community is imprisoned to a large degree by its environment there is intense competition for survival within the limited space, both in seeking food and avoiding capture. Life is largely interdependent, with a mutual advantage for survival offered to each occupant. As an example of this the following food cycle may be cited: plant—tadpole—dragonfly larva—fish—fish-eating bird. Where death occurs the activity of decay bacteria releases mineral salts as food for the plants, and so the cycle is completed.

A study of food chains will show that the diet of predators changes with their age and size. In a pond a newt larva, for example, begins by catching microscopic animals, then changes over to larger food such as water fleas, tadpoles, worms, and so on. When fully adult it can deal with sizeable earthworms and slugs during its land stage.

A general plan of a food-web in a pond is given in Fig. 50. The tendency is for the larger animals to eat the smaller ones, so that in order to maintain the numbers of each species, the smaller kinds tend to produce larger families

FIG. 49. Swimming and feeding devices in pond animals. A, Pond skater on surface film (neuston); B, Gnat larva, a brush feeder; C, Daphnia (plankton); D, Dragon-fly larva, a hidden hunter (nekton); E, *Dytiscus* beetle, a free swimming hunter; F, *Planorbis* snail, a creeping vegetarian; G, Metamorphosing tadpole (vegetarian with swimming tail as larva and carnivore with swimming legs as adult). H, Caddis-fly larva, a scavenger (benthos); I, Mussel, a ciliary filter feeder (benthos).

(see also p. 63). Reproduction is mainly by means of small eggs, a noteworthy exception being the parthenogenetic young of summer daphnia. Eggs which are laid freely are usually demersal, and sink between the stones and water plants (e.g. trout and hydra). Other eggs become attached to plants by means of adhesive coating (e.g. newt, carp). Some animals lay floating eggs (mosquitos and some dragonflies), whereas others pierce plant stems in which to lay them (beetles and water bugs). Frogs and toads produce spawn masses in clumps or in strings. High numbers of eggs are not unusual (about 7,000 for a toad and up to 700,000 for a carp). There is little or no parental care of the young (Fig. 51).

In special and interesting cases where small families are produced some form of protection is given to the eggs and young. The crayfish (*Astacus*) carries the young about for a time. Others provide the shelter of a nest of sorts. The stickleback builds a nest, the silver water beetle (*Hydrous*) makes a floating egg cocoon, and the water spider (*Argyroneta*) fills its web with air to make a diving 'bell'. In each case, whether it is a large unguarded family or a small protected one, the future of the species will be assured so long as a mere handful reach maturity. Water animals which protect their young are of special interest to students of animal behaviour. They can be closely watched with comparative ease and comfort, within the confines of the aquarium. Small animals taken from ponds will readily adapt to confinement. The elaborate ritual of nest-building, courtship and care of the young by a father stickleback is something which every naturalist should witness for himself (see Project 13, p. 179). The classic work done by Niko Tinbergen was carried out in this way.

Winter poses problems in the pond when the water freezes over. Paradoxically a frozen pond

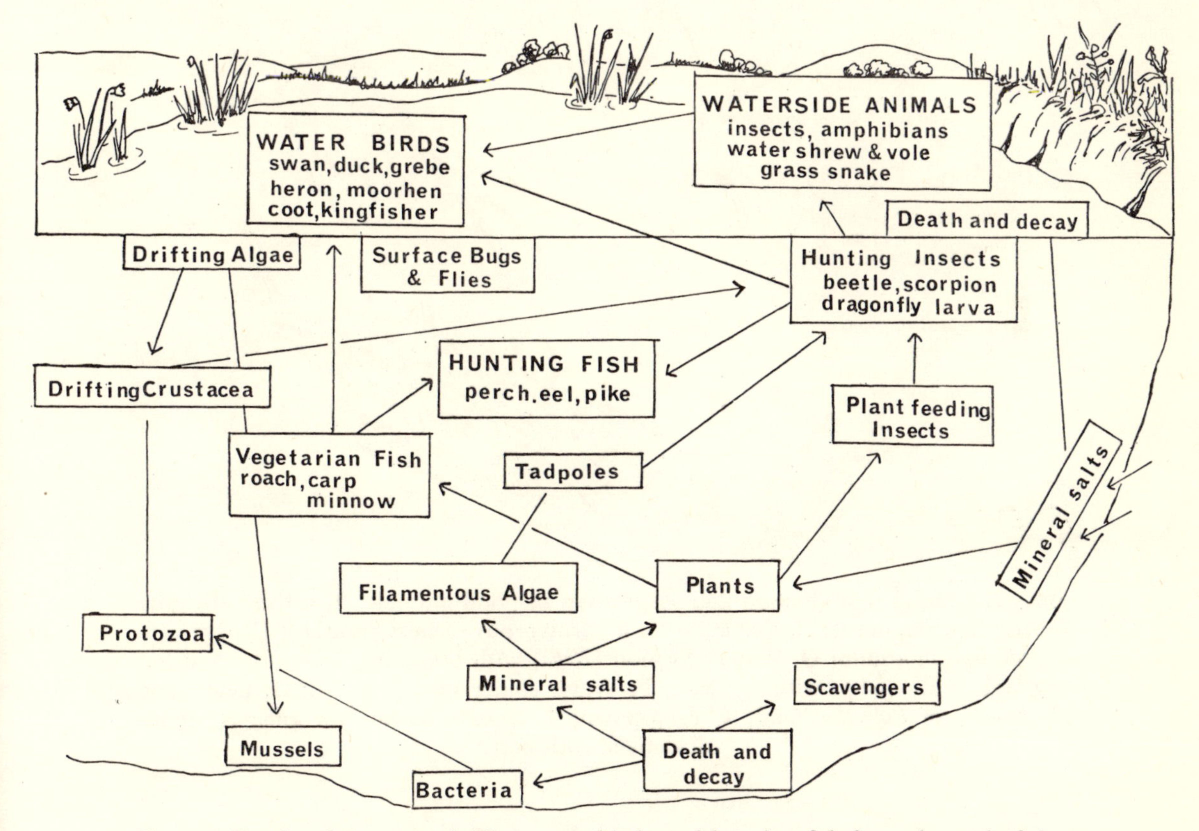

Fig. 50. Food web in a pond. The water birds and hunting fish form the end of the food chains.

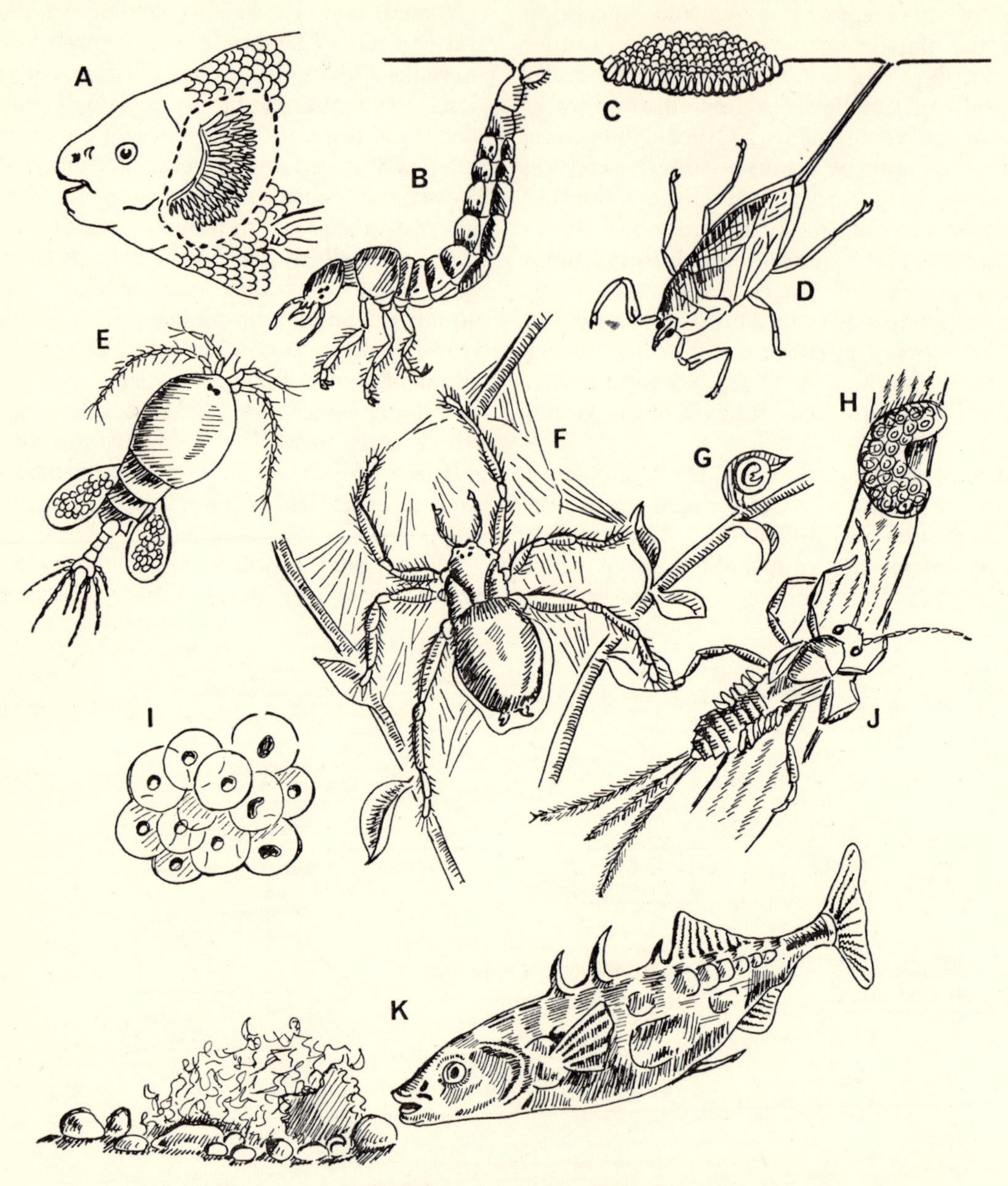

Fig. 51. Animal reproduction and respiration in pond animals. A, a fish with gills (operculum removed); B, Beetle larva, a carnivorous surface breather; C, Egg raft of common mosquito; D, Water scorpion (*Nepa*) with breathing tube, also used for egg-laying; E, *Cyclops* with egg sacs; F, Water spider in diving bell; G, newt egg; H, Spawn of *Planorbis* snail; I, Frogspawn; J, Mayfly larva with gills; K, male stickleback with nest.

becomes a life preserver and a refuge for the inhabitants, rather than a death trap, and this is due to the physical property of water. Down to 4°C it becomes denser and sinks. Below this water expands, and so continues to rise until the whole pond is at zero. The water then freezes from the top downwards. This means that any water below the frozen surface must still be above freezing, even though the air and ground temperatures outside may be down to many degrees of frost. The cold-blooded (*poikilothermous*) pond community remains in a state of hibernation in the pond mud and debris, and will remain so until the spring weather stimulates resumed activity. Hibernation in water also occurs in certain air-breathers such as the amphibians. These can survive the whole winter below water by respiring through their skins (cutaneous), using this after the fashion of a gill for absorbing oxygen from the water. In summer, like aquatic mammals and birds, they surface for air. Water insects also collect air from the surface and tap this with the tracheal system, a series of tubes leading to openings on the body surface (Fig. 48). Pulmonate water snails use air-sacs, and the unique water spider stores air inside its bell. The smaller, soft-bodied animals such as the protozoa, various worms, hydra, etc., take in oxygen over the whole body surface. Fish and crustaceans, also a few insects, use some form of gill apparatus for underwater respiration (Fig. 51). In Appendix 14 (p. 231) is given a list of pond animals found during the survey work done on Baldwin's pond.

Animal Life in Running Water

As a river travels along its course, the volume, purity, speed and temperature of the water, as well as the kind of channel which it cuts, will vary according to the distance from the source and river mouth. This has an effect on the animal and plant life, so that different community zones may be picked out. In each of these there is some peculiar species which may be taken as typical of its zone (Fig. 52).

Headstream or Highland brook. Near the river source the rapid movement of a shallow mountain stream wending its irregular course over bedrock mixed with stones and gravel results in a body of cold and clear water of high purity. Temperature is low and constant throughout the year. Few water plants are found, except for close tufts of mosses and liverworts clinging to the stones. The advantages to the animals living in this reach are the well oxygenated water—the disadvantages a shortage of food and a danger of being swept away. Most of the fauna is benthic. The animals remain hidden by clinging beneath the stones with their flattened bodies. They consist mainly of planarians, insect larvae and molluscs. Some of the insect larvae can actually live in the water film on the exposed rocks. In these mountain streams where an almost arctic temperature is maintained live a number of Ice Age relicts, such as the flatworm, *Planaria alpina* (see also p. 18).

The trout beck. As the headstreams unite to form a steady and more concentrated flow of deeper water which has greater force, a channel is carved out of the bedrock. Small fragments are washed away to settle as beds of silt in the quieter patches between the larger boulders. Little waterfalls are common in the more turbulent stretches. There is more room and opportunity for colonisation by a larger community, but this still remains mostly hidden as a benthic fauna. There are a few exceptions of more powerful necton species, such as the brown trout (*Salmo trutta*). This is the typical 'indicator' of this part of the river system. Under the stones may be found a variety of insect larvae and nymphs, molluscs, leeches and worms. Two characteristic small fishes of trout becks which also hide away, are the loach (*Cobitis*) and the miller's thumb (*Cottus*). The crayfish (*Astacus*) is another hidden dweller, found especially in limestone streams. The benthic fauna exhibits two tropic responses. As lucifuges they immediately seek hiding when exposed to bright light, as when a stone is lifted. They also tend to face the current (stereotropism). As mentioned on page 64, tropisms are also a strong motivating power in some inhabitants of a darkened micro-habitat.

All the animals of this region can endure long periods of starvation. In addition, there is no winter sleep, and an active larval stage may last for a year or two. The best time to collect is in winter, since the various larvae and nymphs will have hatched out by summer (e.g. May-fly).

Minnow reach (*grayling zone on the Continent*). As the gradient slackens off and erosion lessens, more winding occurs around obstacles, and the speed of flow is lessened. Sediments collect in places, and filamentous algae cover the rocks and stones. This tends to hold up drift material which settles and provides food and anchorage for a number of stream plants (e.g. willow moss, *Fontinalis*; milfoil, *Myriophyllum*; and water starwort, *Callitriche*). Around these a soft silt gathers, especially in the backwaters. The life in the trout beck is repeated, except for the absence of stenotherms (low-temperature animals). Large populations gather in the side-streams and pools, especially among the weed-choked shallows and silt beds. Insect larvae and snails forage among the roots and stones, worms burrow in the sediment, beetles swim in the

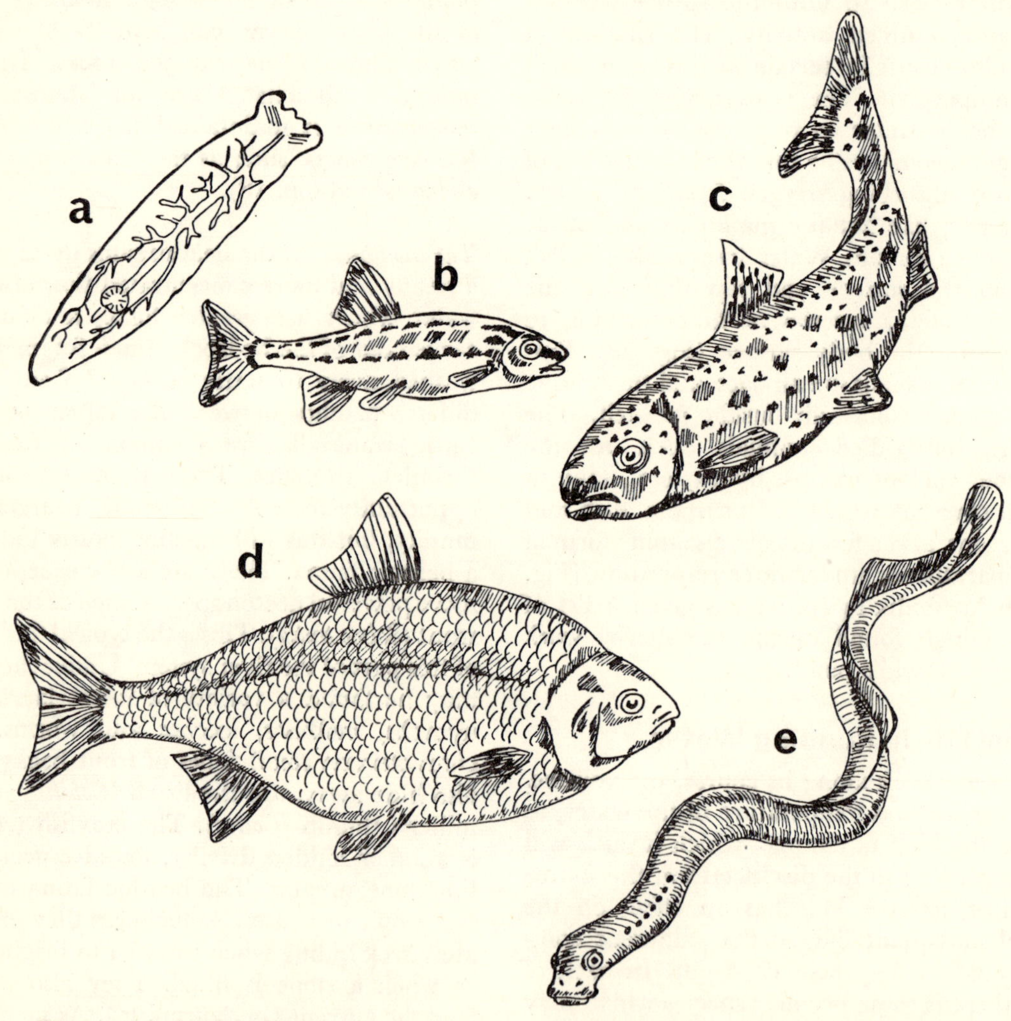

FIG. 52. Animal indicators in river zones. a, *Planaria alpina* in Highland brook; b, Minnow (*Phoxinus*) in Minnow reach; c, Trout (*Salmo trutta*) in Trout beck; d, Roach (*Rutilus rutilus*) in Lowland reach; e, River Lamprey (*Petromyzon*) in Tidal estuary.

pools, and skaters and whirligigs play at the surface. In some respects the community life resembles that of a pond, except that flowing water is near at hand and plankton may be absent. This is the habitat of the brook lamprey (*Lampetra planeri*) and restless little minnow (*Phoxinus*). As in the upper reaches, the water is usually clear and fresh. Where mountains reach down to the sea this may be the final zonal stage of the river system.

The lowland reach (*bream or roach zone*) (Fig. 53). At this stage along the river the greater volume of water, slower current and shallower gradient results in much meandering over a wide valley floor (the flood plain). There is greater contrast in the water depth from one bankside to the other. Shallow shingle banks build up on the inner curves, with a corresponding deep water channel cutting into the steep bank on the outer side. On balance there is more deposition of material than erosion, so that what is lost from one bank is readily made up on another. The water itself is subject to changes in level, low during drought and high in spate, and the surrounding countryside may become flooded after heavy rains. Consequently the river stretch becomes clear at one period and cloudy during another.

Flooding today has been largely overcome in these lowland reaches by the construction of raised river banks. Barriers such as weirs and lock gates affect the river's level and progress, and pollution may be a problem in the neighbourhood of towns and factories. Unlike the mountain tributaries, the lower river fluctuates in temperature, keeping more to that of the atmosphere. All these factors affect the river life and its fringe vegetation of swamp flora and trees. These may be drastically cut back or pollarded, to the detriment of the riverside birds and other animals. In the river itself silt deposition can become a dangerous hazard to the plants, and it may smother the benthic life, causing more harm to life even than pollution (see Sewage Farm, p. 137).

With the exception of the strong-swimming nekton found in mid-stream, mainly fish, most of the animal life is concentrated in the back-waters and inlets, and this compares with that of a pond. River fish include the roving shoals of roach, dace and chub. These vegetarians will often feed at the surface in trout fashion, whereas the hunting pike and perch lurk

FIG. 53. The Lowland Reach (Roach Zone).

among the plants, making use of their camouflage against the reed stems. Carp, tench and bream browse on the bottom, and the smaller gudgeon, minnow and stickleback keep to the shallows over the gravel and sand. During the breeding season in spring and early summer (when there is a close-season for anglers) many river fish make for the shallows to spawn.

Plankton life is composed mainly of algae (diatoms) fed from the backwaters. Sometimes it concentrates in mid-river during warm spells when the water is low and sluggish.

Tidal zone or estuary. Where the river meets the sea and mixes with the salt water a brackish element exists in the tidal ebb and flow between the salt marshes. This is the habitat of the salt tolerant plants which belong to our shores (see p. 101). The animals found here can withstand the changes in salinity and also the rise and fall of the river. Two species, both fishes, can actually penetrate the chemical barrier between salt and freshwater—the eel and the salmon. The mature eel makes for the sea to spawn (*catadromous*) whereas the salmon goes upriver (*anadromous*). At one time the salmon was joined by the sturgeon, the royal fish which spawned in the river mouths, but today no longer enters British rivers.

The river mouth and its mud flats is a favourite hunting ground for many birds. Waders search in the mud for molluscs and worms, and herons fish in the shallows. Geese and duck flight in to feed on the salt pastures. Otter and common seal search for flounders and crabs, sometimes disturbing the nets and lobster pots to the local annoyance of fishermen. Much of the estuarine animal life is hidden in the mud.

Part 2: Section G

Waste ground

A piece of land which, for some reason, has been allowed to go to waste, or has been uncovered by demolition or clearance, will eventually become colonised by a succession of plants and animals, even in built-up areas. A biological vacuum has to be filled in, and such an opportunity arises in places like a demolished building site, a neglected allotment or garden, a new roadside verge, rubbish tip, and so on. In time, whatever cultivated plants are left behind will have to compete, and in most cases give way, to the wild species. The arrival of these invaders will depend upon the agencies of seed dispersal, or perhaps some human activity.

A spectacular invasion of London's bombed sites took place during and after the Blitz of the Second World War (Fig. 54). Here, the colonisation started from nothing, since no soil was present on the burned ruins. From 1946 to 1953, in a unique study of London's ecology, members of the London Natural History Society undertook a bombed-site survey in the devastated area of Cripplegate just behind the Guildhall. A careful record was kept of the gradual build-up of a colony of plants and animals inside the City. Special attention was paid to the ecological factors which this barren environment offered, and to the adaptable characters of each successful species.

The plants arrived first, as was to be expected, followed by the animals which were given facilities for food and shelter by their plant neighbours. The success of each plant species was found to be due to its means of dispersal, then having reached the site, its ability to gain a foothold, to survive and to seed on a very poor plant diet in polluted surroundings with a somewhat unreliable water supply.

The Arrival

Plants which appeared on the bombed sites arrived broadly speaking, in two ways—by air and by human transport (Fig. 55). Among the first to appear were the lower plants, such as algae, mosses and ferns, whose spores are readily airborne. In fact, they are constantly raining down from the sky. Many flowering plants with light seeds having parachute attachments came in on the wind, or in stages with each passing train or large vehicle, swirled along in the vortex of air. There were times when a patch of fruiting thistle, rosebay or ragwort gave the appearance of a summer snowstorm over city streets, and this can often be seen on any piece of waste ground in summer.

The first plants to appear with the help of human agency were from seeds carried in with the large supply of sand required for defence purposes against aerial attack. These included a number of grasses, also maritime species normally seen along the seashore. The author noticed a number of these sprouting from sandbags on a local air-raid shelter even before the air-raids had begun. Such maritime plants can often be seen on the edge of bunkers on a golf-links, even far inland. Fodder plants were introduced with food for horses stabled in London, and from discarded bird-seed and various pet foods. Human food brought in by City workers produced an odd assortment of exotic plants as a result of discarding a half-eaten sandwich, an apple core, tomato, and the stone or pip of a plum, date, cherry, and so on. In the later stages, and after the end of hostilities, many garden flowers appeared on the cleared sites, having been scattered there from seed packets by well-wishers. Fruits with hooked attachments were carried in on clothing, dogs and other animals, possibly a number of them clinging to the fabric of vehicles. The harder, nut-like fruits became attached to the mud on shoes or in the pockets and turn-ups of trousers. Vehicles played their part in picking them up on muddy tyres, and some species of American origin undoubtedly came over with the Allied troops, stuck to the tracks and tyres of war transport and tanks.

Settling In

As mentioned above, the light airborne spores and seeds arrived first, and growths of algae on wet surfaces, mosses on porous stones and in

FIG. 54. A London bombed site shortly after the fire blitz in World War 2. a, Dandelion; b, Rosebay; c, Canadian Fleabane; d, London Rocket; e, Reedmace; f, Oxford Ragwort.

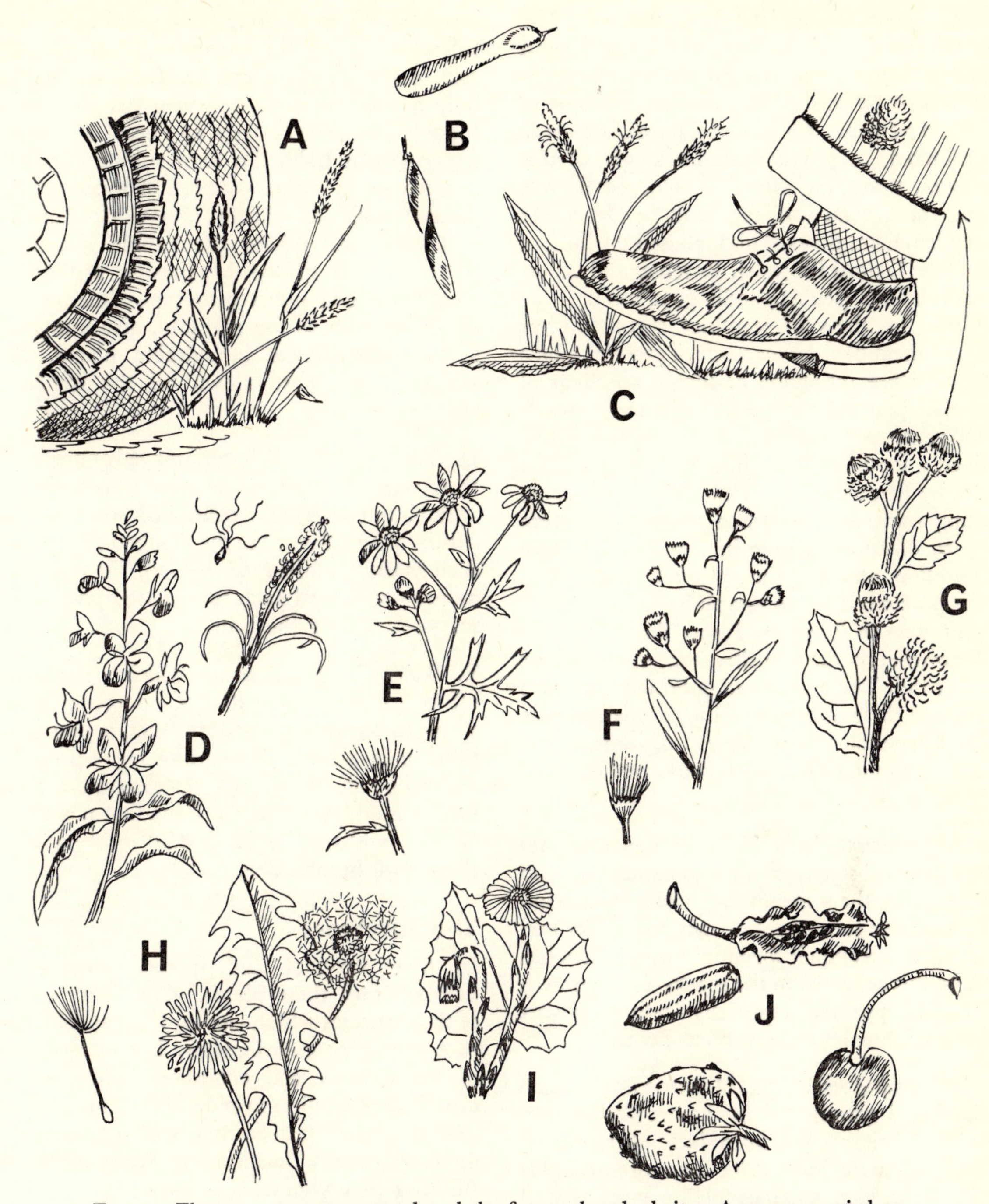

Fig. 55. Flowers on waste ground and the former bombed sites. A, grass, carried on tyres and wind blown; B, winged fruits of sycamore and ash; C, Plantain carried on shoes and in trouser turn-ups; Wind blown fruits of D, Rosebay, E, Oxford Ragwort, F, Canadian fleabane, H, Dandelion and I, Coltsfoot; G, Burdock carried on clothing; J, Discarded food—apple core, cherry, strawberry and date stone.

cracks, were among the first noticed. The humus from their dead remains, mixed with brick dust and powdered rubble, provided the first soil to be washed by rain into the cracks in the pavements. On this the parachute seeds germinated, to produce at the height of this invasion, some fine sheets of coloured flowers along the broken wall tops and in the cracks along basement floors. This included the famous trio—the Oxford Ragwort, Rosebay and Canadian Fleabane (see Appendix 3, p. 221).

Great numbers of seedlings sprouted, but most of them died off for want of growing space and moisture (as many as fifteen seedlings of ragwort per square yard in some places). Plants such as grasses, bracken, and even trees, gained a foothold in the piles of rubble pushed aside by demolition workers. With their penetrating roots and rhyzomes such plants were able to make anchorage in a somewhat loose foundation, and find the moisture level below. Such perennials, once established, have managed to live on for years. Today there are trees still growing on the odd site, some of them over 30 feet tall. The annuals maintained and even increased their numbers as they spread to new areas after each successive crop of seeds.

Five zones of colonisation could be identified on the bombed sites (Fig. 54).

A. *The basement floors.* Slabs of concrete laid directly on subsoil at or below ground level, and sheltered by the surrounding walls from wind and excessive sunshine, lay open to the sky and provided a receptacle for humus and rainwater. Where this collected in the cracks a rich growth of flowers appeared, especially Oxford ragwort which brightened the basement floors.

B. *The standing walls.* Most of the seedlings which lodged and germinated in the cracks between the wall bricks died from lack of moisture. This became the habitat for the more drought-resisting species, such as pellitory-of-the-wall and ivy. In some places, however, especially along the wall tops exposed to sunshine, sheets of handsome rosebay flourished.

C. *The rubble.* Debris of bricks, mortar and broken concrete heaped up in basement corners provided the most productive plant zones, but this varied from site to site in relation to the size of the rubble particles. On loose rubble only a few species with a firm root system or tough rhyzome (e.g. trees and bracken) gained a permanent hold. On finer rubble appeared a much more varied selection of waste-ground plants, such as thistle, mugwort, burdock and coltsfoot. Where the rubble had been flattened and firmed by passing traffic and feet, it was the tough, rosette-type of plant which survived, such as the plantain, also numerous grasses and mosses in suitable places. The winged fruits were blown off the hardened surface.

D. *The city gardens.* Where damaged or neglected these were invaded by the usual weeds of cultivation, but they also contained some garden shrubs of significance to the bombed-site survey, such as the buddleia (see below). Many garden annuals appeared each time a seed packet was scattered among the ruins.

E. *Water.* Deep basements which were sealed off and filled with water as emergency static water-tanks during the Fire Blitz, supported a number of plants. The various algae probably arrived by air, turning the water green in some, but such oddments as the Canadian pondweed (*Elodea*) and reed mace (*Typha*) must have got there with human help.

The Survivors

In their survival and spread through the bombed sites these plants of the Blitz consist of those species which can show great tolerance to a very poor environment. The general lack of water, poor soil conditions and polluted atmosphere would normally defeat most of our native species. Indeed, many of the successful ones were very localised natives before the Blitz, or even aliens.

An acceptance of poor soil, resistance to drought and pollution, rapid turnover of seeds in some, and firm root anchorage in others, coupled with a curious liking for burnt ground, made it possible for these plants to colonise a unique environment. As a war-time legacy and a credit to nature's fighting qualities in reclaiming land which she has lost, it is still hoped that this will never be repeated. Within a year or so the City's scars were clothed in a rich carpet of colour. In a sense the Irishman's dream had come true, for the scene was paved with silver (Canadian fleabane) and gold (Oxford ragwort) (see Appendix 3, p. 221).

The factors with which plants in built-up areas may have to contend are (*a*) reduction of light from screening (smoke, dust and fog), (*b*) choking of leaf pores by tarry deposits, (*c*) souring of the soil and reduction in the soil-bacteria, (*d*) absorption of sulphates by the leaves, causing corrosion, and (*e*) a limited water supply.

Before the London Blitz only one wildflower was still firmly established within the city area, the Pellitory-of-the-Wall (*Parietaria officinalis*), recorded by John Gerard of Holborn in his Herbal of 1597. By 1952 the list of species of wildflowers, grasses and ferns for the bombed sites of London had risen to 269.

Prior to the bombardment few wild animals were seen in the City, notably the London pigeon, starling and rat. By 1953 the record stood at 3 mammals, 31 birds, 56 insects and some 30 other invertebrates. These included snails, slugs, wood-lice, centipedes, flies, aphids and spiders, all finding shelter and food among the flowers and rubble. Insects of special mention were the butterflies and moths. Vanessids such as peacock, small tortoiseshell and red admiral were attracted to the scented blossom of the buddleia, and laid their eggs on the nettle. Many adults hibernated in London, and were seen in early spring flying amid the ruins. A number of species of hawk-moths took advantage of the trees which are the food plants of the caterpillars. To the ubiquitous rat, mouse and cat were added such oddities as hedgehogs, tortoises, lizards and snakes, and other lost or discarded pets. Birds were not in great evidence, as these were mainly visitors in search of food, rather than breeding residents. However one remarkable exception which made ornithological history, was the spectacular arrival of the black redstart, a summer migrant rarely before seen in the whole of Britain. It chose the bombed sites as a suitable nesting habitat. In Europe this bird inhabits rocky terrain and nests among rocks and ruins.

For a list of waste-land plants, including the City bombed sites, see Appendix 2M, page 218.

Part 2: Section H

Hedgerows, roadsides and fields

One of the outstanding features of the British countryside is the patchwork nature of its farmland, a collection of small fields separated by hedgerows in lowland areas, and stone walls in the uplands and moorlands. This change from the old feudal system of open, unenclosed strips of cultivated land, to the age of the Great Enclosures between the sixteenth and nineteenth centuries, had given to the landscape its distinctive atmosphere of homeliness and privacy to the owner and farmer, and what may amount to the most important factor in saving our wildlife from the inroads of a rising human population.

Hedgerows have existed since Saxon times, but it was not really until the Enclosure Act of 1847 that it became lawful to enclose land with hedges and walls in order to mark a permanent boundary. This probably arose from the old custom of dividing the fields with temporary hurdles, along which a row of coarse vegetation and a tree or two would have settled in. Possibly, too, a green stake here and there would have struck root and sprouted. This is especially the case with the hawthorn (*Crataegus*), a name derived from the Anglo-Saxon *haeg*, meaning a hedge.

By the nineteenth century some 180,000 miles of hedgerow had been planted, and today there are about 600,000 of them in Britain, containing some 80 per cent. of farmland. Since the last world war up to a fifth of these have been destroyed in some areas, due to neglect, cost of upkeep, changes in farming practice, the spread of houses and roads, and the dangers of stubble burning which can set alight a hedgerow and destroy it. In mountains and moorland areas the field boundaries are made from the local stone, since this is readily available, and planting in exposed places is not so easy. In the sheltered lowlands hedgerows are the commonplace feature. Many trees have been chosen for this purpose, especially those which plant readily, and can stand the mutilation of pruning, lopping and coppicing. Lopping was once carried out as a commoner's right on crown land such as the Royal Hunting Forests (see Epping Forest Project, p. 146). At intervals during the winter months, branches were lopped at a height of a man and his axe so that in a sense, the tree was beheaded, that is, pollarded. In time new branches will grow up from the crown of the trunk. Such pollards can be seen in woodlands (oak and hornbeam), along river banks (willows) and on the roads in built-up areas (limes, planes and chestnuts).

With coppicing the tree is cut at ground level

from such smaller trees as hazel and willow, in order to obtain the long sticks and poles used in various crafts such as rustic work, fencing and basket-making, and as stakes for plants. Lopped branches were carried away as timber for building, or as firewood.

The kinds of trees which lend themselves to this rough treatment, and can be trained to form a hedge to keep in cattle and horses, are hawthorn, blackthorn, elm, beech, hazel, maple, ash and elder, and on chalky soils, wayfaring tree, dogwood and privet. Today the hedgerow, especially with a ditch, is fast becoming one of the last strongholds of wildlife. It has largely taken the place of the former extensive deciduous woodlands which are a principal habitat for Britain's wildlife. Indeed, there are not many farmland animals which are independent of trees and bushes, and can live entirely in the open fields. Three examples are the hare, skylark and partridge.

It is in arable country where trees abound that the greatest concentration of wildlife may be found, and this is perhaps most obvious in birds. On 100 acres of farmland some 100-200 pairs of birds, of 20-50 different species may be encountered. On moorland there are only about 10-12 pairs of 3-10 species. Hedgerows are now vital to the survival of much of Britain's wildlife, but even so the numbers of plants and animals in a certain strip of hedge can vary considerably, according to its character and upkeep.

The old method of hedging and ditching suits the occupants best. It was done by hand, and at slow speed, so that life was not unduly disturbed. Also, the old craftsmen saw to it that the ditch was cleared of litter from time to time, and that no gaps appeared in the hedgerow. This was achieved by careful layering of the branches, whose side shoots then interlocked as they grew upwards. The modern method is to use speedy mechanical aids, and to trim the hedges somewhat ruthlessly, so that leading shoots are cut off and cannot reach the ground to strike new growth. Bushes become top-heavy, cutting out the light and air, and this holds up the process of decay. As a result the hedgerow bottom and ditch become choked with leaves and debris, and the hedge dies from the bottom upwards.

If this important farming practice were properly carried out, then a rich herb layer would be maintained, attracting a greater number of invertebrate life, and this in turn would bring in the birds and small mammals, followed by the larger predators. This massing of wildlife along the field borders could well be to the advantage of a farmer in the long run.

Hedgerows provide food and shelter to a large variety of animals and plants normally found in woodland. This is also a foraging area for many visitors. Those which divide the fields are usually undisturbed by passers-by, but those which follow the roadways can vary between extremes, such as a quiet, sunken Devon lane, to a wide and busy concrete motorway (Fig. 56). The first is sheltered by trees and a bank. It is cool and moist in summer, and protected from winds and by snowdrifts in winter. This habitat attracts a high number of plants and animals, and if there is a ditch adjoining the hedgerow, so much the better. In rainy spells the lane takes on a similarity to a stream flowing through a wood. Some five zones can be picked out—the lane itself, the verge, the ditch, bankside, and the hedgerow on top. A woodland association can be seen in the flowering herbs which are sheltered by the bushes and trees above. In spring oak blossom and hazel catkins dangle above the primrose, lesser celandine and cuckoo-pint. Ferns and mosses jostle with ivy and honeysuckle to gain a hold on the bank. In the ditch may be found the ragged robin and meadow-sweet. Insectivorous birds which nest in the bank or bushes compete for food with the smaller hedgerow animals such as bank-vole, rabbit and frog. These attract the predators such as fox, weasel, owl and grass-snake. A badger stakes out its underground fortress, with perhaps a rare dormouse nesting above it.

In the verge a small grassland community may flourish, the species depending on the subsoil and humidity, so that foxgloves appear in the acid lane, but cowslips may turn up where there is chalk. A dry stretch contains wild thyme whereas ferns grow in a wet one.

Whatever should happen to grow, it is the plants which largely control the animals. Vegetarians are tied to their food plants, and predators to their prey.

In a 50 yard stretch of Devon lane the author during a holiday once listed the species found there during a sunny spring (see Appendix 4, p. 222).

The lane itself, relatively undisturbed by traffic outside the holiday season (now a serious problem in the West Country), becomes a highway for the hare, fox, badger and stoat. These can often be seen in broad daylight by the quiet watcher. A partridge takes a dustbath and a house-martin collects mud from the puddle for her nest under the nearby cottage eaves. At night a wandering toad or hedgehog may be the only pedestrian.

In contrast, the main road appears like an ecological desert. The broad concrete motorway with its fast day and night traffic is a hazard to animals which cannot know the dangers of jay-walking. From deer to New Forest pony, toad and hedgehog, the many corpses are a silent witness to the heavy nightly toll on our highways, which impartially claims man and animal alike. By early morning our roads are largely cleared, for the scavenging crow and magpie have already taken their pickings (see Appendix 5, p. 222). The broad, roadside verge is monotonous in its repetition of the grasses and herbs which must withstand the attacks of the mower and weed-killer. As with the hedgerow the soil will determine the species of plant, and an observant botanist can tell the kind of countryside he is passing through by the plants which he notices. This search for soil 'indicators' among plants can be a useful and entertaining exercise for passing the time during a long rail or road journey. Between the two extremes of lane and motorway can be found many varied roadside habitats, some more or less undisturbed, but others under constant pressure from cutting and spraying, and passing traffic. In each one can see a similarity of flora and fauna to a more conventional habitat. The Devon lane, for instance, shows the associations of a woodland—bare ground (the

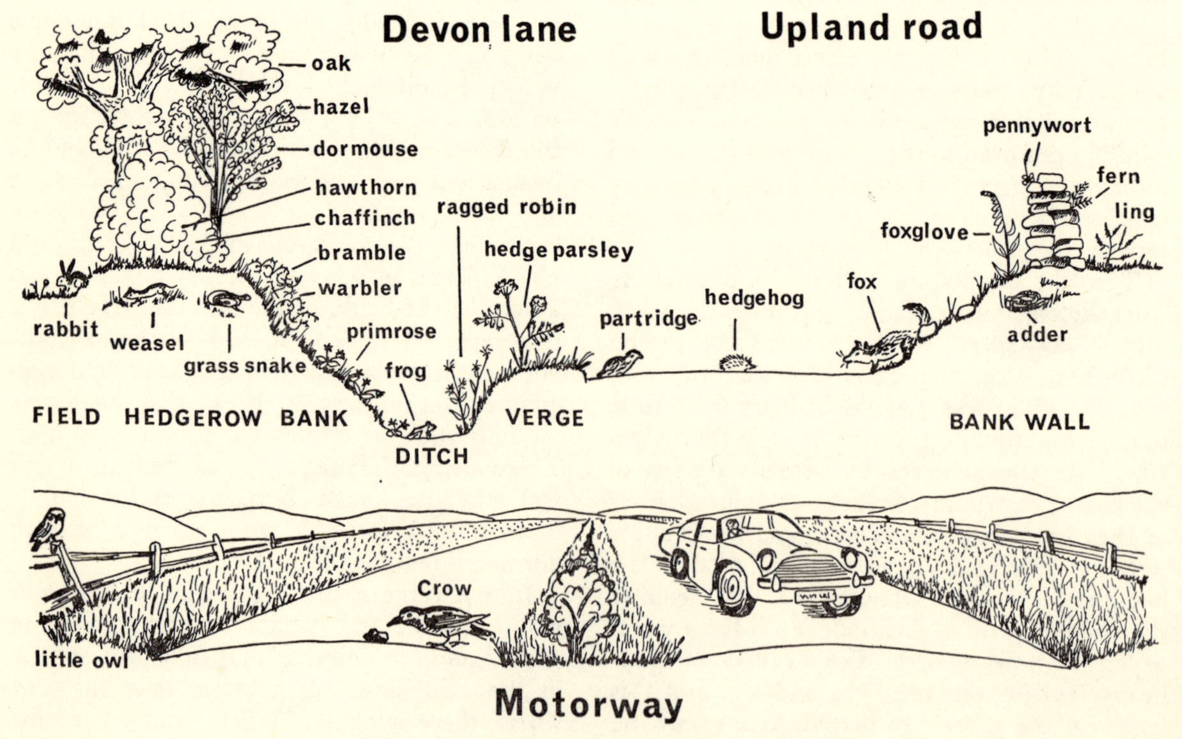

FIG. 56. Examples of habitats along hedgerows and roadways.

lane), herb layer (the verge), shrub layer (the hedgerow) and woodland canopy if there are trees. An upland road crossing moorland supports a verge of heathland plants (heathers and berries), and a stone wall contains a cliff community of mosses, lichens, pennywort and wall-rue fern. A wall over limestone country might resemble a continuous rock-garden of chalk-lovers, with the occasional rarity tucked away in a crevice (Fig. 56).

In a similar way to road verges and hedgerows, the wildlife of our fields will vary according to the subsoil and topography, but probably more so from the biotic factor created by man. Since a field, by definition, is a man-made habitat for the purpose of growing things as food, to support domestic animals, or in a narrower sense, a place for recreation, the wildlife must take second place. No arbitrary distinction can be made about the type of wildlife community to expect in any one field, but in a very broad sense this can be done if the following factors are looked into:

a. *The locality.* A field, say, in the Thames valley, and another in the Welsh hills, will support a number of similar plants and animals which have a wide distribution and tolerance to different surroundings. There may also be some distinctive species in each which are either localised or adapted to valley or hill. In the Welsh field, for example, one might expect, or at least hope, to catch a glimpse of the polecat which appears to be well adapted to hill-farming country, even though it is now somewhat rare. The Thames valley field, on the other hand, might harbour a brown hare or two, a species not usually found above 2,000 feet.

b. *Activity.* A field can be used for growing a crop of cereal, fruit or vegetable, or it may be left as grass pasture. Each will contain some distinctive wildlife which is attracted to it. Taking mammals as an example, one might expect to find harvest mice in the corn-field (a little optimistically, perhaps!), deer among the vegetables, and squirrels raiding the fruit. Field voles would inhabit the rough pasture. The interesting thing here is that these unwelcome visitors change as the crop grows, is harvested, and the earth turned by the plough. House-mice will attack the young corn, foxes will kennel in the growing crop, rooks feed on the stubble, and hares make their winter forms in the ploughed furrows. In fields which are regularly ploughed it is the invading annuals which dominate the wild population. Fields left to pasture will have a higher percentage of perennials. These in turn will attract different species of insect-feeders, including those which attack the human crop. Cabbage white butterflies make for their favourite food-plant, cinnabar moths fly around the ragwort, and some clouded yellow butterflies may produce a second brood in a field of lucerne during a hot summer.

On a recreation field used for cricket or football, or on a golf-links, the grass is kept permanently short, so that there is an open aspect with good view of approaching danger all the year round. Certain birds such as the black-headed gull and fieldfare use it on winter days, and the skylark and swallow may turn up in summer, either for feeding or nesting.

c. *Topography.* This feature, coupled with the climatic factors of temperature and rainfall, and also the nature of the soil, are permanent to the area, no matter how the field is treated. On condition that the wildlife it contains is not eradicated or constantly disturbed, certain characteristic plants and animals may be found. A farmer's field on the South Downs will probably contain a number of chalk-loving plants, such as the horseshoe vetch which is visited by the chalkhill blue butterfly. A crofter's rough mountain pasture field in the Highlands could harbour a pair of wheatears which are nesting in the stone wall close to a clump of insect-catching butterwort. In each case it is the nature of the rock and soil, and to some extent the climate, which determines their presence.

d. *Age.* A field may be young, or many centuries old, and even mentioned in the Domesday Book. Much interesting research has gone into the effect of age on soils, and how the goodness may be retained by careful crop rotation or application of fertiliser. A farmer's care of his field may consequently mask its age, for in the course of time, under natural conditions, an open field originally cut out of a woodland area should return to its former climax. This becomes obvious in a neglected field, where perennial herbs, then shrubs, and finally trees take over. In a properly maintained field this would not occur, but even so an observant naturalist can spot the signs of its age. Things to look for are the condition and age of the hedgerow trees, and whether they shelter any unusual 'relicts' of earlier days (see p. 152).

The diameter of fairy-rings made by mushrooms are indicative of a field's age, and would also suggest that the crop has always been grass.

A list of plants associated with hedgerows and fields is given in Appendices 2O and 2P, pages 219, 220.

Part 2: Section I

Man-made habitats

In a country like Britain, with its high population and two thousand years of civilisation, the countryside no longer presents a natural appearance. Apart from a remote mountain crag or some small part of the coast reclaimed from the sea, signs of man's presence or passing can be seen everywhere, no more so than in his towns.

Earliest traces of man, apart from human remains of bones and teeth, are the more enduring artifacts which he made, such as stone tools, dwelling-sites, burial mounds, monoliths and trackways. These in themselves hardly affect the natural history of our islands—it is what they started at the dawn of civilisation that has led to what we see today. The Neolithic farmers who settled here needed space for their homes, crops and animals. Today we are still in need of these essentials, and have cut more and more into the countryside in order to supply the needs of our ever-increasing population.

Throughout the history of Britain there has been a steady retreat on the part of wildlife, but not a total rout, for there are a number of animal and plant species which, in a sense, have become civilised and have succeeded in taking over man-made habitats. In a modern society the essential needs of man may be listed under buildings, transport and highways, water and food (both grown and stored), disposal of waste, and various amenities for health and leisure. Each of these requires space, and each has been exploited by wildlife which has come to terms with man. It is in such places that a town naturalist, far from envying his country colleague, can pursue some interesting studies in ecology where a constant struggle is going on between himself and nature in the competition for space. Any fair-sized town may be chosen, and London is selected for this chapter as a place for natural history activities within a built-up area.

From what is known of London's geology and prehistory, and also from Roman writings, a fairly clear picture of the Thames valley landscape emerges (Fig. 57). A centurion pacing the walls of his city would have had an uninterrupted view of a placid, clear river, winding up the broad valley, and running with salmon. The sloping banks would have been covered with extensive reed-beds, the home of the otter and bittern. In those days the Thames was probably tidal up to this point, the first fording place up river where a riverside settlement could be built. It was here that general Aulus Plautius in A.D. 54 built his supply port. There were mussels and oysters to be gathered in the waters near by. Later, on the gravel terraces above the flood water level, Roman Londinium was built as a walled city, on two low hills. Today, two of our most historic buildings stand there, having survived both fire and wars—The Tower on Tower Hill and St Paul's Cathedral on Ludgate Hill. In between once flowed the small side-stream (now in the underground sewers beneath Wallbrook) which entered the City through culverts in the northern wall at Moorgate.

Today the tide reaches as high as Teddington, and Roman London lies buried some 20 feet below the present level. At one time much of the terraced slopes, a poor and readily drained gravel soil, supported a heath-like scrub, the home of birds and small game. Beyond this, and as far as the eye could see, the sentry would have gazed on a hostile, dense woodland of oak and hornbeam—a vast extent of trees firmly entrenched on London's principal bedrock, the London Clay—and beech on outcrops of Chalk in the far distance. Here roamed the larger animals, such as deer, wild pig and bear, cattle and lynx; it was a place to avoid in which the Britons laid ambush for the patrolling legions on their marches to and from Colchester and St Albans.

From this vantage point on the battlements an ecologist would have picked out a number of habitats—water where the Thames and its tributaries flowed, marshland where the reed-beds grew, heath and scrub on the gravels, and woodland on the clay and chalk. Today, after two thousand years of occupation, the Thames valley only shows traces of these old habitats for wildlife. The polluted river is now a highway for ships, and the salmon have long since gone. The reclaimed marshes and reed-beds, used in the Middle Ages as rich farmland for sheep-rearing and fruit growing, have now been built upon, or opened up as dockland and reservoirs.

Glimpses of the ancient heaths on the terraces may be seen in the few open spaces preserved for recreation, such as the parks and commons. Higher up the valley slopes fragments of the old woodlands preserved by the Normans can

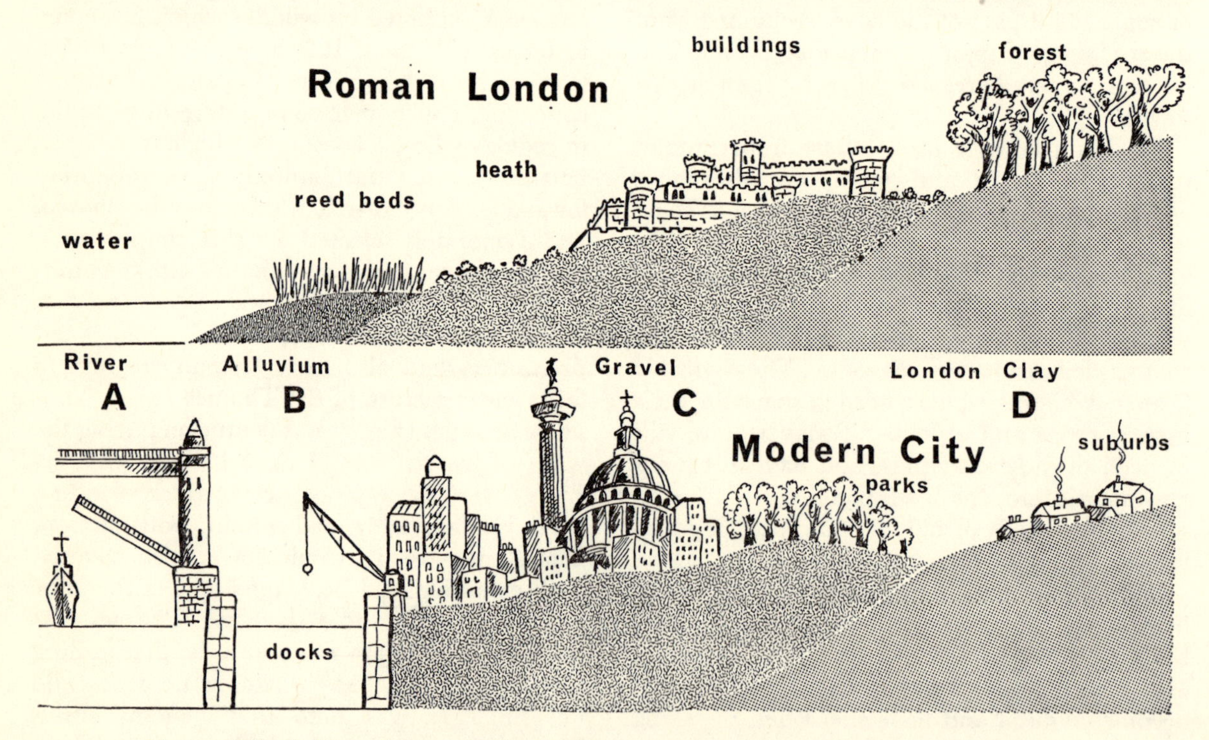

FIG. 57. The Built-up Area. Modern London compared with Roman London showing changes in the former habitats (for explanation see text).

be seen in the pitiful remains at Epping (the Forest of Essex), Ken Wood (Middlesex Forest) and at Richmond Park, once a Royal Chase. Apart from these precious relics, most of London's woodland life has to be sought in the suburban parks and gardens.

Over all this now sprawls the brick and concrete jungle of Greater London, some 30 miles in diameter, with the original 'square mile' of Roman London at its centre.

The Buildings

From cave to modern flat seems a long step, but in either extreme it seems reasonable to assume that the animals with which man has come into closest contact are those which live on his doorstep, or on his person. As true parasites the human flea, bed-bug and louse, and the occasional nematode and tapeworm, are still with us, even in the most sophisticated of homes. So are the 'hangers-on' or commensals which share our buildings and food, such as the rat, mouse and sparrow (Latin *mensa*—a table). Neglected buildings and unguarded food are always a target for attention by such enterprising animals and plants, and have been a problem since the first homes and crops were built and grown. The presence of these unwelcome invaders will depend largely on the relaxed vigilance of the housewife, storekeeper and sanitary official. Houses, shops, offices, public buildings, warehouses, hotels, and even hospitals, are all vulnerable, and the economic or social importance of this problem is in direct relation to the damage or nuisance which is caused.

A careful survey of his home might reveal to the surprised reader a large number of uninvited guests which he has harboured during a single year. The smaller invertebrates, in particular those which find their way in through open doors and windows, down chimneys, up drainpipes, even through cracks in the roof, often arrive quietly and unnoticed. Apart from these wild occupants, which may range from protozoans in a bowl of neglected flowers, to bats in the loft or rats in the basement, there are also the large numbers of pets kept in Britain, and animals for education or research. A census taken of London's pet population would make an interesting sociological exercise into human likes and dislikes where animals are concerned, for this ranges from the conventional family poodle or budgerigar to an alligator in the bathroom. Generally, these have little effect on the community as a whole, if kept under proper control, If not, they may prove only a nuisance value of local importance, such as a barking dog, neglected rabbit hutch (which can attract rats), or an escaped pet snake. A predatory cat can be a serious menace to garden birds, especially at nesting time, and a pond of goldfish might attract the early heron. In large towns such occurrences are usually localised and passing. They may even be welcomed and publicised, if overdramatised, by the popular press as useful copy to fill out the news.

Plant invaders of buildings are less in evidence, and more a gardener's problem, but mention should be made of the ubiquitous fungi. Minute spores, carried far on the air currents, somehow reach their destination, so that neglected food, clothing and building material become a target for attack. Mould appears in damp, overlooked places, and from a tiny patch of pin-mould (*Mucor*) on some jam, to an alarming spread of dry-rot (*Merulius*) in the staircase, the fungi keep us on constant guard against damage and decay. This is perhaps also a timely reminder that decay is an essential link between life and death.

A list of commensals and invaders of buildings is given in Appendix 2Q, page 220.

Gardens

The wildlife seen in our gardens can be quite varied and numerous, more so in country districts where there is ready access from outside. Usually these outsiders are not tolerated. With the possible exception of birds, and such useful allies as the hedgehog, lady-bird, spider and centipede, all of which are useful predators of insects, slugs, and so on, the wildlife must face the attack of hoe, weed-killer and insecticide. A neglected garden, however, will soon attract the attention of a host of so-called plant 'weeds' and animal 'pests', to use the gardener's terms. Taken as a whole they form a community which is normally part of the habitat found in a woodland or hedgerow where trees and bushes give cover and food.

Every town and village has its gardens, and around London these form a broad belt which separates the inner built-up area from the open countryside. From the air this dormitory area looks not unlike a circular strip of open woodland in which the dividing walls and fences are hardly noticed. This is the bird's-eye view, and that of insect and wind-blown seed, and each can readily spread from one garden to the next, if given the chance. The fruit trees and ornamental shrubs lend a woodland atmosphere to this artificial habitat (Fig. 58). The soil is enriched with each leaf-fall, and birds find the necessary food and sites in which to build their nests.

That such a man-made habitat will ultimately revert to a woodland was put to the test by the author when he moved to his present home, in a London suburban district only eight miles from the City. The end portion of the garden is covered by old fruit-trees, the remains of a former orchard. On a triangular plot, measur-

Fig. 58. The Garden habitat for birds. A, loose tile (starling, house-sparrow); B, eaves (house martin); C, tree (mistle thrush, wood pigeon, jay); D, nest box (blue and great tit, spotted fly-catcher); E, climber (hedge sparrow, chaffinch, goldfinch, wren); F, bush (song thrush, blackbird); G, pond (for drinking and bathing); H, bird table (food-crumbs, seed, fat, nuts); I, rubbish dump (food, slugs, worms, snails); J, lawn (worms); K, shed—open door (robin). All these birds have appeared in the author's garden. (See page 223).

ing about 100 feet on each of the three sides, a miniature woodland was encouraged to grow. The heavy growth of garden escapes and invading waste-ground weeds which had grown up was first cleared away, and the rich soil turned over. On the bared ground a number of common woodland herbs (bluebell, anemone, ground-ivy, lesser celandine, etc.) were introduced, as well as a shrub layer of young woodland trees and bushes. Here and there tree logs were laid down to offer micro-habitats to woodland invertebrates, and nestboxes erected to attract the birds. A small pool was also built nearby to ensure a permanent supply of water, and to maintain a small colony of pond-life.

Since this little plot is mainly surrounded by treeless gardens, and is sheltered from the noise and disturbance of traffic (and children!) it has become a miniature oasis for wildlife within a built-up area of roads and houses. These, however, are flanked on one side by a golf-links and suburban park, and on the other by part of Epping Forest. No doubt these open spaces act as reservoirs of wildlife which constantly spills over into this small garden 'woodland'. Ground travellers such as hedgehog, frog and toad, airborne insect and birds, and the occasional wildflower, now regularly enter this sanctuary. This experiment was started in 1954. By 1963 the small plot had settled down to a well-established climax of woodland flora and fauna (see list in Appendix 6, p. 223).

In this man-made or rather man-controlled woodland, the old apple and pear trees form the plant dominants. From time to time a badger, fox, squirrel and tawny owl has represented the animal dominant (very tame and kept in cages during my absence).

Obviously the main controlling factor is a biotic one—man.

In return the environment has a strong amenity influence on its creator. A secluded shed in pleasant surroundings makes an ideal retreat for writing and quiet thinking, a pleasant antidote to daily commuter living.

It was noticed during the early years that a number of the sturdier plants of waste-land kept a tenacious hold on the ground, and had to be dug out repeatedly, especially some of the grasses. In time, however, as the shade canopy increased with the growth of the shrub layer, these disappeared. Today only the odd cultivated escape puts in an appearance. One interesting discovery was that the lower plants of the Bryophyte layer, such as mosses and fungi, were among the most difficult to establish. Soil and dead wood containing mosses or toadstool mycelia from the nearby Forest were placed down frequently, but met with failure. These only appeared on the logs and brick path already in position.

Reservoirs and Sewage Farms

These two essential services—the supply of purified water and the disposal of waste material—are modern methods used to safeguard the health of the community. As a result some very unusual communities of wildlife have now become our permanent neighbours, for in both these services certain organisms are exploited by use in the control of disease. This form of biological control has improved over the years, from crude methods of filtration and sewerage control, to efficient uses of bacteria and algae. These in turn have attracted further wildlife, as we shall see.

The sewage farm

Apart from drinking, washing and other domestic needs, water is extensively used in the transport of waste. At one time rainwater simply washed away rubbish and excrement from the roadside, or it was simply dumped into rivers and pits. In mediaeval times the marshy waste at Moorfields just north of the City was used as an open sewer, and the Thames suffered from serious pollution due to the discharge of untreated excreta. Latrine facilities were little better, although some early civilisations tried methods of controlled sewerage disposal. The Romans, for instance, used pipes. Other methods were to distribute the liquefied waste over the land. On porous soils this worked fairly well for a time, as did the first contact beds consisting of pits filled with stones. Clogging was the chief problem, and it was not until the turn of the present century that an efficient system came into practice in which the bacteria which live in the so-called filter beds render the sewerage harmless.

The stages in this process are broadly as follows (Fig. 59). Crude town sewerage from the streets and houses flows into the *Grid Pit.* Stones, grit and other heavy objects sink to the bottom, and such material as paper, rags, fruit and vegetable peelings are caught on a screen. Excrement is broken up. A turbulent liquid passes on to the *Sedimentation Pit.* In this, every twelve hours or so, the settled sludge is removed and dried, and makes a useful fertiliser. This stage controls the rush of sewerage during peak periods, and allows the mixture to acquire a good oxygen level. Effluent from the Sedimentation Pit is then sprinkled onto the *Percolating Filter Beds*, which are about six feet deep and filled with loose stones or clinker, of the kind which does not crack during frost. Sprinklers rotate above, or move backwards and forwards, so that the tank receives a steady flow of liquid. Each stone becomes soaked in effluent, and collects a massive colony

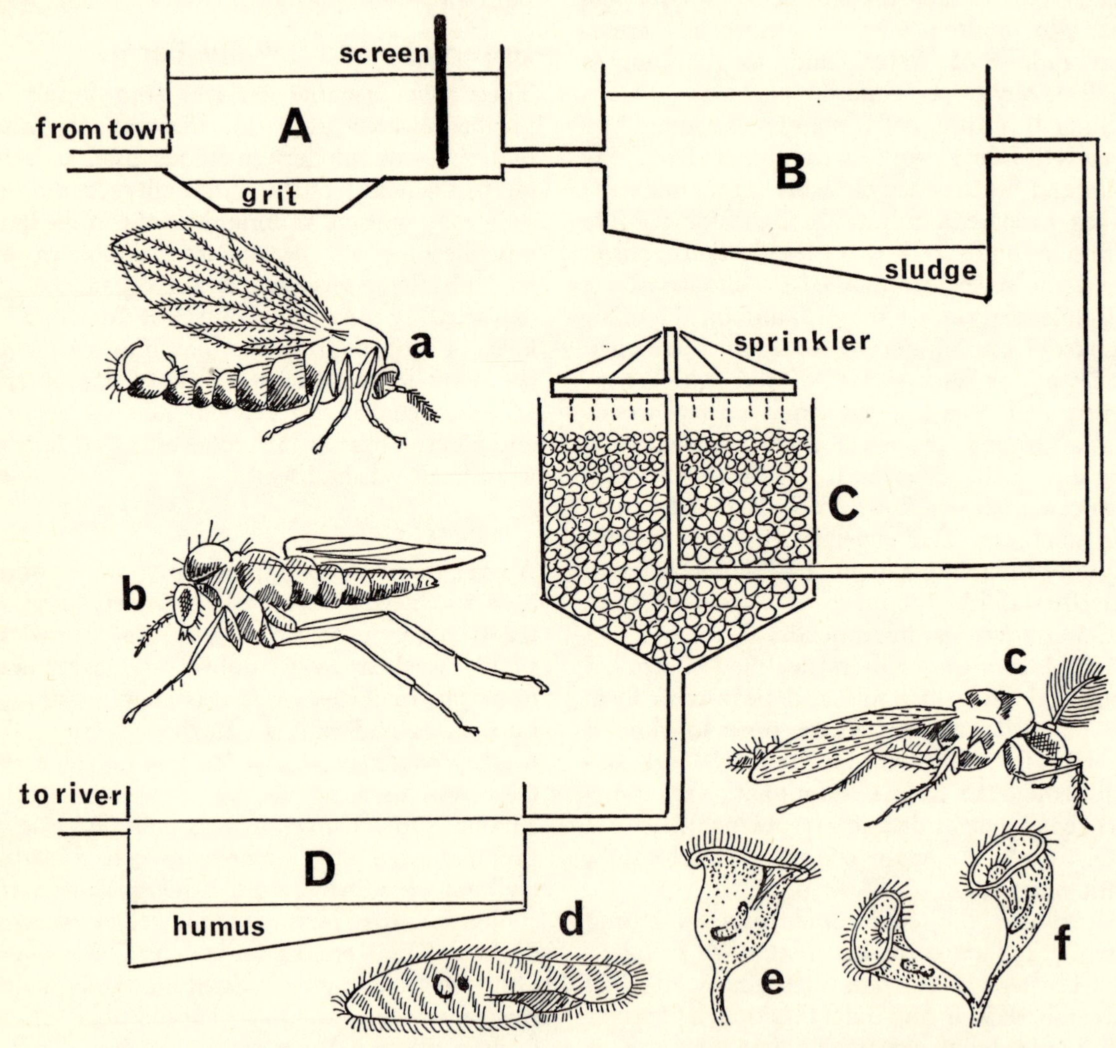

FIG. 59. The Sewage Works. A, Grit pit receiving crude sewage; B, Sedimentation tank; C, Percolating filter bed; D, Humus tank. a, Moth fly (*Psychoda*); b, Window fly (*Anisopus*); c, Sewage midge (*Metriocnemus*); d, *Paramoecium*; e, *Vorticella*; f, *Epistylis*.

of bacteria and algae to form a slimy film. This in turn provides food for a host of micro-fauna, such as the ciliated protozoons and various fungi. In time this might choke the filter system were it not for masses of worms and insect larvae which now feed on the rich meal. It is estimated that in early summer there are some 40 million flies busy on each acre of the filter beds. These in turn attract numbers of insectivorous birds, such as wagtails, fly-catchers, warblers and starlings. Waders, mainly as winter visitors, feeds on the worms. At intervals this food chain breaks down, and there is a heavy death rate among the lower organisms which become starved of food. The surface film to each stone is shed, in what is called the 'vernal slough'. This is carried to the *Humus Tank*, where a fine broken-down sediment collects. When drained off and dried this can also be spread over the land, or carried away for disposal. If taken to sea it should be dumped well away from the river mouth, otherwise it is simply swept back with the returning tide. Such a fine sediment can result in even greater harm to life than pollution, since it can choke out all river life, including the plants.

Life found on the sewage farm can be compared to that which lives on the estuary mud-flats (see p. 122).

The reservoir and waterworks

Water supplied for drinking or for personal use must be of high purity to ensure good health, since a number of disease organisms can live in water. At one time rivers, wells and springs supplied Britain's needs, and more often than not such water was taken and used direct from its source. This is always risky, since pollution or disease can sometimes seep in from underground, or may be put in from above, either accidentally or deliberately. Today the water passes through a series of stages in order to eliminate all traces of disease organisms. London's water comes from the Thames and its tributary, the river Lea, once a favoured fishing river of Izaak Walton. Subject to the Water Board's regulations governing a river's flow and level, below which stagnation might become a health hazard, the pumping stations divert the river water into large storage reservoirs. Water purification begins here (Fig. 60).

Sedimentation of much of the river's contents takes place because the water is no longer in movement. Larger forms of life or floating objects can be checked by screening, and any likely entry of polluted water becomes diluted as it spreads out through the reservoir. This has a large surface area exposed to the open air and sunshine, and this encourages high concentrations of plankton life, such as bacteria, unicellular algae, and various kinds of small crustaceans and insect larvae. In such an illuminated area, in water which has a constant low temperature, even in summer, any likely disease organism usually has a short life. This is because most water-borne germs live best in the intestines where it is warm and dark (e.g. the causes of typhoid, cholera and dysentery). The algae help in water purification by manufacturing large quantities of oxygen. As a result they are buoyed up to the surface by the bubbles, a common sight among pond-scums in any pond.

In a reservoir the large rooting plants normally found in the marsh and swamp zones of lake-sides are absent because of the sloping concrete sides and lack of shore mud. Swarms of water-borne insects hatch out from eggs laid on this (e.g. *Chironomus* flies). The larva, called a blood-worm, sometimes colours the mud by the water's edge of ponds and river backwaters. Of the larger forms of animal life it is the fish and water birds which make a reservoir their temporary or permanent home. Fish are sometimes deliberately introduced as a sporting amenity, such as pike, perch, roach and trout. This attracts the fishing birds such as heron, grebe and cormorant. At present the largest heronry within the London area is situated on an island in one of the Lea valley reservoirs near Walthamstow. Occasional otters have been known to visit this spot.

The small list of larger animals, and the masses of planktonic flora and fauna with little in between, show a similarity to the larger lake or Highland loch community (see p. 109). As a winter retreat such a large and quiet inland water, even close to large towns, attracts many winter visiting waders, duck, geese and swans, so that this habitat has become an attraction for ornithologists. London's reservoirs, as well as sewage farms are under regular observation by enthusiastic amateurs who are given permission by tolerant officials to carry out their counts and surveys among the surprising numbers of winter visitors. One of these is the Black-headed Gull, the so-called London Gull in these parts. Over the past eighty or so years it has taken to wintering inland in the large towns all over Northern Europe. In London, birds ringed on the Thames Embankment and in St James's Park go as far as the Baltic to breed. During winter they scavenge by day on the rubbish dumps and marshalling yards, and receive their daily ration of bread and cheese from their human friends. Towards dusk they quietly fly away to roost on the reservoirs. When in London this gull takes on the role once carried out by the ancient kite, as a kind of dustman.

From the reservoir, where a regular supply of water can usually be relied upon, the screened water enters the filter-beds at the waterworks (Fig. 60). These are a series of elongated tanks whose floors are covered with a graded series of layers, from sand down to stones, and through which the water must pass on its way out. This traps most of the solid particles, including plant and animal, but not the micro-organisms where disease may lurk. These are

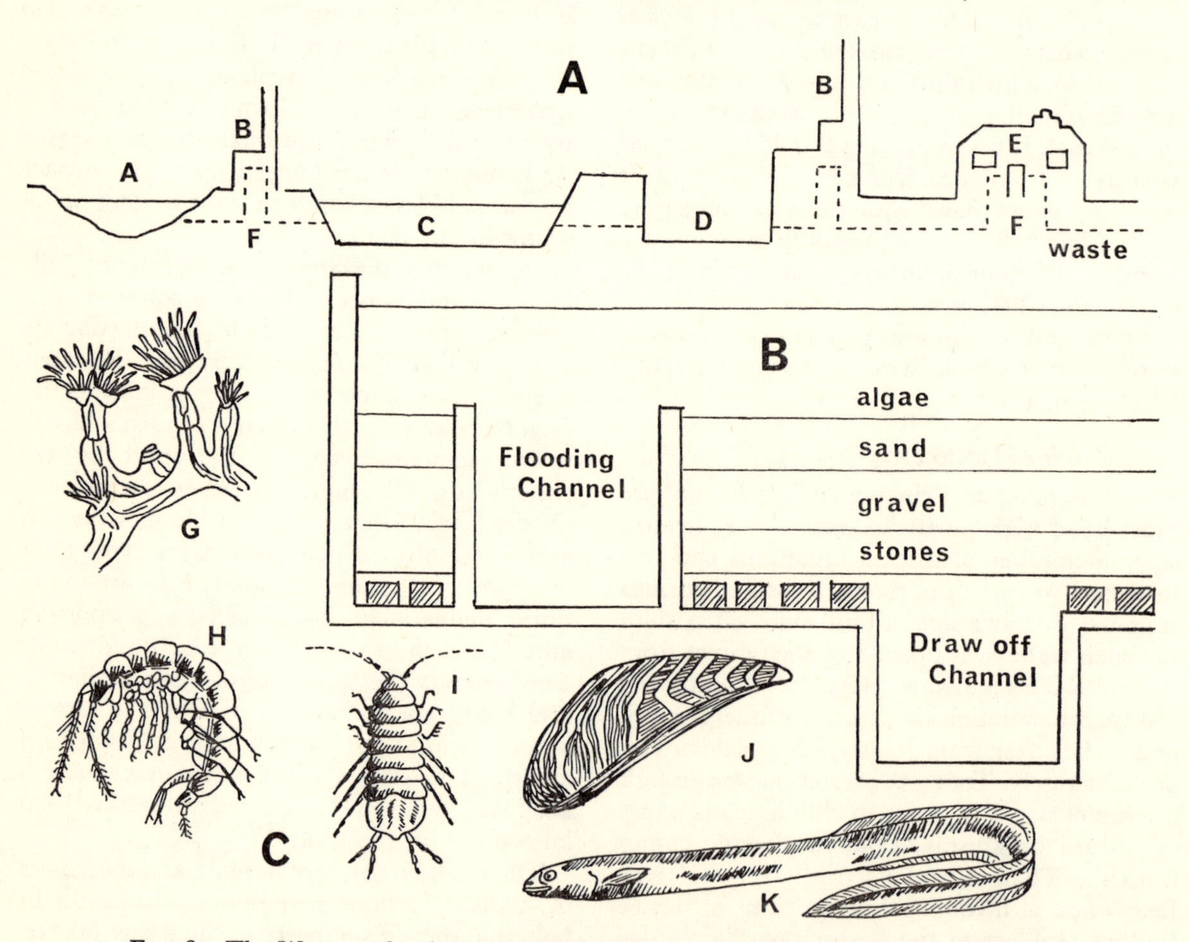

FIG. 60. The Waterworks. A. water supply from water source to consumer. A, source (river); B, pumping stations; C, reservoir; D, filter bed; E, Consumer; F, pipes; B. section of filter bed; G, a polyzoon colony; H, freshwater shrimp (*Gammarus*); I, freshwater louse (*Asellus*); J, Zebra mussel (*Dreissensia*); K, Eel.

checked by a growth of filamentous algae which spread over the sand at the bottom of the filter. Even bacteria fail to penetrate this delicate growth, although samples of water are constantly checked. It is assumed that if *Bacterium coli*, a common and harmless intestinal parasite of man, is below a certain concentration then there is little danger of harmful germs being present. A final precaution is the chlorination plant through which the water passes on its last stage to the consumer. The life of the algal film is limited, and from time to time each filter bed is emptied in turn in order to remove the old film. Using rakes the workmen 'roll up the carpet'. Meanwhile a cleaned and refilled bed is brought into operation.

One of the former problems of a waterworks was the constant clearing of the supply pipes which became choked with so-called Pipe fauna. An early account from Hamburg tells of the untreated water from the Elbe which was drawn from the river, then passed on direct to the water mains. Domestic filters failed to stop the entry of disease bacteria, and in 1892 a heavy cholera epidemic killed some 80,000 people. A search of the pipes revealed a remarkable colony of animals living in darkness and clinging to the sides of the pipes (see list in Appendix 7, p. 224). Today the filter system prevents this colonisation, and when it was introduced into Hamburg the Pipe fauna died off from starvation.

Plants are of minor importance owing to the absence of light, but certain bacteria and algae can still become a nuisance. Iron bacteria cause 'rusting', and together with certain blue-green algae may cause discolouring and unpleasant smells. Sulphate of copper can be used as a control.

In describing some of the man-made habitats found in built-up areas such as London, sufficient has been said to suggest that field natural history need not be confined to the open countryside. The subject may be pursued almost anywhere. A common complaint among many town naturalists, is that leisure time is all too short for visits outside the home territory. Even so, a little time spent daily, or at weekends, on a nearby rubbish tip, waste plot, sewage farm, golf-links or other undisturbed place, should not be outside the scope of any serious naturalist. Even in such unlikely places many ecological problems still wait to be solved.

Part Three

Field Work

Part 3

Field work

A LONG-TERM PROJECT

Over a period which could last for a number of years a team of naturalists can work at a long-term project in ecology within a chosen area. By making regular visits over the seasons, during which records and observations are made of the plant and animal life, and the factors which affect their lives, a useful exercise can be carried out to show the ever-changing pattern of nature which is going on around us. Apart from any educational value and training in field lore which this gives, the results should provide a useful record for the future. The area may not always remain the same. Also, some new discovery could add to our store of knowledge.

Many amateur naturalists, apart from scientific workers, college departments, museums and field centres, are now undertaking regular surveys and studies in their respective areas, and there is no finer way of getting to know the countryside and its wild life. Before setting out on such a venture a number of points should be considered, and a plan worked out for the operation:

1. Choose an area which contains a number of different habitats, if possible on undulating ground, some of it exposed and some sheltered. This will give the maximum scope to the exercise.

2. Ascertain that the area will remain undisturbed for a number of years, and that no major change is likely to occur in the foreseeable future. Human interference on a minor scale is almost certain to occur, but this will add interest to the work (i.e. the biotic factor of man).

3. As a basis for records prepare a large-scale map of the area (using a 6-inch O.S. map) and mark on this:

a. the geological details of solid rock, including any drift material should this be present,
b. the topography (contours),
c. the habitat areas. Where possible use such convenient demarcations as a ditch, hedgerow, wood border, road, etc.

4. Prepare separate habitat maps on which can be recorded the zones and associations of animals and plants. These maps will also come in handy for any short-term project likely to be carried out from time to time. If these habitat maps are run off in quantity they can each be used for a different survey, or a sequence of surveys on the same topic, thereby providing a useful reference over a period.

5. Prepare charts of the essential seasonal variations in rainfall, sunshine, winds, temperature (day and night). Figures can be supplied by the Meteorological Office. With the use of simple apparatus more exact and localised figures can be recorded (see Weather, p. 67).

6. Build up a collection of specimens (except rarities) for each habitat. This should include rock and soil samples as well as animal and plant, and makes a tangible record which can be supplemented with field sketches and photographs, both of specimens and of habitat scenery.

7. Field notes, made *on the spot*, also correct labelling of specimens at the time of discovery, are essential. One should never rely on memory.

8. At intervals carry out the standard field activities, such as quadrat counts, transects, sampling, etc., so as to retain a picture of the ebb and flow of life within the habitat.

9. Keep regular observations on a selected number of species, especially those which appear to be dying out, or are new arrivals. This may be due to some important new factor within the area, and might make a valuable record for the future. Note also the effect on the animals and plants of any extreme seasonal variation (e.g. a severe drought or hard winter, or some sudden human interference such as a fire, tree-felling, a building operation, and so on).

10. Do not overlook interesting micro-habitats which may appear from time to time under a fallen tree, a rubbish tip, deserted building, etc.

11. Observe the Country Code and local by-laws (see Appendix 8, p. 225); also, show the expected courtesy towards the owner or authority in whose area you are working, by asking permission of entry. Always notify the owner when the area is being trapped, specimens collected, or when any digging or staking is necessary.

THE EPPING FOREST SURVEY

The area here selected covers a number of habitats of more or less permanent stability within the boundaries of Epping Forest, a semi-natural and protected woodland lying to the north-east of London, and well known to the author. It is chosen for this book because the woodland habitat is the most easily accessible throughout Britain, and happens to be the climax community for these islands.

Apart from the natural factors at work in the Forest, there has been, and still is, a strong biotic influence at work, namely that caused by man. Formerly a Royal Hunting Forest maintained almost exclusively for harbouring royal game, such as deer, the Forest has been preserved over the centuries, since Norman times. During this long period certain rights have been enjoyed by the Forest dwellers, or Commoners, such as free grazing of their cattle and pigs, and the lopping of firewood. This has left its marks on the trees and undergrowth. Since the 1878 Epping Forest Act, all common rights except that of cattle grazing have ceased, and the Forest is now preserved and maintained as a pleasure ground under the trusteeship of the City of London Corporation. The duty of the elected members of the Common Council, the Verderers, the Superintendent and his keepers and woodmen, is to 'preserve the Forest in all its natural aspects'.

Forestry in the economic sense is not carried out, since the trees are not grown for timber, but to give an amenity value and natural beauty to the area. Any felling or clearance which takes place is aimed at preserving the scenic variety, by keeping the open spaces intact, and by encouraging a rotation of tree growth in the wooded parts. The occasional felling, pond clearance, and the control of such recognised harmful animals as the grey squirrel, wood pigeon and jay, form part of the duties of the keepers and woodmen, a deliberate conservation programme designed to preserve the natural features and wildlife of the Forest. No hunting, shooting, trapping or collecting of animals or plants is permitted, except by special permission. The Forest is also guarded from encroachment, and remains open at all times. There are no enclosures or fences.

The Forest habitats include closed woodland,

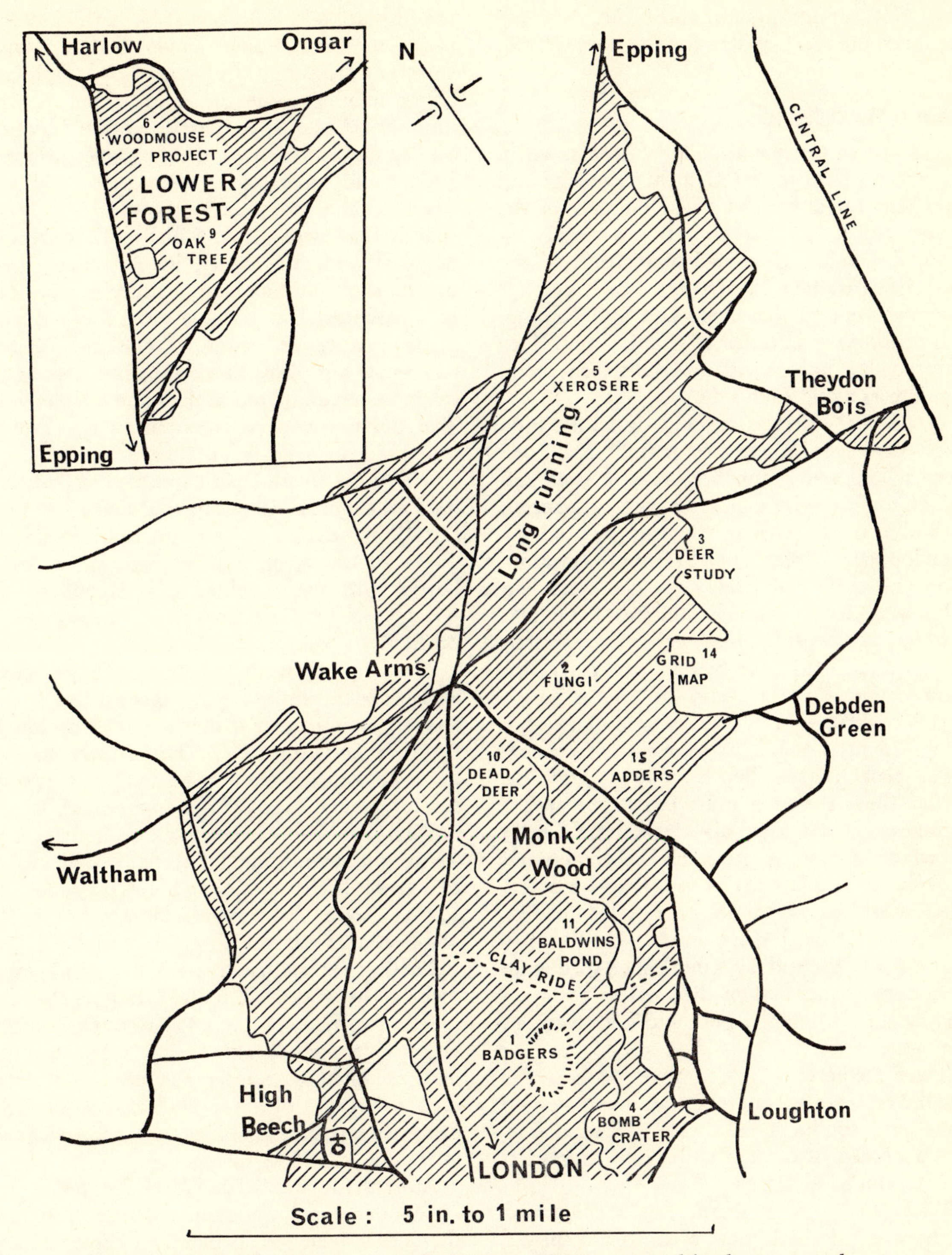

Fig. 61. Map of Epping Forest showing the habitat areas used in the survey, also a number of sites of the various projects.

open glades, ponds, heath and scrub, of which the following were chosen for this survey (Fig. 61).

Beech Woodland

Lying just to the north of the Forest town of Loughton (close to the Central Underground line from London) is the deepest part of the Forest, called Monk Wood. It lies mainly on exposed London Clay and forms part of the watershed drained by streams which run off east and west to the rivers Lea and Roding, two northern tributaries of the Thames. Two forest roads running north and south form useful boundaries along each side.

Much of the area is covered by a false high forest of beech. These are mainly pollarded trees whose new branches have grown sufficiently tall to form a closed canopy overhead. Pollarded trees are those whose branches have been lopped at a man's height from the ground. Here and there are a few coppiced trees (cut at ground level) whose branches have also regrown to a considerable height. On undulating ground deeply cut by old streams which meander between the trees, the area is now a closed woodland in a state of secondary high forest. Light is the most important factor on plant growth (see Beechwood, p. 76). In earlier times the area must have presented a strange sight. At intervals of twenty years or so whole groves of trees had their crowns decapitated (pollarded) so as to leave a somewhat stark battleground of isolated trunks. This would have let in considerable light, so that a varied ground flora must have sprung up from time to time, then died off as the trees regrew their branches, once more cutting out the light.

Today, wherever a clearance takes place, and old trees are cut out to reduce overcrowding something similar takes place. Light enters, and a process of recolonisation and succession of flora can be observed.

With beech as a dominant tree, Monk Wood contains a poor selection of flowering plants in its ground flora. Lack of light is mainly the cause, but there is also the factor of the cold clay soil which inhibits bacterial activity, so that the formation of humus is slow. A thick carpet of mummified beech leaves is the result. Constant pressure of feet caused by people and horses moving along the rides has killed off a number of plants, in particular the bluebell. Showy flowers such as the foxglove, primrose and orchid have largely disappeared, mainly due to vandalism.

It is the lower plants which still flourish in Monk Wood. The fungi, in particular, grow in profusion, and as many as fifty species may be encountered in a single day's foray during a good autumn season. Woodland mosses, liverworts and ferns also occur, but here again the ferns have suffered at the hand of collectors. The polypody, once common in the crowns of the old pollards, is no longer found.

There is a shrub layer composed of scattered oak and hornbeam. The oaks are in poor condition, except where an odd specimen grows in the open, and are fighting a losing battle with the beeches. This is one of the puzzles of the Forest, since the heavy soil is particularly suited to oak. Perhaps the heavy demand for oak timber, once needed for houses and wooden ships, has weakened the Forest oaks to such an extent that they can no longer compete with the beech. This sturdy grower, more at home on well drained calcareous soils, has now taken over the role of tree dominant.

The hornbeam, a famous Forest tree, is in a subordinate position, but steps are being taken to preserve it, so that Epping Forest may continue to produce the finest stand of hornbeams in the whole country, if not throughout Europe.

Further secondary trees of the shrub layer include the wild crab, holly, bramble, sloe and hawthorn. The crab was formerly encouraged to provide food for the fallow deer, and the holly planted as winter shelter. Apart from their usual diet of grass, buds and acorns, the deer will eat the apples in autumn, and gnaw at the holly bark in spring.

Animal life in Monk Wood is typical of a closed, deciduous woodland. Mammals include the fallow deer, fox, badger, grey squirrel and woodmouse. There are a few small rabbit warrens, and with the decimations caused by myxomatosis, an increase in hares. Birds are limited to those usually found in the deeper parts of woodlands, such as woodpeckers, nuthatch, tree-creeper, wood-pigeon, crow, various tits and finches. Winter visitors include the brambling, goldcrest, redwing, woodcock, and rare waxwing. In summer one notices the redstart in the pollarded hornbeams, the three leaf-warblers, the cuckoo and turtle dove. Birds-of-prey, now sadly decimated throughout Britain, include the occasional sparrow-hawk and kestrel, and the tawny owl.

Insect life abounds, and occupies many niches such as the trees (on leaves and fruit, bark and roots), undergrowth and leaf-litter (see Project 9, p. 171). The litter fauna is high, and plays a similar role to that of the fungi in breaking down the leaves and fallen branches into humus.

Oak Woodland

To the north of Epping town lies a piece of isolated woodland, called the Lower Forest. This consists of mixed oak and hornbeam, with a scattering of birch. It lies on London Clay overlaid by some Boulder Clay, and holds surface moisture most years. After heavy rain it becomes waterlogged. Due partly to this, also its greater distance from London, the Lower Forest is far less disturbed by humans, and supports the greatest variety of wildlife in the whole Forest area. This is due in part to the higher incidence of light, the richer humus layer of oak leaf-mould, and to less disturbance and vandalism.

The author's list of flowering plants for Monk Wood is about thirty species, whereas for the Lower Forest it exceeds a hundred. Much of this is typical of lowland oakwood, and includes the expected primrose, wood-anemone, bluebell, violet, celandine, early purple orchid, foxglove, and so on. In turn the animal range is much higher. Among the mammals, only here in the entire Forest have the red squirrel, shy dormouse, and pygmy shrew been seen in recent years.

Birch Woodland

To the north and west of Monk Wood, around High Beech church and in the vicinity of the cross-roads by the Wake Arms Inn, are a number of pure groves of silver birch. These are situated mainly on glacial gravels and boulder clay, on what was probably a heathland in former days. A hundred years ago the birch was recorded as a rarity for the Forest. Today this area shows the striking success of this coloniser. Wherever land is cleared by fire or felling, the birch often takes over, especially on the poorer, more sandy soils. Patches of birch may now be seen all over the Forest, in places where bombs have fallen during the Second World War.

If allowed to grow unchecked, the birch will completely cover an area where once there was an open common or heath. There are strong signs of this happening in the Wake Arms locality. The author can remember flourishing communities of heath plants in this area, such as ling and heather mixed with gorse and moor-grass. The birch, bracken and coarser hair-grass (*Deschampsia*) have now taken over. Typical of the birch are the two well-known fungi, the birch polypore and the fly agaric. In summer the redpolls search the trees for seeds, and the nightjar (now rarely heard) nests between the bracken. The high growth of bracken is a favourite retreat for fallow deer, and a secluded place to which the does retire to drop their fawns.

Open Commonland

Beyond the birch groves to the north of the Wake Arms lies a stretch of open ground on gravelly soil, called Long Running. The silver birch which grows in clumps here and there is a dominant tree, but since the area is frequently subject to forest fires, has not yet succeeded in closing up the area. Soil analysis gives a high acid content, and pollen analysis has revealed the earlier presence of a number of heath plants and moorland grasses. Today the area is almost entirely overgrown with the hair-grass (*Deschampsia*) which has taken the place of the former ling (*Calluna*), commonly found about thirty years ago. The reason for this change from heather to grass may be connected with the interference factor of fires, after which the stronger plant only can survive.

Long Running is of especial interest in the study of colonisation. After a heath fire has died down the ground is bare of surface vegetation. In severe fires even sizeable birches are destroyed. A new community springs up in succession, beginning with filamentous algae on the bare and damp peat-soil. Mosses and liverworts take over in a year or so, followed by grasses and herbs, and the masses of birch seedlings. The tough bracken and hair-grass already in the area usually manage to push their way through the burnt ground, whereas the all-too-rare ling suffers in a losing battle against fire and competitor. Small plants of ling planted by the author in safe places have managed to hold their own against the grasses. This suggests that fire is the chief controlling factor (see Project 5, p. 161).

Insects and other invertebrates peculiar to open acid soils are found on Long Running. This is also the home for lizard and adder, vole and hare. Birds include the tree-pipit, yellow-hammer, occasional summer whinchat and wood-lark, and a ranging kestrel.

The Grass and Scrub

Scattered along the roadsides where there are open glades are a number of small places which are kept open by periodic clearance by the woodmen, and by occasional fires. As a safety experiment the Conservators have cut back the road verges in parts, so as to give motorists a clearer view and warning of cattle or deer which stray on to the highway. A mixed scrub of young birch, thorn, bramble and briar usually builds up in such places, and is then cut back every ten years or so. The ground flora is mainly of hair-grass (*Deschampsia*) mixed with various sedges, rushes and acid-loving heath plants such as the tormentil, heath orchid and heath bedstraw.

These glades and verges support isolated populations of reptiles, such as adder, viviparous lizard, slow-worm and grass-snake. There is ample cover and retreat for these animals under the impenetrable thorn bushes. The adder hunts the field vole and takes the occasional lizard and bird's egg from a ground nest. The grass-snake hunts the frogs and toads which wander into these areas from the Forest ponds after their breeding season. Lizards pursue spiders, grass-hoppers and other insects which live in the dry tussocks of grass. In these glades, surrounded by Forest trees, birds abound. There is ample food, cover and nest sites. Typical garden birds such as robin, thrush and chaffinch share these places with their shyer migrant cousins such as the warblers, flycatcher and nightingale. These are good spots in which to hear the Dawn Chorus.

A Forest Pond

Most of the Forest ponds occur along roadsides or near forest rides. Originally man-made, they are the result of gravel-digging operations during the last century when material was extracted for lining the roadways before tarmacadam came into use. Some ponds lie on high ground, and are drowned gravel pits worked from the hills and old river terraces where gravel caps the London Clay. Other ponds were created almost accidentally where a road or ride happens to cross a stream, and has been raised on a piled up bed of earth. This has resulted in a kind of dam across the stream's course, so that running water has piled up against the up-stream side of the roadbank, to form a pond. Such a pond, called Baldwin's Pond, was formed about seventy years ago when a ride, called Clay's Path, was built across a stream near Baldwin's Hill. Due to the surrounding trees and the constant invasion of the reed-bed, this pond might have filled in long ago, were it not for the foresight of the Conservators. Every few years this pond is dragged to remove as much of the natural debris and advancing reed-bed as possible, so as to retain the open water. Were it not for this check, then Baldwin's Pond would have reached the final stage of a hydrosere long ago, and become a shallow marsh, if not dry land. By holding back the seral succession towards this climax, the pond community has been preserved, giving the author the opportunity to study and keep records of its pond-life over the past thirty years (see Project 11, p. 175).

Hedgerows and Fields

Much of the Forest land, although unenclosed by any fences or walls in accordance with the law of commonland, is now hemmed in by built-up areas, especially in the southern parts. To the north there is still plenty of open country, in the form of fields, which meet the Forest edge. The particular area of fields and hedgerows kept under observation lies within the Green Belt, along the north-east side between Debden Green and Theydon Bois. It is possible that the fields hereabouts have been under cultivation for a great number of years. It is recorded that 'written record goes back to 1066 when we find in a charter of Edward the Confessor his confirmation of the gift of Tippendene (Debden) along with several other estates from Earl Harold to endow his new monastery at Waltham. It is recorded in Domesday book as a flourishing manor. So it continued as a typical rural community of some 30 households, retaining its own economic identity, governed by the often harsh Forest laws and paying obeisance to the Lord of the Manor, the Abbot, until late mediaeval times. Then, it appears, possibly by succession, to have become part of the Manor of Loughton.'

When the author first visited the area some thirty years ago the fields were in use as meadowland by the two farms, at Debden Green and at Birch Hall. Cattle and horses were a common sight. There was also a piggery near one corner of the Forest. Camping was permitted on some of the fields, and restricted to a few recognised bodies, such as the Boy Scouts, Girl Guides and the Camping Association, each with a permanent camp-site in its own field. Equipment was stored in the farm outhouses, and fresh water and dairy produce obtained from the farmer. Wood for camp fires was gathered from the Forest, but only as dead or fallen timber.

The author used to camp with his Boy Scout Troop, and in those days, during the 'thirties, the night sounds of moving and grazing animals, owl calls and nightingale song were a common experience. Fallow deer could be seen from the tent doorway, emerging at dawn to feed on the meadow grass. An odd encounter with a prowling badger, hedgehog or fox after dark added interest to a young naturalist's week-end camping. Pigs were set free to roam beneath the Forest oaks in search of acorns. The pig farm close by was probably one of the last in the Forest area to continue the practice of the ancient right of pannage (i.e. allowing pigs on common land to forage for acorns and beech mast).

In those days the fields and hedgerows on the farmland and neighbouring estates were little affected by man, but more so by animals. Many trees along the Forest border had curious umbrella shapes, with flat undersides where the farm animals and deer had reached up to browse the leaves at head height. The soil here and there was much churned up by rooting pigs, a kind of natural hoeing process which can uncover the buried tree fruits, and so produced a good crop of oak, hornbeam and beech seedlings in the following spring. Today when an area is cleared, the woodmen harrow the ground for the same reason, to break up the bared leaf-mould and encourage regeneration. It is quite probable in days gone by that the ancient herds of deer, wild cattle and wild pigs played an important role in breaking up and fertilising the soil.

In the fields mushrooms in late summer and autumn were a common sight, and benefited from the horse droppings and cow-manure, and we enjoyed a tasty breakfast. As campers we ourselves were responsible for the fine crops of tomatoes which used to grow along the ditch in our latrine area! By early summer the meadows were knee deep in grasses and flowers. Rabbits would come out to feed and play at dusk, giving us fine opportunities to try out stalking techniques. Along the hedgerows little touched by cutting machines stood many fine Forest trees which had spread into the open, such as oak, hornbeam, beech and maple, with thick growths of briar and bramble around their trunks. Sloe and hawthorn blossom appeared in spring, followed by rose, blackberry and honeysuckle in summer. The bushes were alive with birds, and in one week-end count along a hundred yards stretch some thirty nests were noted.

Today, much of this is but a memory. Farming days are gone, and many of the old camping fields are now covered by a housing estate. There are no more cattle, horses or pigs. Fortunately some of the area is safe from building. The old farm-house, cottages, barns and a number of the fields at Debden Green were acquired by an educational authority, so as to provide facilities for week-end recreation and education in pleasant surroundings close to London. The fields are now used for camping, and for games ranging from football and cricket, to archery and field sports. The house runs residential courses on a variety of subjects, including natural history, both for adults and schoolchildren.

This highly commendable venture, from the human standpoint, has meant that a strong biotic (human) factor is now at work in the area. The fields are regularly mown to keep them trim for campers and players, and the meadow flowers and mushrooms have now gone. The hedgerows are cut back, not by cattle and horses, but by machines. Nesting birds are far less common. Deer and rabbits seldom show themselves, except at night, or by day during hard winters. All told, there is too much disturbance caused by the presence and activities of humans, even with the best of intentions.

Possibly the least changed places in this field and hedgerow habitat are to be seen along the ditches under the hedgerows. These have remained undisturbed, and here one can still find some survivors of a former wealth of plant and animal life when more natural factors were at work. In secret places known to the author can still be found the odd primrose and orchid. Grass-snakes occupy one of the ditches in a quiet corner, almost under the eyes of unsuspecting campers. Goldfinches still visit a neglected corner where thistles grow. In one undisturbed spot grows the strange fungus called the earth-star (*Geaster*).

With some knowledge of what the area looked like thirty years ago one can see the changes brought about by the actions of a human community. It is such records which prove valuable when one looks back. Like the 'relict' species which have survived in our mountains and dales since the last Ice Age, this 'ditch' community of our hedgerows is a reminder of earlier days. It compares with the ancient buildings which survive in our modern towns.

FIELD PROJECTS IN EPPING FOREST

In this chapter are given some of the short-term projects which have been carried out during the long-term Epping Forest survey. Most of them were completed during a succession of short visits, each lasting an hour or so, by one or two persons. They have helped to build up a general picture of the various habitats kept under observation, and each was carried out for a specific purpose. There is little point in undertaking a field project unless it can provide some useful information, even of a negative nature, which will add to the store of knowledge and understanding of our countryside and its wild life.

With plants which are more closely bound to the soil, the more convenient approach can be made on quantitative lines, that is, with the aid of transects and quadrats, and by plotting zones of the differing plant species. These relatively straightforward but sometimes tedious exercises may produce some interesting information, such as the increase or decrease of a particular plant, or a new arrival. This will lead to a more critical examination, in which varying factors may have to be examined, to find out the cause for the change.

With animals this type of 'static' project is not so easy, since animals come and go. A search for food, shelter, a mate, or the avoidance of enemies may explain their presence or absence. Within limits one can count and assess a population, noting any absentees or new arrivals, by direct observation (very time-consuming) or by using traps. The result can then be related to the available food supply, some interference factor or weather change, or some other cause. Shorter rhythms of activity and changes in population are always going on, whatever the immediate change, and these occur at dawn and dusk, as regularly as they do between high and low tides along our coasts. Mating activities and family affairs, hibernation and migration, will also swell or decrease the species and population numbers.

The general picture which emerges from this should give an idea of how a community of plants and animals exist in an environment subject to fluctuations in weather, food supply, human interference, and so on. A wise ecologist will take note of all these factors in the area which he studies, and try to find out the reasons for each change.

In tackling such a major undertaking as the Epping Forest Survey, all ecological aspects have been considered, as well as every major group, from algae to trees, and protozoa to mammals. This could well become a lifetime study. Ecology of this kind, in which the community is taken as a whole, is called Synecology. The alternative approach, termed Autoecology, has as its target a single species or group, and tries to find out its relation with the rest of the community. The first can be compared to an observation made on the teamwork of a football eleven taken as a whole. Autoecology puts a single member of the team under focus, to see how he fits into the game, and what effect he has on his team mates.

An example of this for Epping Forest would be a study of the ancient herd of fallow deer, in an attempt to find out their role in the community, and incidentally the reason for their rapid decline in recent years. Such single species studies are usually carried out on animals of economic importance, as in the case of the control of the alien grey squirrel, or the pursuit of the elusive but valuable food fish, the herring.

In the following pages are given a selection of exercises carried out in the different Forest habitats, together with some of the conclusions drawn from them.

Project 1

BADGER WATCHING

As an exercise in patience and quiet observation, two essential qualifications of a good field naturalist, the badger makes a worthy subject (Fig. 12). It can only be studied on its own home-ground after dark, at times with which many of us are unfamiliar. This new experience for many can bring about an entirely different understanding of the wildlife of the countryside—the nature-by-night which is all too little known. Also, quite apart from watching this interesting creature going about its homely affairs, there is much to learn from the different signs and clues which it leaves about. Badger study around the home territory and at the actual sett makes an excellent introduction to the fascinating detective work which no field-worker should ignore.

A sett in Monk Wood, known to the author since observations first started in 1935, has been occupied more or less continuously by badgers, with foxes as occasional temporary residents. On occasions both badger and fox have been living side by side, rearing cubs in neighbouring holes. Twice the different youngsters were seen from up a tree, playing almost side by side, and once the crime of vulpicide was committed by the badgers. Mutilated bodies of fox cubs were found one morning near the sett.

This particular sett has been tunnelled into the embankment of an early Iron Age encampment, and over the years the various entrances have shifted over an area of some 50 yards square (Fig. 62). The position of some old doorways, long abandoned and filled in, can now only be picked out by the hummocks of excavated soil, a mixture of hardened clay and sand forming a contour with the ground.

The overhead cover consists of mixed beech and hornbeam, with some scattered holly and thorn, making observation fairly easy since there is little ground cover to obscure the view. On the nearby hill-slope stands a grove of birch with an open area of bracken, used as a playground and latrine area.

Wherever a badger sett is left undisturbed, that is, where no digging, trapping or poisoning takes place, as here in the Forest, successive families may continue to use it for a great number of years. The age of this particular sett may be judged from the number of holes (at one time there were 15) and the distance between them. In one case a tunnel ran for

40 yards, and was confirmed with the help of an obliging ferret which was played out on a line and negotiated the entire tunnel.

Signs left by badgers which may be looked for include:

a. The five-toed, plantigrade footprint, with a broad, kidney-shaped ball-pad and long claw marks of the front paws.

b. The easily identified stiff hairs sticking to the soil and roots around a fresh entrance—about 3 inches long, greyish in colour with a dark band just behind the tip. These are also found on wire and fences where the animal has pushed its way through.

c. Well-worn runways leading to and from the holes, playground and latrine area. In some overgrown and undisturbed woods these may be the only pathways.

d. An occupied entrance, often triangular in outline to fit the shape of the badger's body, with a pile of fresh earth in front. Flies and ground beetles may be seen, and melted snow around the doorway in winter-time. Also, water vapour may rise from an occupied hole due to condensation from the warm body of a sleeping animal.

e. A play area not far away with trodden vegetation, and which may include a play-tree or hillock. There may also be a rubbing or scratching tree with smoothed bark marked with claw scratches, and with mud or hair attached.

f. Ventilation holes. Small shafts or 'chimneys' leading down to a tunnel where the badger has broken through to the surface. This may be done deliberately, or caused accidentally.

g. Scrapes or shallow diggings in the soil where roots, bulbs or grubs have been sought after.

h. The latrines. A collection of shallow pits in the neighbourhood, some containing dung. This can be analysed by teasing out dried droppings softened with water, or by making a microscopical smear. Contents should reveal such material as plant fibres, hard seeds, beetle chitin, hair, worm setae, etc., giving evidence of the badger's omnivorous diet.

i. Material at the entrance. This includes discarded bedding such as grass, bracken and leaves which are raked out from time to time, also fresh bedding dropped on the way in. Any skeletal remains, such as rabbit, chicken,

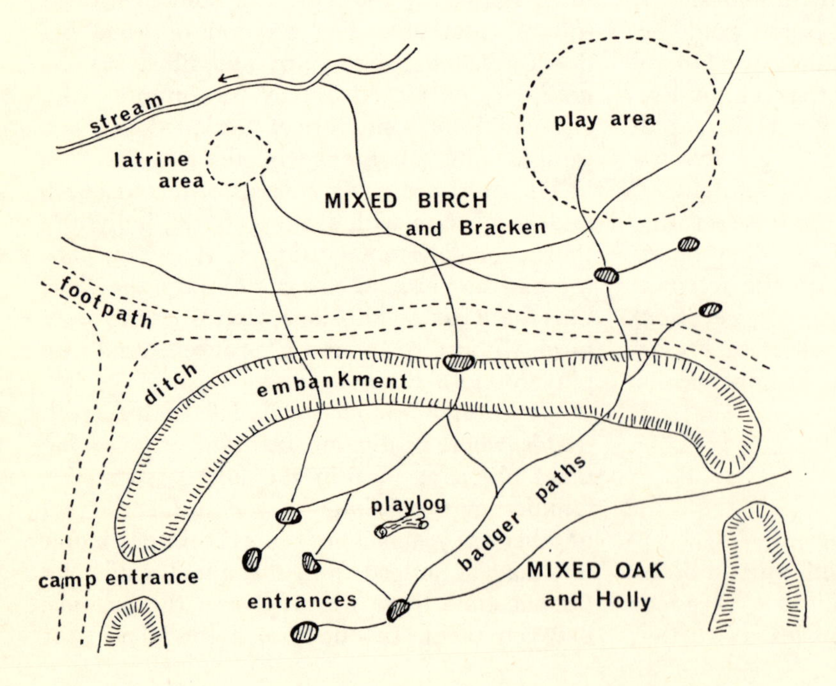

FIG. 62. Plan of badger sett in Iron Age camp, Monk Wood, Epping Forest.

etc., pushed out by badgers are probably the meal remnants of a fox in previous occupation. Foxes will use badger setts, taking food back to their earth, whereas badgers usually eat their meals on the spot. Where badger remains are found among the debris, then an animal may have died below ground. It has been stated that badgers will inter their dead by pulling them below ground.

In order to watch badgers it is usually necessary to arrive at the sett at least half an hour before dusk. Choose a freshly cleaned entrance, test the wind by wetting the finger or watching cigarette smoke, and take up a comfortable position to windward, some 6 to 10 yards away. Some watchers prefer to sit in a nearby tree, others to sit on the ground using the tree-trunk as a back rest. In either position avoid making a silhouette, as any strange object can make a badger suspicious. Badgers have moderate eyesight, good hearing, and an acute sense of smell. Watching them is a matter of patience, silence and slow movements. Wear old clothing of neutral colour and which does not rustle, and come prepared for the painful attacks of midges and gnats during summer watches. A torch with a red filter is useful, and does not appear to worry the animals.

The pleasure of badger watching, and the thrill of meeting this ancient Briton who is still well entrenched over much of our country, is its own reward, apart from all that one can learn about its secretive ways. Food placed near the entrance, especially anything sweet such as honey or treacle, when the cubs are about can bring these inquisitive youngsters right up to one's hand.

Some delightful scenes can be witnessed. On one occasion a cub at the Monk Wood sett had thoroughly covered itself with treacle, to which twigs and leaves became stuck, when mother took it in hand. During its protests and struggles to avoid her licks (cubs make delightful sounds) it received a box about the ears! A tape recording on such occasions would have been valuable. A lot has yet to be learned about the badger's vocabulary and secret ways.

Brock is inoffensive and rarely harmful to man. The excuse to go badger-digging on the grounds that badgers take domestic stock or damage property and gardens can seldom be justified. The even more senseless and cruel practice of badger-baiting, now illegal in England, is but a reflection on man's inhumanity. The use in our language of the word badgering, meaning to torment, is a legacy from former days when, like bear and bull baiting, this was a common British sport. The noble and ancient badger surely deserves better treatment than this.

Project 2

FUNGI

A seasonal activity now regularly carried out by natural history groups and mycologists, and called a fungus foray, is made with the purpose of adding to the records in a given district. Dating back to the 1890s this kind of survey has been a popular autumnal attraction for visitors to Epping Forest. Accumulating finds now run well into a thousand different species. In a good season a small party can record as many as eighty species in a single day's foray. An example is given in Appendix 9, p. 225.

The nature of fungi, where chlorophyll is missing, is to feed on organic substances, and this can be done in dark places. Most species are saprophytic, that is, they subsist on dead material such as fallen leaves, dead branches and tree stumps, animal droppings, helping in the process of decay. Some are parasitic, attacking living things both plant and animal.

The permanent fungus, made up of penetrating threads or hyphae forms a network growth called a mycelium (e.g. mushroom 'spawn'). This is usually hidden away inside the food medium such as the leaf-mould or dead wood. During the fruiting season, which in Britain coincides mainly with the mild and damp autumn weather before the frosts begin, a fruiting body or sporophore emerges. This takes on the characteristic shape of a toad-stool, puff-ball, stinkhorn, fairy club, bracket, and so on, by which the fungus may be recognised. With a little experience, and the aid of a field-guide, one can soon learn to identify the family to which the fungus belongs, and in many cases run it down to its genus, even the species.

Records kept in this way make interesting comparisons over the years, especially where the field notes include the weather conditions of rainfall and temperature for the season. There are good and bad years for fungi in fruit, as well as periods of rest such as one sees in fruit trees.

Certain qualities may be recognised in the different species, how one prefers a certain growth medium or habitat, where another species will not normally occur (Fig. 63). For

FIG. 63. Fungi in Epping Forest. A selection of species from various families of larger fungi. Agaricaceae—A, Parasol (*Lepiota procera*); B, Death Cap (*Amanita phalloides*); C, Field Mushroom (*Agaricus campestris*); D, Sulphur tuft (*Hypholoma fasciculare*); E, Oyster mushroom (*Pleurotus ostreatus*); F, Lawyer's Wig (*Coprinus comatus*); G, Fly Agaric (*Amanita muscaria*); Boletaceae—H, Penny bun (*Boletus edulis*); Polyporaceae—I, Birch bracket (*Polyporus betulinus*); Phalloidaceae—J, Common stinkhorn (*Phallus impudicus*); Sclerodermaceae—K, Earth ball (*Scleroderma aurantium*); Hydnaceae—L, Wood hedgehog (*Hydnum repandum*); Lycoperdaceae—M, Bird's nest (*Crucibulum vulgare*); N, Earth star (*Geastrum*); Pezizaceae—O, Orange-peel (*Peziza aurantia*); P, Fairy Club (*Clavaria*).

instance, in Monk Wood, the following places can be searched:

a. In the leaf-mould under the trees, which include oak, beech, hornbeam, birch and elder. There are no conifer woods in the Forest.

b. Tree stumps, fallen trees and dead branches.

c. In grass.

d. Under bushes or in bracken thickets.

e. On acid, neutral or burnt soil.

f. On animal droppings.

In the case of *a*, the association of a certain fungus species with a particular tree (e.g. the fly agaric with the birch) is of interest. In such cases the fungus becomes attached to the tree roots by a so-called mycorrhyza (fungus-root), and a patch of fly agaric will fruit yearly in the same spot. It is possible that a kind of mutual advantage is gained by tree and fungus through this association (symbiosis). Experiments in forestry have shown that transplanted conifer seedlings will do much better in a new area, if some of the original soil is transplanted with them, so that the fungus is also present.

Any felled tree, or one which shows signs of dying, can be visited regularly in order to keep a record of the successive fungus species which grow on it, and finally turn it into wood powder. In 1938 a beech tree was found with fruiting bodies of the honey fungus, called the 'Foresters' curse' (*Armillaria mellea*). This parasitic agaric was doing its deadly work, and within a year or so a number of branches were in a dying condition, producing fine crops of the beautiful white beech tuft (*Armillaria mucida*). The tree was felled by the woodmen, and in the ensuing years some twenty species of fungi, from minute slime-fungi (myxomycetes) to large brackets, fed on the rotting wood. By 1946 the tree was reduced to a small heap of powdered wood scattered among the leaf mould.

Apart from the more serious recording work done on our fungi, a number of people derive pleasure from collecting them for eating purposes. This pursuit, a traditional activity in many Continental countries, has always been viewed with suspicion in Britain, partly because of the mistaken idea that all fungi, apart from mushrooms, are poisonous. In actual fact many of them make good eating, although this is a matter of taste. With the exception of the *Amanita* toadstools, in particular the death cap (*A. phalloides*), there is little risk of coming to an untimely end.

With the rising popularity in colour photography one can find many attractive subjects for the camera among the fungi in their autumn setting, even winning first prize in amateur competitions, as once happened to the author.

Project 3

BEHAVIOUR STUDY—THE FLIGHT DISTANCE

There are innumerable observations and tests which can be carried out in the field in order to try and understand why animals do what they do. Much of their behaviour is governed by their needs, especially in survival, and this study can become a real challenge to the field worker. Deer study, for example, can become a test in skill and patience for anyone who is prepared to come to terms with one of our wariest of animals. The approach can be made either in the traditional manner of a deer stalker, by following signs until the herd is in view, then making a stalk upwind, or by waiting in hiding near a deer sanctuary or feeding area. In various parts of the country, especially in forestry plantations, deer observation platforms have been erected for this purpose.

There are more deer roaming wild in the British Isles than is generally realised. Some like the red and roe deer are indigenous, others are introductions which have escaped from parks and zoos. In Epping Forest the black fallow deer are in a unique position, being of dark colouring and great antiquity. They have never known captivity, and have roamed the Forest since Norman times. The exact range of this species (*Dama dama*) is uncertain, but it is thought to be indigenous to the eastern Mediterranean countries, and an introduction to the rest of Europe. Introduced into Epping

Forest for the purpose of hunting, this herd has survived the native red deer which were finally rounded up and transferred to Windsor Park in the early nineteenth century when hunting rights began to wane.

The black fallow are by no means easy to approach, and their dark colouring makes them inconspicuous among the Forest trees. Unlike their tame park cousins they are ever on the alert for danger, and a certain amount of guile is necessary in order to make a close contact, as will become evident from the experiments which were carried out in this exercise.

Observations took place during the 1930s in the Debden area, along the Forest border, which used to be a favourite feeding place at dawn and dusk. In those pre-war days the total herd numbered some 200 animals, and small parties were a common sight in various parts of the Forest. The spot chosen for observation was in a dry ditch behind a tree, from which the Forest as well as the open fields could be watched without being seen.

Although long familiar with humans, the deer always display a degree of nervousness, and their flight reaction can be measured under different situations. Flight distance is the distance between the animal and the source of attraction, below which it will not approach. Familiarity, curiosity, hunger or some common danger like fire will reduce the flight distance. A painful experience, sudden alarm, or a danger signal may lengthen it. The following notes are some of the examples which the author experienced, mainly from the above-mentioned hide:

a. Mid-day, January. A hard winter with snow. A pile of food placed down near the hide—hay, potato, apples and bread some days before. Tracks show that the food has already been visited. A small herd approaches with caution, out in the open. A buck catches my scent and barks in alarm. The deer run back into the Forest, then return to the food. This is eaten under tension until gone, then the herd moves away. Conclusion—a strong hunger drive which overcomes natural caution alerted by the danger smell. Actual sight of my presence, i.e. a visual danger sign, might have kept the herd at a distance. A keeper friend informs me that deer regularly enter his yard where food such as hay is put out, as for sheep and cattle on farms, and the deer show no signs of fear. Here is a good example of 'familiarity breeds contempt'.

b. Dusk, mid-summer. Seen from a hide in the ditch. Deer emerge from the Forest to graze in the field, moving slowly towards me. Heads are alerted at 30 yards (my scent?) and the herd moves off in orderly fashion, showing no panic. Conclusion—a normal flight reaction from a safe distance.

c. Dawn—late spring. A similar situation to *b*, but with wind blowing towards hide. When some 30 yards off I wave a handkerchief. Immediate alertness and curiosity brings herd to within 10 yards. One yearling sniffs at the hide, then backs in alarm, and the herd flee into the Forest. Conclusion—flight distance reduced by curiosity, then broken by danger sign. This illustrates a clash between an approach and a flight reaction, one stimulated by curiosity and the other by fear.

d. Rutting time in October, late afternoon in Monk Wood. Bark of calling buck heard, and I make careful approach upwind to a clearing. Two bucks are sparring with does watching close by. I show myself with slow movements 30 yards off. The does move off in alarm (one calls), but the two bucks continue to engage. The does return, although obviously nervous, so I quietly retreat, leaving them to their nuptials. Conclusion—a strong herd instinct accentuated by the mating ritual which overcomes normal flight reaction. Had I come closer the herd would no doubt have moved off. Even so, some animals during rut appear quite oblivious to outside danger. The 'mad March hare' is a good example. So are the young courting couples who continue to spoon in public, even as the policeman approaches!

e. A summer night in the Forest—about 2 a.m. with a bright full moon. I hear the tapping feet of approaching deer through the dry beech leaves, and stand perfectly still, in full view of the moon. The herd stops abruptly on sight of my silhouette. There is no wind or sound to give me away. The puzzled deer approach slowly to within 10 yards, then retreat in confusion. Conclusion—scent and sound have failed, but natural curiosity at my outline brings the herd into closer sighting distance. One cannot interpret the visual acuity of deer, but presumably eyesight would detect a motionless human more readily by day than by night. Night, however, is a much 'safer' period when humans are not expected; also a motionless human could be mistaken for a tree. A fisherman, outdoor artist, motionless game-keeper or poacher will see far more than a moving person, and this dodge of 'freezing' at the approach of a shy animal is a useful tactic to adopt.

f. Mid-day in spring. Deer are feeding in the stubble of the field near the Forest border. A farmer is operating his noisy tractor for late ploughing at the other corner. I join him on his machine as it slowly crosses the field towards the herd. The deer move casually aside as we pass within 30 yards. Conclusion—from experience the deer know that the machine is noisy but harmless, and appears to have no connection with humans. This is a common experience with horse riders and people in cars, and an everyday occurrence in game parks. Petrol fumes also help to mask a human scent. The farmer told me that the deer only run off when he leaves the tractor, that is, becomes a separated object.

g. Late June—afternoon. A doe and her fawn are discovered in a dense patch of bracken on Long Running. At my approach the doe backs away, but stays some 30 yards off, stamping her feet and barking. I move some way off and she returns to continue licking her newborn. By now it has struggled to its feet in order to suckle. When I actually found it and stroked the wet body, the mother at first stayed close by. Conclusion—a strong and quite understandable mother instinct to protect the young, in spite of the equally strong danger stimulus of man's presence. Cases of domestic pigs and cows attacking a familiar farm-hand near the piglets or calf are known to occur. In game parks a mother antelope, zebra or giraffe has been known to drive off an attacking lion in defence of its young.

There is much variety of behaviour in this kind of study even within a single species or individual, and this can be calculated in terms of the particular situation and stimulus which controls the flight distance. For instance, wood-pigeons and wild duck which flight into London each day, especially in hard winters, are almost hand-tame in the parks, but become extremely wary if met in the woods or on mud-flats outside London. Deer seen in parks are approachable, even taking food from the hand, but should any escape and go wild they become extremely wary of man. By day many wild animals are fairly tolerant towards the presence of man, provided that he does not come too close, but show extreme caution if the meeting takes place at night. Humans are not normally night prowlers, and any naturalist abroad after dark is well aware of the tension both in himself and the animals he encounters. By rights he should be in bed!

Project 4

COLONISATION OF A HYDROSERE

As already explained (p. 59), a biological vacuum occurs in nature where a piece of ground is bared as the result of a fire, building operation, ploughing, tree-felling, or some other human activity. On this exposed ground will appear a succession of animals and plants in a series of stages called seres, leading ultimately to a climax of community life for that particular area. Where this occurs on dry land the climax stage is called a xerosere (see p. 59). Colonisation may commence at an earlier stage, where the area to be occupied is under water. In this case the water is first colonised, and slowly fills with humus to become land, so that the stages in the land succession can then follow on to the climax. This is called a hydrosere.

An opportunity to follow such a succession occurred during the winter of 1940 at the height of the London Blitz. The author discovered, a week after the incident, where a parachute mine had exploded at the edge of a glade in Monk Wood, leaving a depression in the London Clay some 10 yards across and 6 feet deep. The crater was already filling with rain-water seepage. The undergrowth and oak trees within a radius of 20 yards around the crater had been blasted, and much of the ground bared. Large lumps of clay lay scattered around in all directions. Some of the oaks died that winter.

A survey of colonisation of this 'pond' began in 1941. Regular visits were made during periods of leave from the Armed Forces, up till 1944, when the author went overseas. On return the survey continued in 1946, up till 1955 when the work ended.

Over the years this small patch of static water has passed through a hydrosere, and today it supports part of a woodland community. Where there was once a community of pond-life now grows a 12-foot oak tree with its attendant ground flora and animal life. Of the sketch-plans made on the spot at about six-monthly intervals, four are given in Fig. 64. The first was made in the year following the explosion, and shows the crater with blasted earth around it, lumps of exposed clay, and the position of the neighbouring oak trees. Some were already dead (1941). Water in the crater already contained microscopic life.

By 1946 grass had spread out across the bared soil to within a few yards of the crater, mostly of a woodland species, the Hairy Brome (*Bromus*). Mixed with this were a number of oak seedlings, clumps of woodland cushion-moss (*Leucobryum*) and patches of woodrush (*Luzula*). The crater was fringed with a growth of bulrush (*Scirpus*) and water iris. A flourishing growth of pondweed (*Potamogeton*) covered the centre of the pool, with Canadian pondweed

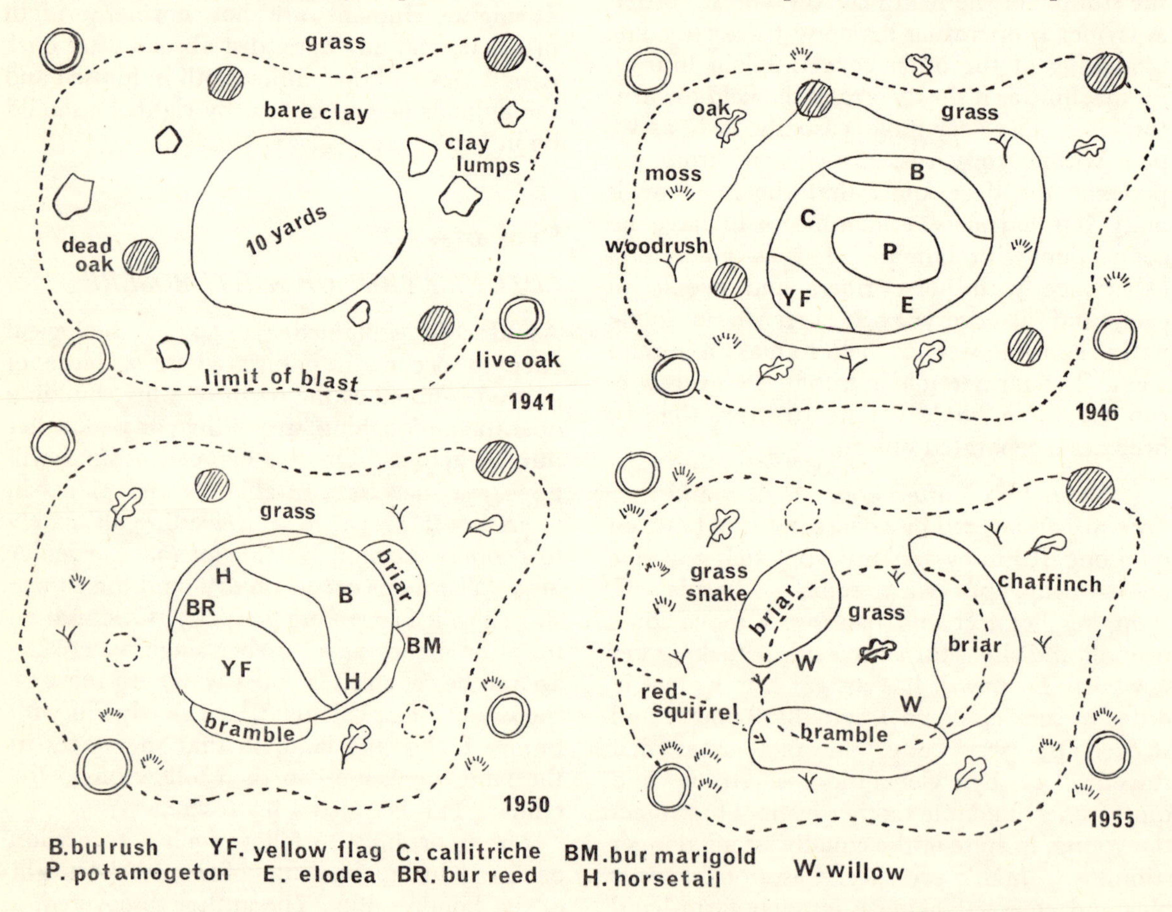

FIG. 64. A Hydrosere. Colonisation of a bomb crater in Epping Forest (see text).

(*Elodea*) and water starwort (*Callitriche*) filling in the gaps. About twenty species of water insects and crustaceans were found, and visiting dragonflies noted. There were two species of newt, also a number of toads, seen mainly during the breeding season. Some sticklebacks were captured during one netting visit, and on another occasion a pair of mallard were disturbed. It was not possible to ascertain how all this life reached the crater, since the author was away during the previous years.

By 1950 little bared ground was left, except for some patches to mark the edge of the crater. On it grew a number of waterside species, including the bur-reed (*Sparganium*), bur-marigold (*Bidens*) and some horsetails (*Equisetum*). The crater was now almost covered with bulrush and iris, with pondweed struggling to grow in the shallow water. No newts were recorded for that year, and only one toad was found. The sticklebacks had disappeared. The crater had reached a marsh stage, and two willows were growing out of the centre. The grassland surround was well into a scrub stage, with a number of young oaks growing here and there, also patches of bramble and briar and many grassland herbs. The expected invasion of birch, a common feature seen today around old bomb-craters in the Forest, did not materialise. Possibly the heavy clay soil at this spot prevented this.

In 1955 the crater contained hardened soil on which one could walk, and had merged with the surrounding ground to make a shallow depression. Flanked by two thick growths of briar and some bramble, it contained a turf of grass, two small willow trees and a bushy, 6-foot oak.

After some fifteen years of colonisation in this hydrosere, a rain-filled bomb crater had finally emerged as a modest little woodland climax containing all its growth layers—the tree canopy, shrub layer and ground flora.

In one of the briars a chaffinch had her nest. Under the other lived a grass-snake. My last record reads 'a red squirrel, one of the few left in the Forest, ran across the grass to feed on the blackberries by the old crater. Any passer-by watching this normal Forest activity would never know that it all began with a war-time incident, now buried and forgotten.'

Project 5

COLONISATION OF A XEROSERE

Severe fires occurred in the Forest during the very dry summer of 1959, especially on the open commons and in the glades. Lack of available water made the work of the fire brigades difficult, so that some of these fires continued to burn for many days. One of these, on Long Running, completely destroyed all surface vegetation, including a number of birch trees, some over 25 feet tall. It was the rainfall in late September which finally put out this fire, which was still smouldering underground in the peaty soil.

In the following winter a patch of bared soil 10 yards square, was marked out with corner stakes, and a survey of colonisation commenced in the spring of 1960. On each visit the plot was marked out with tape to form squares at one-yard intervals. The resulting grid pattern was drawn to scale on field-cards. The idea was to mark out on successive cards for each visit, the appearance, spread, and in some cases, disappearance of each species of plant. Each visit took about an hour or so to complete, and four of these plots are given here (Fig. 65).

The first was made in the summer of 1960, a year after the fire. Heavy rain during this season had thoroughly soaked the bare ground, and the map shows a large spread of filamentous algae. Here and there a few heads of bracken (*Pteris*), rush (*Juncus*) and a hair grass (*Deschampsia*) were found pushing out of the ground. These were survivors from the fire, whose rhyzomes or roots had escaped. Before the fire a birch tree was growing in the plot. The tree, killed by fire, now lay along the ground.

The second map made in the summer of 1961 (two years after the fire), shows a spread of bracken, grass and rush, also a heavy crop of birch seedlings, as many as twenty growing in some squares. Also, on some bare patches flourishing colonies of bryophytes had appeared, in particular the two mosses *Polytrichum* and *Funaria*, and a liverwort (*Marchantia*). These are commonly found as colonizers on burnt, acid soils.

By 1963 the liverwort had almost disappeared, partly due to dry weather, and the mosses had retreated to the shelter of the fallen birch. Rush was now established on two shallow, damp hollows. Grass and bracken were advancing towards each other, and one bare spot held a small clump of rosebay, a ready coloniser of burnt ground (see Bombed Sites, p. 123). Only a few of the birch seedlings now survived, and among these a single oak seedling was noted. The nearest oak to this plot is some 30 yards away. Presumably the acorn was carried and dropped, or buried, by a passing squirrel.

By 1965 the grass and bracken had met across the middle of the plot, and there were two patches of rush still holding their own in the corners. The fallen birch its bark peeling off, lay half hidden under the bracken. Some fine brackets of *Polyporus betulinus* were growing from the dead wood. The mosses, liverworts and rosebay had almost disappeared, and only three birches had survived, now nearly 5 feet tall. So had the young 2-foot oak.

After five years from the date of the fire this plot is now in the same seral stage as it was before its destruction, and will probably remain so with bracken, grass and rush in a balanced competition, that is, until the next fire breaks out. If none occurs, then the trees may eventually overgrow the ground flora. It will then become a struggle between the three birches and the oak, so that one or other will one day dominate this little piece of ground.

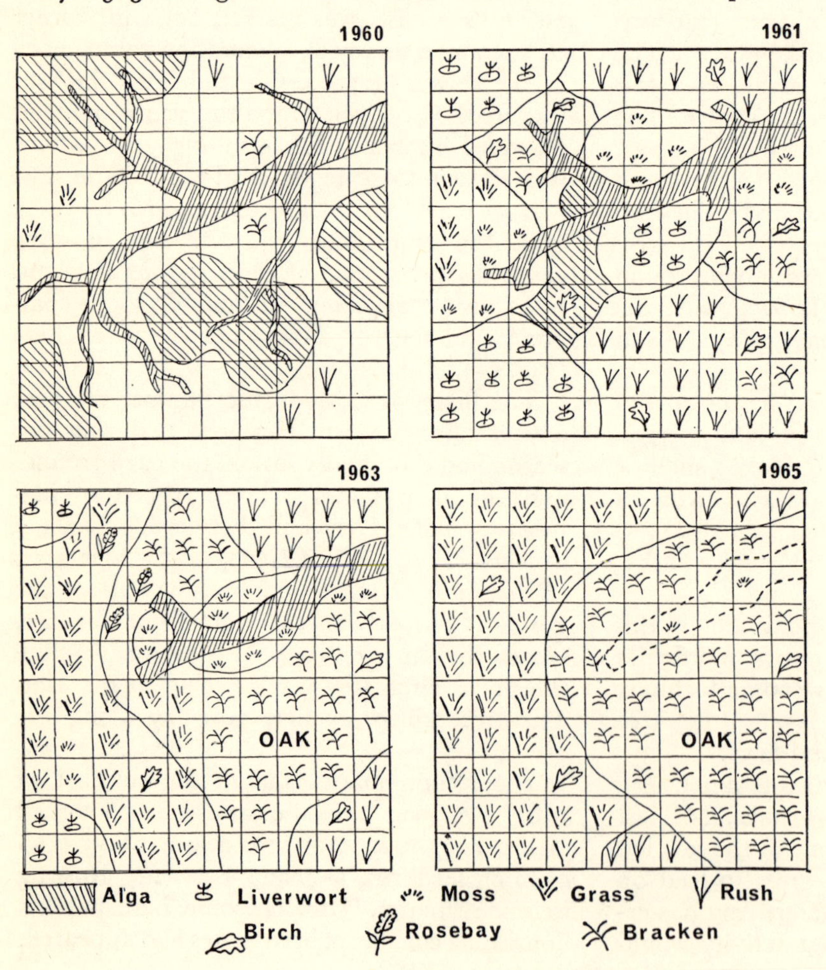

FIG. 65. Four stages in the succession of plant invasion on a burnt patch of heath, Epping Forest. (For details see text.)

Project 6

THE WOODMOUSE—a study of small mammals

As a special project in animal competition the Woodmouse (*Apodemus sylvaticus*) was selected in order to see how it fitted into community life in the Forest. With the mouse as main objective, attention was also paid to any likely predators in the area, also competitors which might occupy the same food niche or habitat.

The area chosen, some 300 yards square, is situated in a corner of the Lower Forest just beyond Epping Town (Fig. 61). It was divided into six fairly well defined zones:

A. An overgrown wood of mixed oak and beech, on heavy and often waterlogged clay soil supporting a dense undergrowth of bramble, ground flora, mosses and ferns.

B. A builder and decorator's yard and works containing tool sheds and stores, piles of planking and dumps of rubble, heavily weed-grown.

C. A row of houses and gardens.

D. An open wood of oak and hornbeam, with a good evergreen shrub layer of holly, and light undergrowth.

E. A birch grove mixed with a few small oaks and heavy undergrowth of grass and bracken, on a light dry soil.

F. A scrub area of small oak, about 12 feet tall, with thick ground cover of grass and heather—a succession stage after tree felling some years before.

G. A steep bank with thick hedgerow separating the builder's yard from the wooded areas.

The woodmouse project was commenced in 1936, when the area was mapped into the above sections and the distribution of the mouse, its competitors and predators, marked in. Information was obtained in the following ways: (*a*) by direct observation of animals seen, (*b*) by looking for tracks and signs of feeding activities, nests, etc, (*c*) by live trapping (captures then released), (*d*) by examining food pellets of owls in the area, droppings of badger and fox, and prey caught by cats, (*e*) by examining discarded bottles and tins in which animals may have died, and (*f*) by questioning residents, workmen at the yard, the Forest keeper and woodmen.

In making any survey of this nature it helps to know in advance something about the kind of habitat the animals occupy, what they feed on, their breeding potential, and so on. The following notes may therefore help the reader to anticipate what was discovered during this project. Only those species relevant to this survey, and actually present in this part of the Forest, were taken into account.

THE WOODMOUSE (*Apodemus sylvaticus*), main object of this survey.

Habitat. Normally a species of woods under open cover of bushes and trees, also hedgerows, gardens and allotments. Makes extensive tunnels in and below leaf-litter. An active leaper and climber. May occupy outhouses and sheds in absence of house mouse. Nocturnal, living in loose colonies, often sharing burrows.
Food. Seeds, nuts, fruit, buds, and a high proportion of insects and snails.
Breeding: Five to six young in about five litters, from March to October.
Nest. A ball of grass and leaves at the end of a burrow.
Predators. Weasel, stoat, fox, badger, adder and owls. (Chief Forest predator probably the tawny owl.)

The following competitors were all found in the area:

THE HOUSE MOUSE (*Mus musculus*).

Habitat. A commensal of man, occupying buildings, stores and corn ricks. Tends to move into open in summer, into fields, hedgerows and walls. Sociable and mainly nocturnal. Uses existing tunnels.
Food. Omnivorous—all kinds of food in houses and towns. On farmland feeds mainly on grain and grass seeds.
Breeding. Five to six young, of up to five litters at all seasons, even higher numbers in sheltered places with regular food-supply, e.g. corn-rick.
Nest. A rough ball of any available material, somewhere out of sight in a hole, under floor-boards, in a corn-rick, etc.
Predators. Cat (in buildings and on farms),

weasel (along hedgerows) and man. (Chief predator in this area probably cats.)

FIELD VOLE (*Microtus agrestis*)

Habitat. Normally in rough grass of undisturbed pastures, woodland, glades, moorland, reed-beds, dunes, scrub areas and young plantations. Active day and night, under ground cover of grass, bracken, etc. A surface dweller, making tunnels through grass stems. Rarely jumps or climbs. Lives in loose colonies.
Food. Mainly grass stems, also bulbs, roots, bark and seeds. A fairly strict vegetarian.
Breeding. Three to six young, in three to four litters from April to September.
Nest. A ball of grass and other plant material at end of surface tunnel situated at base of a grass tussock or under a log or stone.
Predators. Kestrel, buzzard, barn and short-eared owls, fox and adder. (Chief Forest predator probably fox.)

BANK VOLE (*Clethrionomus glareolus*)

Habitat. Thick undergrowth, more so than woodmouse, along woodland borders, hedgerows and scrub areas, especially in banks and walls. An active climber and burrower. Largely diurnal. Lives in loose colonies.
Food. Mainly green plants, also berries, seeds and bark. Prefers softer food to that of woodmouse.
Breeding. Two to four young, in four to five litters per year.
Nest. A ball of grass and herbage at end of more or less vertical tunnel in bankside.
Predators. Mainly weasel and tawny owl, also little owl.

COMMON OR BROWN RAT (*Rattus norvegicus*)

A strong commensal of man—in buildings, stores, yards, sewers, rubbish dumps, farmland and corn-ricks, spreading outdoors in summer to river and canal banks, along hedgerows and into woodland. An active burrower, living mainly below ground. Mainly nocturnal but often abroad by day in quiet places, coming well into open. Sociable. Uses existing tunnels.
Food. Omnivorous. All kinds of food in buildings, mainly corn on farmland. In hedgerows and along river banks and seashore mainly a scavenger. Will catch small animals and eat carrion.
Breeding. Six to ten young of up to five litters in all seasons, more so in sheltered places with plenty of food, e.g. in corn-ricks and stores.
Nest. A loose and untidy nest of available material, somewhere out of sight in a hole, under a building, in a sewer, corn-rick, etc.
Predators. Fox, stoat and man. (Chief enemy man).

COMMON SHREW (*Sorex araneus*)

Habitat. A wide range, but usually under dense cover of grass, scrub, bushes, bracken, etc. Lives in runways made in and above leaf-litter, also uses existing tunnels. Climbs well. Restless and aggressive, somewhat solitary.
Food. Most invertebrates and other small animals found in leaf-litter (beetles, worms, spiders, grubs), also carrion.
Breeding. Six to eight young in variable numbers of litters, from May to August, after which the adults die off.
Nest. A loose ball of grass in leaf-litter, or at surface under some cover, e.g. a log.
Predators. Mainly cats and owls, especially barn owl, also adder.

A few rabbits existed in a small colony in the birch area *E*. Grey squirrels were frequently seen, and the concentration of summer greys suggested that a small colony of about a dozen or so animals was centred in the oak wood *A*, making food forays into *C* and *D*. Since rabbit and squirrel were not considered to be serious competitors of the woodmouse, they were ignored during this survey. Attention, however was paid to the following predators:
Fox. Occasionally seen by day in the area, and heard barking at night. Tracks found in snow. Once observed, digging in *A*, possibly after woodmice. Also seen twice in *F*, stalking and pouncing in between grass tussocks, as if in pursuit of field voles. Frequently reported after poultry in *C*. Nearest earth about a half-mile away, where remains of rats and chickens were found.
Badger. An occasional visitor. Tracks found along muddy rides in the area, also in snow. Signs of

digging in *A* and *D* (searching for grubs or mice?). Nearest sett one mile away.
Weasel. Seldom seen. One shot by Forest keeper, and sometimes seen foraging along bank and hedgerow *G*, also in builder's yard *B*.
Tawny owl. Permanently in the area. A number of roosts found, and large number of pellets examined. Found to contain a high proportion of woodmouse skulls and teeth, also one or two shrews and house mice.
Barn owl. Not common. Reported by keeper and local residents (a 'white owl which screamed'). A lucky find of a temporary roost occupied for a few weeks in an empty shed in builder's yard *B*. Pellets examined contained bones and teeth of shrew, woodmouse and house mouse.
Cats. Often encountered abroad in wooded area *A*, also *B* and *C*. Householders report capture by rats of mice (house or wood mouse?) and 'a mouse with short tail', i.e. vole, or 'a mouse with long nose', i.e., shrew. The cats appeared to have dined well on a diet of mouse, vole and shrew, although the last may not have been eaten.
Adder. Usually present in bracken in *E*. A number caught with leather gloves at different times. Regurgitated meals included field vole, woodmouse, lizards and odd nestlings.
Grass-snake. Occupied the ditch and hedgerow *G*. An occasional woodmouse regurgitated, which was understandable because of a total absence of frogs and toads, the more normal prey. This species normally feeds on cold-blooded prey, e.g. fish and amphibians.

By 1939 sufficient material was accumulated for a review of the woodmouse situation, and this is shown in Map 1 (Fig. 66). It sums up as follows:
Woodmouse. A main concentration in wood *A*, with overspills into the builder's yard *B*, the gardens *C*, and the oak-holly section *D*. Nests found mainly in *B*, under planks and between rubbish covered by overgrown weeds. Numerous pellets found at tawny owl roosts in *A* and *D*. Woodmouse trails were common in snow in *A*, seen as a series of hops (voles and shrews run over the ground). Here and there a trail would disappear into the snow where a mouse had dug through the snow crust, either to reach its nest or to forage in the hidden leaf-litter. A carpet of snow makes a protective blanket, covering the woodland harvest of seeds, fruit and nuts in a kind of deep-freeze, at the same time giving protection and insulation to the hidden mice. In such conditions many small burrowing mammals can survive the hardest of winters.
Housemouse. Noted mainly in *B* and *C*, but not recorded for the wooded areas. Occasionally trapped by householders and caught by cats. Nests found in shed of *B*, where live-traps were set. Enquiries among house residents at *C* produced varied results. Some occupiers responded with interest, others curiosity, and one or two with righteous indignation!
Brown rat. Areas *B* and *C* only. Specimens seen both in builder's yard and in gardens, usually hanging around the poultry pen. A rodent operative called on one occasion after complaints from the neighbours. One rat seen during summer, foraging along the stream and bank *G*.
Field vole. Mainly concentrated in the grass at *F*, and in bracken at *E*. Vole runs common among grass tussocks, where a number of surface nests were noted. A few vole remains found in owl pellets.
Bank vole. The only specimens found, by live trapping, were along the bank and hedgerow *G*. No field voles ever turned up there. A few remains in owl pellets.
Common shrew. Found by trapping, also discovering nests, in all areas. A few remains in tawny owl pellets, but mainly in pellets of barn owl. 'Mice with long noses' reported by householders, either caught by cats or 'just found dead in the garden', presumably killed by cats and left uneaten. Since the shrew is an animal feeder, its presence would not seriously affect the rodent population.

Map 1 shows an interesting share-out of territory over the whole area. Although there was some overlap, especially during summer movements, each species kept pretty well to its own habitat—woodmouse in wooded places (*A* and *D*), field vole in grass and bracken (*E* and *F*), bank vole in hedgerow and bank *G*, and rat and mouse in buildings and yard

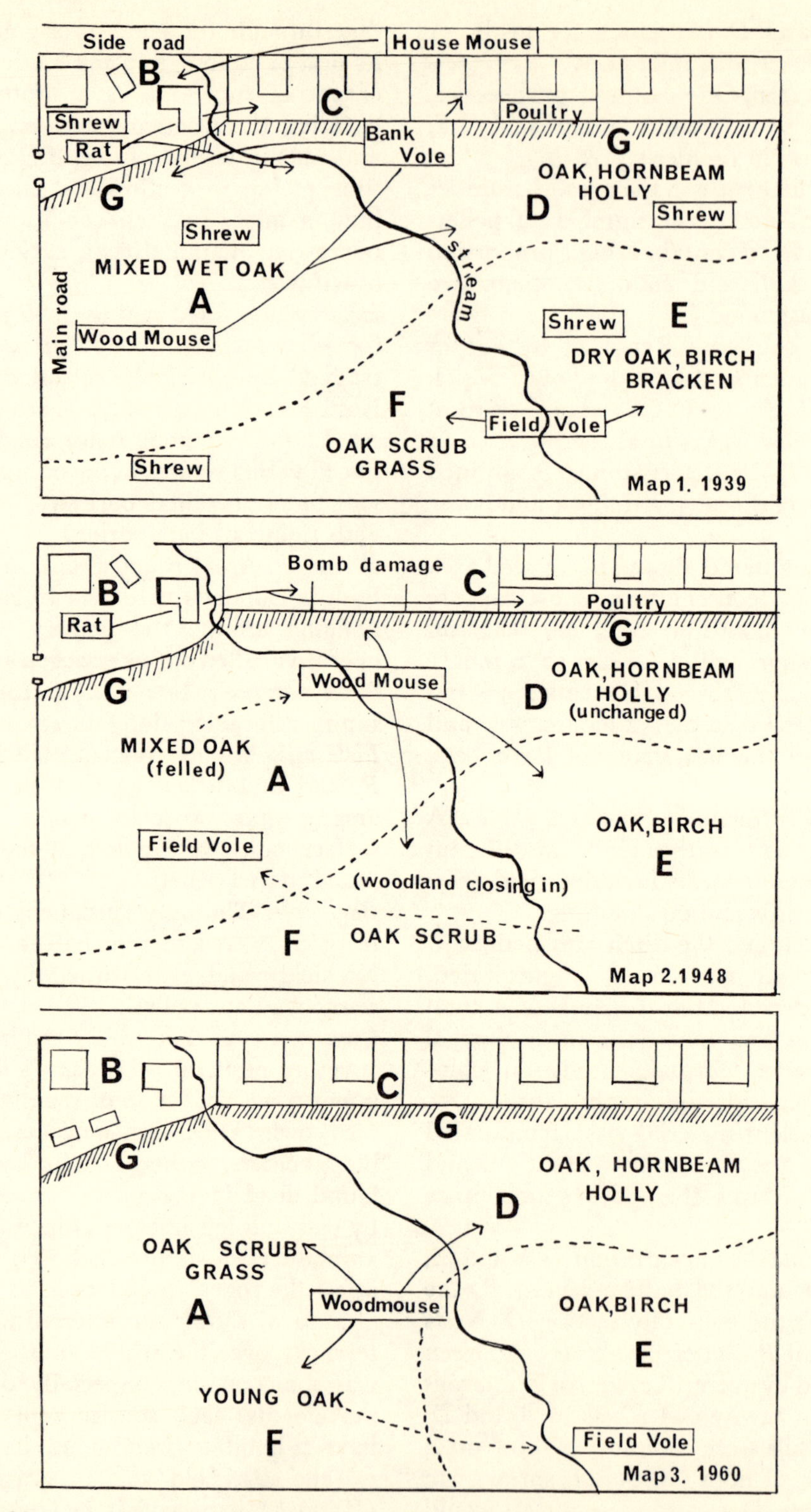

FIG. 66. A survey of small mammal distribution in Epping Forest, showing three stages (1939, 1948 and 1960).

(*B* and *C*). Only the shrew appeared to turn up in most areas.

In 1941, during the author's absence on war service, some drastic biotic changes took place. The oak area *A* was heavily thinned out by the woodmen, the builder's yard *B* closed down, and three houses in *C* were severely damaged by a bomb. The poultry pen was unscathed and had been enlarged as a war-time food measure. Areas *E* and *F* were closing in to give more tree cover. This situation was found during 1948-49 when the woodmouse survey was repeated. A shift in the species distribution is shown in Map 2.

Woodmouse. The felled wood *A* was now entirely deserted, and woodmice appeared to have shifted into the undisturbed area *D*, and were beginning to invade vole territory in *E* and *F*.

Field vole. A few left in *E* only. A definite shift into the former woodmouse territory in the felled wood *A*, where grass was now making heavy undergrowth.

Rat. Now in sole command of the deserted buildings in *B*, also in the ruins of the bombed site *C* with frequent raids on the poultry houses.

House mouse. Much less in numbers. A few recorded in the occupied houses, presumably losing a battle for territory with the rats. These two species normally live apart.

Bank vole. Appeared to have deserted the bank *G*. Possible pressure from woodmice may have been the cause.

Common shrew. No sign anywhere, except in occasional owl pellets at a roost in *D*. The reason for this disappearance was never determined.

In 1960 a third and final survey was made. By now the builder's yard was again in use by another firm, and the place kept in tidier condition. In *C* the bomb damage was fully repaired and the new houses occupied. The poultry yard had gone, and the hedgerow along the bank *G* severely cut back. In the wooded areas *A*, *D* and *F* the trees had more or less joined up in a continuous cover. *E* retained its character of birches and bracken. Intensive trapping with the new Longworth trap, by then in regular use by field mammalogists, revealed a further shift in populations from the previous two surveys. The woodland *A*, *D* and *F* was almost entirely dominated by woodmice. Only in the birch area in *E* did field voles still turn up. Rats and mice were absent from *B*, *C* and *G*, and the householders and yard appeared to be free from these unwelcome commensals. A very few shrews were found in *A*, which had by now advanced well into a scrub stage.

The shifting populations of our small mammals, as revealed by this survey, is probably a normal pattern in most of Britain. A high human population with all its attendant activities is bound to disturb the countryside sooner or later. Whether or not this may be for better or worse, so far as we are concerned, it will always be exploited by nature in one way or another.

Project 7

TRACKS AND SIGNS

Searching for tracks and signs which animals leave about can add interest to a field project, apart from helping to trace their whereabouts. Such information provides the clues where any direct observation is difficult. This applies particularly to mammals. Shy and elusive, well camouflaged, largely silent and often nocturnal, the mammals of our countryside betray their presence in, broadly, one of four ways—by their footprints, feeding activities, droppings, and nests or burrows.

Tracks. Soft ground such as mud, wet sand or clay, and especially snow, will show clear prints even down to claw marks. These can be looked for along pathways, borders of ditches, ponds and lakes, by the riverside and along the seashore, particularly on mud-flats and in estuaries, and at the entrance to earths (Fig. 67).

Usually the space of ground where tracks may occur is limited to a few yards, so that continuous trails are not visible. An exception would be in snow, in a sand dune, or at low tide along the shore where an animal's journey can sometimes be followed for long distances. From this one can learn the mode of locomotion, and how the feet are placed along the trail. Speeds and movements can be worked out, and a good tracker will learn much about the animal he is following, even before he sees it.

Records of tracks and trails can be copied into the field notebook with the aid of a measuring tape, so that they can be drawn to scale. Another method is to take photographs. Best results are obtained by angling the camera so that shadows heighten the contours. These show up best in slanting sunlight during early morning or late afternoon.

Permanent records of tracks are made with the aid of plaster-of-Paris (see Appendix 10, p. 228).

Feeding Activities. With a little experience one gets to know the various signs of feeding activities which mammals leave about. Teeth marks on bark, twigs, nuts, cones, acorns, bone, and such objects as the dropped antlers of deer may suggest the work of some small mammal, a squirrel, rabbit, or even a deer. These vegetarians also have characteristic ways of dealing with grass, leaves and growing twigs. Some bite off their food neatly, others tear it off, or strip away pieces.

The work of carnivores may also be recognised. For instance, a half-eaten bird with feathers torn out is probably the work of a bird-of-prey. If the feathers are bitten through, then perhaps a fox has made a kill. The empty skin of a hedgehog, a torn-out wasp's nest or rabbit ncst is what one expects from a badger. Gnawed bones of a picked carcase could be the work of a rat. In the case of bark damage, the height to which this reaches from the ground may help to determine whether a rabbit, squirrel or deer has been busy in the area.

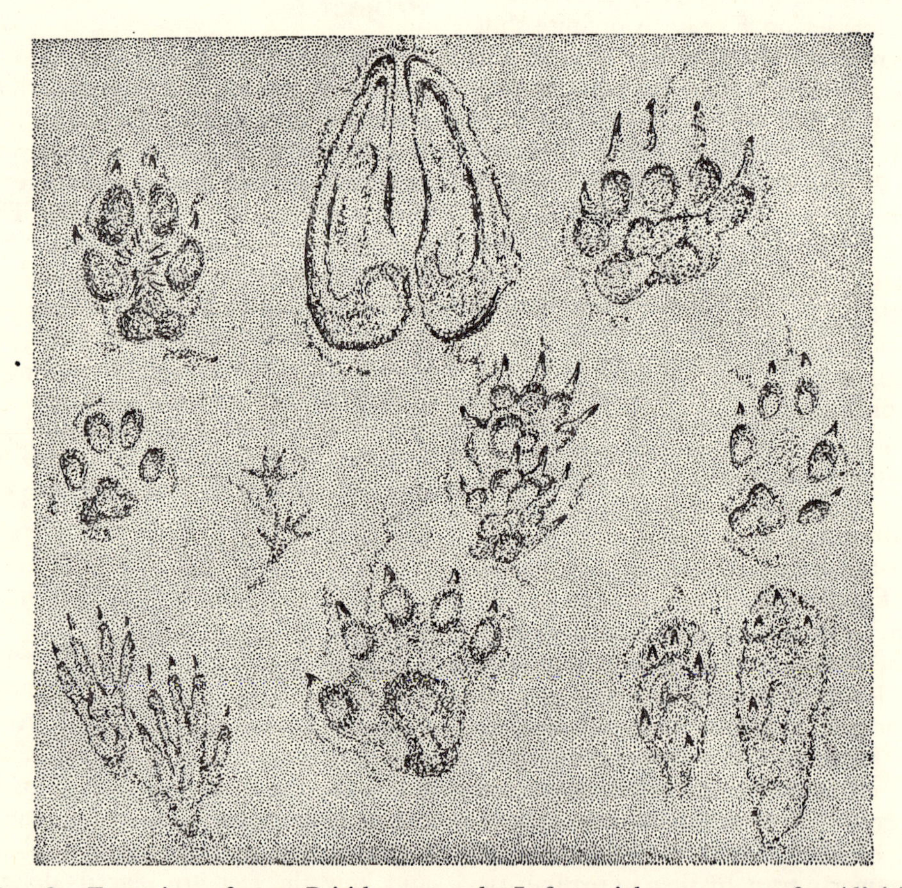

Fig. 67. Footprints of some British mammals. Left to right; top row—fox (digitigrade); fallow deer (unguligrade); badger (plantigrade); middle row—cat; wood mouse, fore and hind; hedgehog, fore and hind; pine marten; bottom row—grey squirrel, fore and hind; otter; rabbit, fore and hind.

Droppings or scats. Many mammal and bird droppings, even those of reptiles and amphibians, and many invertebrates, have a characteristic shape and size, especially when fresh. Colour, texture, contents, and sometimes odour, are helpful in identity. The subject of scatology involves a study of these waste products, and much can be learned about an animal's feeding habits. The specimen can be treated as a microscopic smear, or by teasing out the contents. The presence of some small mammal or invertebrate unknown to the locality can be detected by this method.

Typical examples of food remains are the hard parts of beetles found in the scat of a hedgehog or toad. A microscopic search of the dung of a badger will show the setae or bristles of earthworms which they eat. Otter contains a high proportion of fish scales, hair is largely present in fox, and so on. Equally important are the regurgitated food-pellets of a number of birds such as owl, hawk, heron, gull and crow. In these can be found quantities of teeth, skulls and bones of small birds and mammals even the occasional reptile or amphibian, and insect. Bird-of-prey pellets are particularly useful in tracing the whereabouts of small mammals, as happened in the Woodmouse Project (p. 163).

Sometimes the discovery of the contents of a scat or pellet can reveal surprises, especially when there is a shortage of natural prey. This occurred with the buzzard and fox when the rabbit was hit by myxomatosis. During one severe winter the author examined the droppings of a fox, and found them full of starch grains. The starving animal had been feeding on potato.

Nests and burrows. A mammal has to rest sometimes, or remain in an area to rear a family. As a result some kind of home or shelter is prepared, and here again useful clues can be searched for, especially if the home is occupied. An observant naturalist will constantly shift his eyes from ground level to the trees above, and from ditch to hedgerow. An exposed nest may first be spotted as a silhouette against the light, in a bush or tree. Otherwise, there may be a tell-tale opening in the ground. Its location, size and shape will help to identify the occupant, and a closer examination of the nest material may help to confirm the first suspicion.

As in a criminal investigation all these signs must be taken together so as to build up firm evidence as to which mammal (or bird) has been responsible. Sometimes the obvious signs may be misleading. At a badger sett well known to the author there were signs of occupation at one hole. However, the musty smell, remains of rabbit and a fresh heap of dung nearby showed that a fox was in residence. Meanwhile, the fastidious and clean badgers had moved to a fresh hole further along the bank. Piles of fresh earth had been dug out, tracks were identified, and badger's hairs found stuck to the roots above the doorway.

The value of this kind of detective work is suggested in the Badger and Woodmouse projects (Projects 1 and 6).

Project 8

DAY AND NIGHT ACTIVITIES

Anyone who has taken up badger watching, gone camping, worked at an outdoor night-shift, or simply taken a walk after dark in the countryside, must have noticed the different sounds and movements among the animals after the sun goes down. The hoot of owls and churring of nightjars take the place of the song of day birds and the drumming of woodpeckers. The yap of fox and yell of badger are heard in place of a squirrel's chattering or bark of deer. Moths and beetles take over when the bees and butterflies go to rest, and bats are on the wing in place of swallows.

All these animals are busily searching for food, each part of a food chain or food web. In some cases, as with many small invertebrates, feeding may be continuous, but with larger animals there is usually a period of rest between meals. Their rhythm of activity is linked with night or day, so that two distinct feeding cycles are going on in succession within the same habitat area. The night shift, as it were, takes over from the day feeders.

A walk through a woodland by day, and also by night, making recordings of the activities of animals and what they feed on, soon makes it clear that food chains are flexible. In some cases quite different species may occupy the same food niche, by feeding at different times. For instance, sky hunters like swallow and dragonfly pursue the flying insects by day, whereas bat and nightjar take over after dark. On the ground birds like thrush and robin forage for worms, molluscs and wood-lice in the leaf-litter. At night this becomes the role of badger and hedgehog. Some hunters share the same food. Kestrel, weasel and grass-snake catch the voles by day, whereas fox, owl and adder find them by night.

Frequent excursions by the author into the Forest by day and night in all weathers throughout the year have made it possible to form a general picture of what is going on above and below the trees in a woodland community (Fig. 68). It is possible from observations and counts to build up a pyramid of numbers to make a food chain (see p. 62). The 'key' animals on which much of the animal population depends are the caterpillars, aphids, woodlice and other small invertebrates found on the leaves and in the leaf-litter. These occur in greatest numbers. The 'apex' animals at

FIG. 68. Day and night activities in Epping Forest. Arrows denote some of the food chains. A, grey squirrel; B, dragonfly; C, fly; D, spider; E, bark beetle grub; F, Great spotted woodpecker; G, fallow deer; H, little owl; I, weasel; J, bank vole; K, earthworm; L, song thrush; M, frog; N, grass snake; O, snail; P, tawny owl; Q, bat; R, nightjar; S, moth; T, beetle; U, badger; V, rabbit; W, fox; X, woodmouse; Y, hedgehog; Z, adder.

the top of the food pyramid are the fewest in numbers, such as the owl, hawk, weasel and fox. Key animals usually remain all their short lives on one spot—on the leaves, on bark, in the leaf-litter, and so on. Apex animals roam about more freely throughout the habitat area, even leaving it at intervals, depending on the availability of food, search for a mate, or the competition for territory. A hawk will range the entire woodland, whereas a caterpillar stays on a single leaf (see Fig. 21).

An area was selected in Monk Wood, and records kept of the animals seen or heard during a number of visits between 1947 and 1955. From these a number of species have been chosen in order to give an overall picture of the food web which holds the animal community together (Appendix 11, p. 229). This is very much a matter of supply and demand—so much food for so many diners, and the more diners from the same table, the more selective are their dishes. This will become clear from the next project.

Project 9

FOOD NICHES—THE OAK TREE

The sharing out of food is obviously essential to an animal community if it is to survive as a harmonious unit, even though there may be a high sacrifice in life. To preserve the species at a proper level the losses should be in relation to the birth rate (see example on p. 63). For every single hawk in a woodland there are thousands of insects. These small 'key' animals far outnumber their predators. How do these enormous numbers of small animals manage to share out the plant-food which they need?

To find out something about this a single food source was chosen—an old and pollarded oak-tree. In Epping Forest, as in any British oak wood, this well-known and long-lived tree is the home, shelter and food source of a large number and variety of animals, big and small. Each occupies its own food niche.

The tree in question used to stand in the Lower Forest, the least disturbed part of Epping Forest (see p. 149). When first examined, in 1935, it appeared in a healthy and vigorous condition, in spite of its high population of small animals and seasonal attacks from defoliators. By 1939 there were signs of fungal attack and rotting in the pollarded crown due to rain. In 1947 a winter gale blew it down, and by 1955 the remains had broken up apart from a branch or two and a mouldering stump. In 1960 there was no sign of a tree ever having grown on the spot.

Many visits and observations were made over the years. The oak became, one might say, as well known to me as a second home, in which almost every branch, leaf and bark surface were as familiar as the furniture at home. Every available nook and cranny seemed to provide a home for one or other species of animal or plant. Not all the life was permanent. Some made a seasonal appearance, others came as casual visitors. It was used for food, shelter, or as a breeding place. Many species were born, grew up, and died on the old oak. It was their only home.

To find many of the smaller animals which might be overlooked, various collecting techniques were used (see p. 201). Three selected lists from separate visits are also given, showing the shift and change in the oak tree community. One is for 1936 when the tree was in good health. The 1948 count was made after the tree had fallen (in 1947) and the third list made in 1955 from the tree stump and fallen branches (Appendix 12, p. 230).

In the complicated food web at work on this tree it was noticed that all parts came under attention from the animals, such as leaves, bark, buds, roots and so on. Of these the leaves were the most acceptable, and here the competition for a food niche was intense. Each leaf tenant appeared to have its own preference, so that on occasions many different species were feeding on the leaves at the same time (Fig. 69). Leaf miners tunnelled under the epidermis, aphids pierced the leaves for the juice, and the grubs of gall insects remained on the spot where the eggs were laid. Caterpillars wandered about more, biting off portions of a leaf with their jaws. Squirrels nibbled off the buds, and occasional deer browsed from the lower leaves.

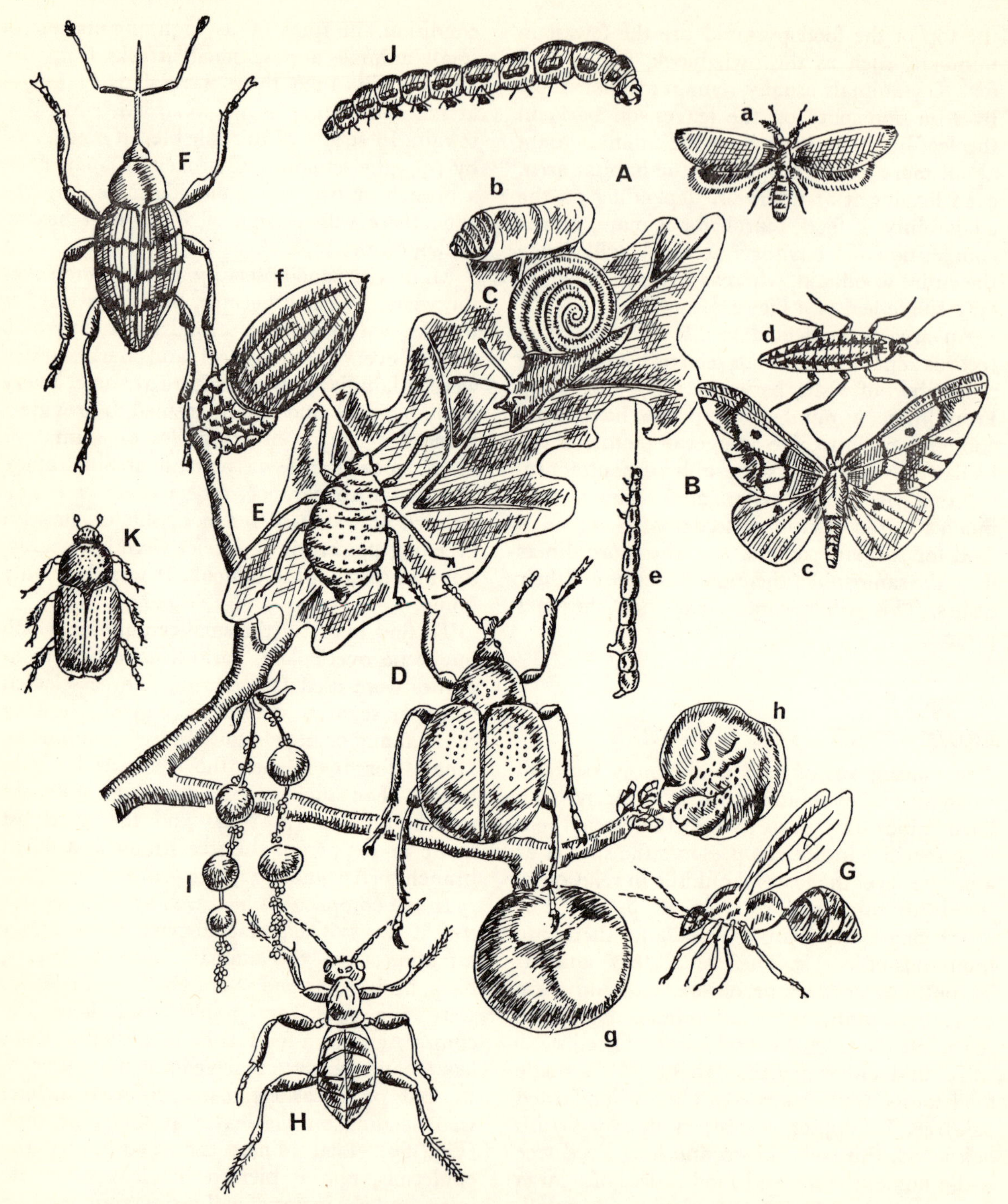

Fig. 69. Some animals and their food niches on the Oak. A, Oak roller moth (*Tortrix viridana*)—a, adult and b, pupa (a defoliator); B. winter moth (*Cheimotobia brunata*) c, male, d, female and e, larva. (a defoliator); C, a snail; D, a weevil (*Attelabus curculionoides*); E, an aphis (sap feeder); F, nut weevil (*Balaninus* (= *Curculio*) *nucum*) on acorn f; G, gall wasp *Andricus kollari* (female) on marble gall g; H, gall wasp *Biorhiza pallida* (agamic form) on oak apple h; I, currant galls on oak catkins formed by *Neuoterus quercus-baccarum;* J, goat moth caterpillar (*Cossus cossus*); K, oak bark beetle (*Scolytus*).

Some of these leaf feeders worked at night, and others by day. Mildew fungi also took their share. Even when the leaves fell in autumn their use was not finished. Here a new army of feeders took over, among them, woodlice, litter grubs and saprophytic fungi.

Project 10

MICRO-HABITATS

Scattered around the countryside are numbers of hidden places, often overlooked, which are sheltered and enclosed in such a way as to contain a set of factors different from those of the surroundings. In these confined spaces, called micro-habitats, live communities of small animals and occasional plants which rarely stray far from the boundaries of their small world. Three of these places, commonly found in the field and usually ignored, were studied by the author, and are given here—a log, a carcase, and a cow-pat.

A. A LOG

In the author's small sanctuary, described on page 135, a log was placed down on the leaf-litter below the shrub layer. During summer it was mostly hidden by the ferns and other ground flora which grew up around it. The intention was to provide a micro-habitat within a woodland environment for different kinds of animals and plants which make use of such a place as shelter and a source of food. The log had been cut from a felled beech tree, and was about 6 feet long and 2 feet in diameter. In the following year, 1954, little turned up apart from an occasional stray invertebrate such as a woodlouse, spider, centipede or two, and a cockchafer grub. By 1955 a small colony of woodlice of two species had settled in, mostly under the bark which was beginning to peel away. More centipedes, spiders, also earthworms were noticed. Fruiting bodies of the honey fungus (*Armillaria mellea*) also appeared along one side of the log, where the mycelium had penetrated. The presence of this parasitic fungus, the so-called Forester's Curse, was the cause of the death of this tree, and the reason why it had been felled.

The third year saw a considerable build-up of the colony. The honey fungus continued to fruit, and was joined by some attractive brackets of *Polystictus versicolor*. Much of the bark had now broken away, and on the bared wood in wet weather appeared one or two species of Slime fungi (*Myxomycetes*). Various seeds caught up on the loose bark and in cracks produced a grass or two, a beech and hornbeam seedling, and some wood sorrel. Two or three small ferns grew up in that year. These plants were already established in the garden woodland around the log, although colonisation of the log was artificially encouraged by deliberate scattering of seeds from time to time.

In 1960 a list of some dozen species of invertebrates was recorded by examining the log. This meant slowly rolling it over and carefully replacing it so as to disturb the colony as little as possible. Vertebrates which used the log included an occasional visiting woodpecker which left behind it marks on the wood and chippings on the ground, a woodmouse which nested below and reared its family, and a toad which returned regularly during one summer to shelter by day in a hollow underneath. Newts from the nearby pond would hibernate in company with snails, slugs and spiders. One winter a queen wasp slept there, and the following spring built her nest close by.

As an experiment some plants with epiphytic tendencies were deliberately placed on the log, such as mosses, ferns and fungi, the latter placed cap downwards in the hope that the spores would settle and germinate. Some toadstools did succeed and produced fruiting bodies, such as the oyster mushroom (*Pleurotus ostreatus*).

A particularly successful experiment was the introduction of some stag beetle larvae (*Dorcus*) found in a rotting tree-stump in the local park. By now the wood of the log was beginning to soften (1957) so that holes could easily be bored, and the grubs placed one to each hole. An adult was seen on the log the next year, followed by more, young, grubs, so presumably the adults had mated. The colony lasted for six years, but by 1964 the log was in a mouldering condition and breaking up. That winter a blackbird and robin were seen foraging in the soft pulp, in search of any grubs. By 1966 there was nothing to show for this flourishing little woodland colony which had come to an end. The total list of those visitors and colonisers of this log which were seen and recorded over the ten years is given in Appendix 13 (p. 230).

B. AN ANIMAL CARCASE

Anything dead immediately becomes a target for the agencies of decay, or the food for scavengers. In situations where some means of preservative is available, however, the animal remains may last long enough to become transformed into fossils. Apart from such unusual circumstances as mammoth frozen in the Arctic soil, insects trapped in amber, and pollen grains preserved in peat, most fossils are found in sedimentary rocks originally formed on ancient sea-beds. On the land surface exposed to the air, rain, wind and other weathering agencies, quite apart from being eaten by scavengers as mentioned above, a carcase soon disintegrates.

An opportunity to find out what happens to a dead animal was presented when the body of a dead fallow buck was found in Monk Wood, Epping Forest, in the summer of 1938. The animal had obviously been poached, the legs and haunches having been severed, leaving behind the torso and head. The head was removed by the author, and carefully cached in a nearby ditch filled with leaves. The intention was to recover the cleaned skull and antlers at a later date. This is now in the author's collection. The body was left where it died, lying on the leaf-litter under some beech trees. To hide it from passers-by some fallen branches and bracken were placed over it.

Regular observations were made on both remains, to see what kind of scavengers the decaying meat would attract. Within two days the belly of the torso had been broken into and the entrails removed. Nearby signs and a not too distant sett suggested that badgers had been busy. In the week that followed, when the carcase was still fresh, observations were made from a nearby hide, using binoculars. Carrion feeders such as carrion crow, magpie and fox were seen at times near or on the carcase. The crow was noticed to peck at the deer's eyes. This is often the first part of a dead animal to attract attention.

Flies soon gathered on the decaying flesh, and maggots appeared. A number of specimens were collected for identification. Maggots were reared in bottles with muslin tops, and containing sawdust and pieces of deer meat. Various other invertebrates appeared, mostly under the carcase which was used more for shelter, i.e. a dark and humid micro-habitat. These in turn attracted various birds, drawn to the site in search of an unexpected meal. A blackbird was seen to poke around the remains, and one morning a flycatcher was seen hawking after flying insects hovering around the carcase.

In the general analysis of what was seen and collected, it appeared that the various visitors to this spot were attracted for one or other of four reasons—to feed on the remains, to use it as a breeding ground, to use it as a shelter, or to prey on the smaller animals already attracted to the spot. A list of animals recorded over a period of six weeks, by which time the bones were picked clean, is given in Appendix 13 (p. 230).

During the same period the deer's head in the ditch, which was dry at the time, was also examined. This was done very carefully, by gently brushing aside the covering bed of leaves, turning over the head to look underneath then replacing everything as it was found. Being hidden, the head was not found by any of the larger animals, surprisingly enough, since a badger or fox with keen sense of smell might have been expected to locate it with little trouble. As it so happened, only the litter fauna turned up, such as worms, slugs, centipedes and beetles. With the exception of some burying beetles (*Sylphidae*), it was never fully decided whether these invertebrates had gathered to feed on the head, or merely for shelter. Some so-called vegetarian animals are known to feed on flesh at times, and situations like this make one realise the vagueness with which the descriptive feeding terms of 'carnivore' and 'vegetarian' are often applied.

C. A COWPAT

This was selected from one of the fields at Debden, where cattle were put out to graze. The one chosen had dried out so that it could be lifted and turned over for examination. Beetles, earthworms and dipterous maggots were found sheltering and feeding underneath, and these in turn attracted birds such as the flycatcher and starling, as well as toad (found after dark) and hedgehog (also at night). The droppings of the toad and hedgehog both revealed beetle remains. The presence of a green woodpecker was incidental, and almost certainly due to the ants which nested in the mole hummocks in the area (see list in Appendix 13, p. 230).

Project 11

POND LIFE

Much of the author's work in this sphere has been carried out at Baldwin's Pond mentioned on page 150. Some idea of the animal and plant life which it contains may be judged from the list given in Appendix 14 on page 231. This is by no means complete, but is typical of what has been found over the years. It shows the large numbers of groups which contain freshwater species, in itself a good reason for safeguarding our ponds and using them as easily visited sites for ecological research and educational studies. This can become a life's work of never ending interest and discovery.

As explained on page 115, the interrelation between animals which goes on in such a community can best be studied by identifying the occupants, finding out what kind of food they are after, and fitting them into their separate niches in the food web.

With the plants which have a common food supply, and are bathed in it, life is more a competition for space. Being static they can be mapped out into different zones. The great variety of plants and animals found in a pond, in which they are largely imprisoned, can be used to illustrate these niches and zones (Fig. 50 and 70). The plants in Baldwin's Pond are shown on a surface map and in a profile transect which passes from the woodland border across to open water. The zones of plants which this reveals can be clearly seen by the trained eye as one approaches a pond. In the case of the animals, however, much goes on below the surface in a constant search for food, so that it becomes necessary to make observations on captive specimens. To gain this information the author has collected and kept many pond creatures over the years.

Most specimens can be kept in small containers or jam-jars, in some cases for a year or two, for making close studies into movement, feeding habits, reproduction and life histories (see Project 13, p. 179). One can easily find space near a window to hold a row of containers, one for a beetle, another for a snail, a water-bug, and so on. It is best to use rain or pond water, rather than tap water, to which is added a little portion of water plant (e.g. *Elodea*). Also, some pond-mud or sand may be placed on the bottom. In other words, one is providing a kind of micro-pond for the inhabitants. One can, in comfort and leisure, learn a great deal from this indoor pond study, and if handy with a camera can take some very rewarding close-up studies of pond life.

In the confined space of a jam-jar it soon becomes very apparent which animal is the aggressive predator and which the harmless vegetarian or scavenger. For obvious reasons each must be kept apart from the other. In assessing the position of each species along the food chain, some understanding of each animal's habits is needed. Is it bold and active, shy or retiring? Does it keep to the surface or is it a bottom dweller? These are only broad distinctions, for it does not necessarily follow that a hunter will pursue a harmless species. Size and age make a difference. A small predator would not normally tackle a large vegetarian. Also food preferences may change with age. A tadpole is an underwater plant feeder at first, then changes to a meat diet when legs appear. As an adult it catches live insects and worms on land. An adult beetle might attack a small fish, whereas its larva would leave it alone.

A typical collecting trip should be carried out with some planning. The equipment to use is suggested in the list given on page 198. The pond-side is approached cautiously, and the ground and vegetation examined for any resting or basking animals which may be there, such as dragonfly, frog, water bird or grass-snake. Next, the waterside is quietly approached taking care to avoid vibrations from heavy footsteps, and making shadows over the water. First, look for surface animals such as spring-tails, water skaters, rising bugs and beetles, newts, perhaps a colony of whirligig beetles or a basking fish. Plankton swarms will be detected in the sunny patches just below the surface. If the water is clear and not too deep, various wandering beetles, snails, caddis larvae or tadpoles may be seen on the plants or on the bottom. There may also be a stickleback guarding his nest in one corner, a mussel slowly cruising over the bottom, or a colony of tubifex worms waving over a mud patch.

Collecting will depend on what is needed, i.e. for identification or for home study. Starting at the surface, a few gentle sweeps of the plankton net will gather in the small drifting

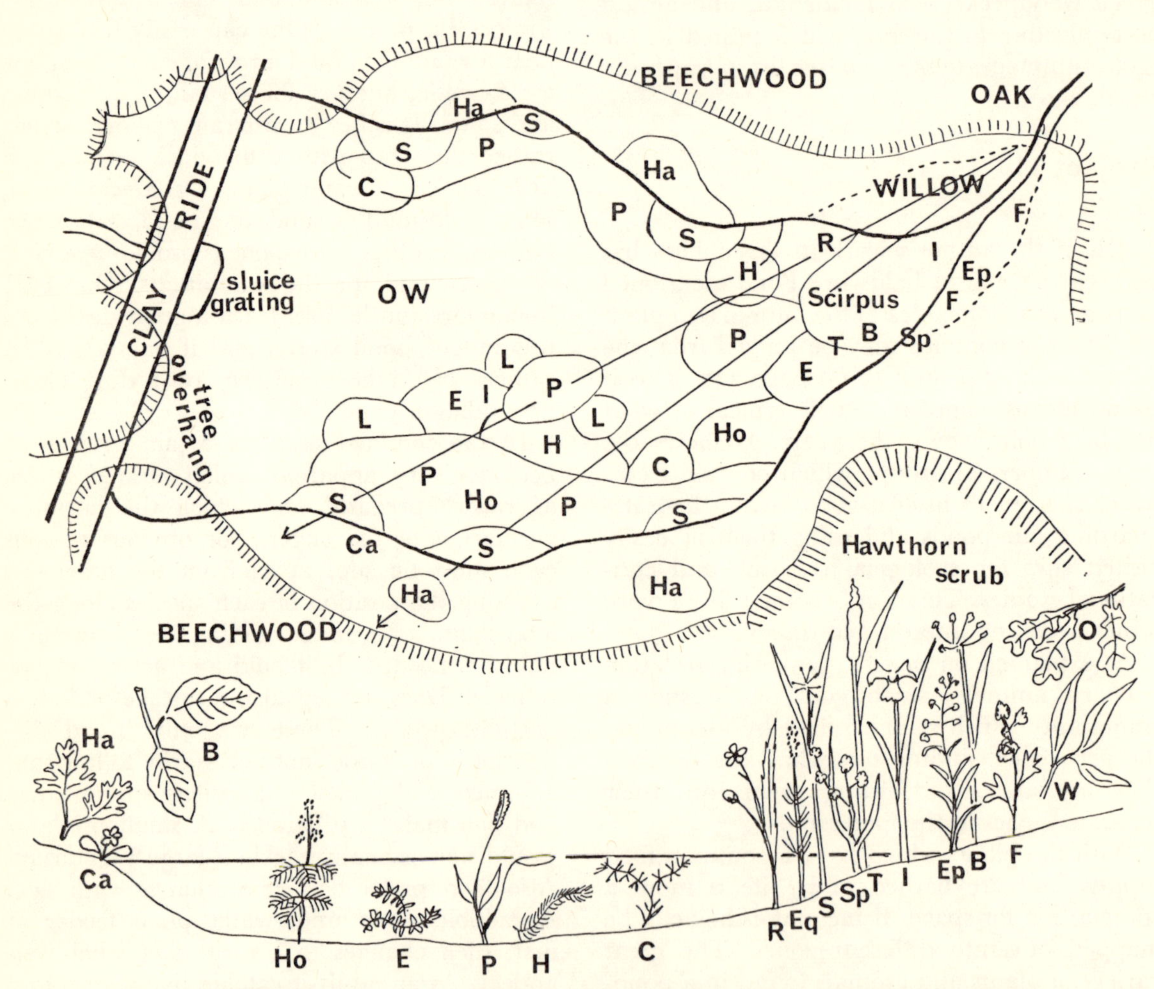

FIG. 70. Surface map and profile of Baldwin's Pond, Epping Forest, showing distribution of main shore plants and aquatics. Ha, Hawthorn; Be, Beech; O, Oak; W, Willow; OW open water; C, *Callitriche*; S, *Scirpus*; L, *Lemna*; E, *Elodea*; P, *Potamogeton*; H, Hornwort; Ho, *Hottonia*; I, Iris; R, *Ranunculus* (spearwort) B, *Butomus*; T, *Typha*; Sp. *Sparganum*; F, Meadowsweet; Ep, *Epilobium*; Eq, *Equisetum*; Ca, *Caltha*. (The profile transect below was sighted between the two arrowed lines).

animals, and a more rapid thrust with the sweep net will capture a rising newt or beetle. There is a pause between each catch, to allow things to settle, then the net is pushed steadily through the clumps of underwater vegetation, where the hidden dragonfly larva, snail or water scorpion may be found. Finally, the bottom is slowly dredged with the trawl net, and the contents tipped out to look for the detritus feeders, such as worms, caddis larvae, water louse and mussel.

To avoid overcollecting each captive should be examined, and if possible identified in its container. For the smaller specimens the pond-side 'aquarium' or a white dish is useful for close inspection, and where a hand lens can be used (see p. 201).

Life in a pond is always fluctuating, and this is due to the changes of seasons which affect such things as temperature, oxygen and carbon dioxide contents, also the concentration of mineral salts, all of which are necessary to life. During a single year a chart was compiled to show this for Baldwin's Pond (Fig. 47). Readings were taken and water samples collected at approximately fortnightly intervals.

Mapping a Pond. Preparing a scale map of a survey area by means of the grid system is explained in the Debden project on page 185. Laying out a grid is not possible in such places as a rocky seashore, or across open water. In this case a sighting technique is adopted, as was used for plotting the outline of Baldwin's Pond (Fig. 71).

A drawing-board or some similar flat surface is fixed at its centre by means of a clamp which pivots on a swivel attached to a stand. A camera tripod would be ideal. The tripod is positioned over the required spot so that it is vertical (by using a plumb line). The board is then adjusted with the aid of a spirit level so that it is horizontal in all directions.

Drawing-paper is fixed to the board, using

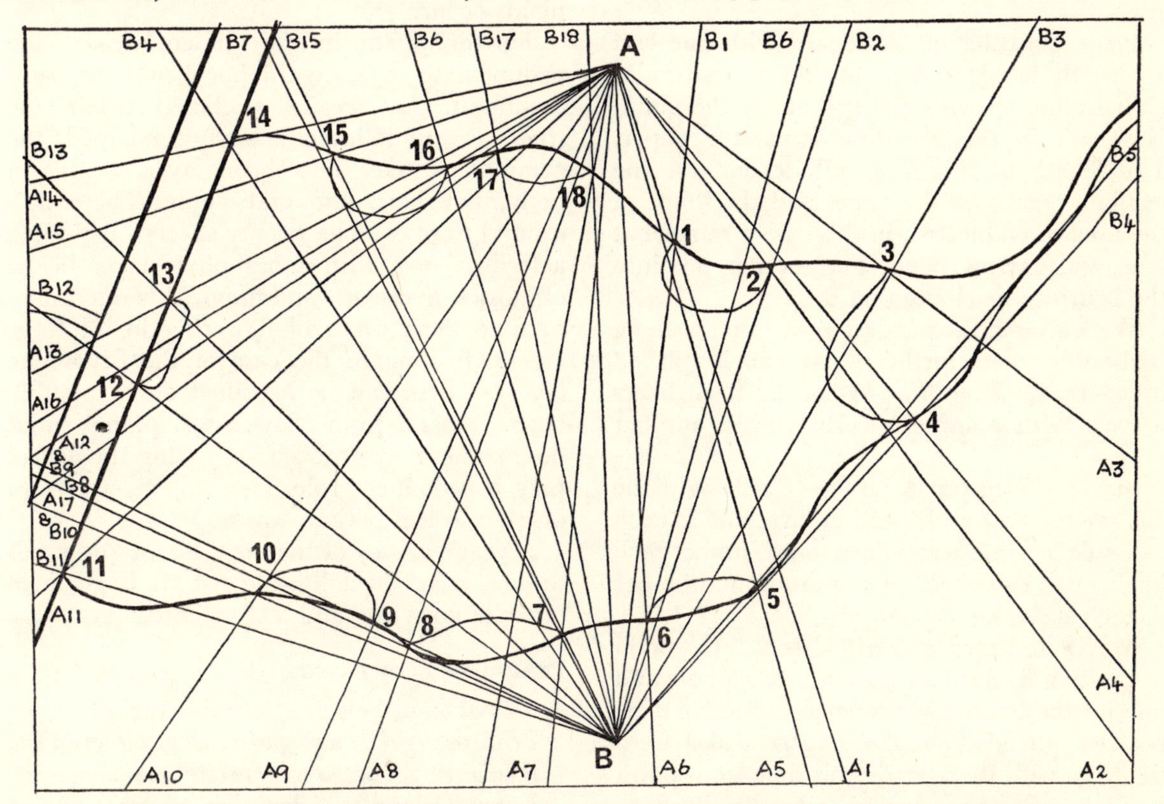

Fig. 71. Mapping Baldwin's Pond. The alidade is sighted from points A and B and the pond outline drawn through the numbered points where the sightings intersect.

drawing-pins so that it does not blow away. A base line is next selected which will cover the entire area to be mapped, in this case Baldwin's Pond (Fig. 71). Because of its elongated shape this line was taken across the narrower width, from A to B. Pickets were driven into the ground at these two spots. A was then removed, and a short peg placed in the hole. The tripod was placed directly over this, and the board clamped horizontally in position. The N.-S. bearings were marked in one corner with a box compass thereby fixing the position of the board.

The plotting now proceeds as follows. Fix a pin into the paper on the board to mark position A. From this pin various points along the pondside will be sighted and marked onto the paper. For this an alidade is required. This usually has a telescopic sight, but a much cheaper instrument which gives good results at short distances, is a strip of planed wood, or even a ruler. This has two small nails driven in at each end, placed carefully in line with the edge.

Place the ruler on the board with one end against the table pin A. Sight along the two ruler nails to line up with picket B across the pond. Now rule in the base line across the paper. The length of this line will depend on the available space on the paper and the shape of the area to be plotted, in this case a somewhat long and narrow pond. Fix another pin into the board to mark position B.

Walk around the pond, selecting points along its border where further stakes can be placed in positions visible from A and B. Each bears a card with a large, clearly visible number on it.

Starting from pin A on the board, sight the alidade towards each stake in turn, and in each case rule in a line across the paper. Number each line as it is ruled with the appropriate number of each stake, i.e. A1, A2, A3, etc.

Move the board and tripod round the pond to position B. Replace picket A in its hole and line up the board by reference to the compass bearing already on the paper. Take fresh sightings with the alidade this time from point B. These will be marked in as B1, B2, B3, etc., and will intersect the lines taken from A. By connecting these intersections an outline of the pond will be produced.

Project 12

WOODLAND FLORA

As already explained in the chapter on Woodlands (p. 74), the richness of its flora, especially in an oak wood, is due to the beneficial factors which are present in such a habitat. The heavy carpeting of the ground flora, and the jostle for space among the shrubs and trees above, would suggest a fierce struggle for existence in the competition for light and living room. Much of this apparent overcrowding is made possible by the layering of the root systems and height of the shoots, and also by the succession in growth and activity of the leaves and flowers. To observe this it is necessary to make repeated visits to the same area over a whole year. This was tried out in the Lower Forest where a good number of woodland plants occur.

The dominant in the chosen area is the pedunculate oak, with hornbeam as sub-dominant. An area was chosen where the trees are unpollarded, having escaped the commoners' axe. The shrub layer consists of holly, hawthorn and crab-apple. There is a ground flora of some twenty species, including a fern or two, with a bryophyte layer below of mosses, liverworts and fungi. It is interesting to note the periods of flowering in all three layers (fruiting in the case of the bryophyte layer). These can be classified on a monthly basis. Apart from flowering, plants need nourishment to grow, and in this the leaves play a prominent role. Here again, a certain overlap in leaf periods was noticed.

A year's survey of the area, about 100 yards square, gave the following list. It is given in order of flowering and leaf activity.

PERIOD OF FLOWERING

Pre-vernal (mid-February to mid-April)

Primrose (*Primula vulgaris*), Dog Violet (*Viola riviniana*), Wood Anemone (*Anemone nemorosa*), Lesser Celandine (*Ranunculus ficaria*), Sallow (*Salix caprea*).

Vernal (mid-April to end May)
Oak (*Quercus robur*), Hornbeam (*Carpinus betulus*), Bluebell (*Endymion non-scripta*), Bugle (*Ajuga reptans*), Ground Ivy (*Glechoma hederacea*), Cuckoo-pint (*Arum maculatum*), Pendulous Sedge (*Carex pendula*), Early Purple Orchid (*Orchis mascula*).

Aestival (end May to end August)
Enchanter's Nightshade (*Circaea lutetiana*), Foxglove (*Digitalis purpurea*), Rosebay (*Chamaenerion angustifolium*), Honeysuckle (*Lonicera periclymenum*), Black Bryony (*Tamus communis*), Male Fern (*Dryopteris felix-mas*).

Autumnal (September to November)
Various fungi in fruit. Various mosses in fruit.

Hiemal (December to mid-February)
Hazel (*Corylus avellana*), Dog's Mercury (*Mercurialis perennis*).

PERIOD OF LEAF ACTIVITY

Vernal Green (February to June)
Bluebell (*Hyacinthoides non-scripta*), Lesser Celandine (*Ranunculus ficaria*), Cuckoo pint (*Arum maculatum*), Wood anemone (*Anemone nemorosa*), Dog violet (*Viola riviniana*), Wood sorrel (*Oxalis acetosella*), Primrose (*Primula vulgaris*), Ground Ivy (*Glechoma hederacea*).

Summer Green (June to September)
Oak (*Quercus robur*), Dog's mercury (*Mercurialis perennis*), Honeysuckle (*Lonicera periclymenum*), Enchanter's Nightshade (*Circaea lutetiana*), Burdock (*Arctium minus*), Foxglove (*Digitalis purpuraea*), Sanicle (*Sanicula europaea*), Yellow pimpernel (*Lysimachia nemorum*).

Winter Green (September to February)
Bugle (*Ajuga reptans*), Ground Ivy (*Glechoma hederacea*), Primrose (*Primula vulgaris*).

Evergreen (all year round)
Ivy (*Hedera helix*), Holly (*Ilex aquifolium*), Butcher's Broom (*Ruscus aculeatus*).

NOTE. The species listed above are placed in the period of year when they show maximum leaf activity. In the case of primrose and ground ivy there is a return to leaf growth after the autumn leaf-fall.

Project 13

KEEPING ANIMALS AND PLANTS

One advantage of keeping animals or plants in confinement, even for a short while, is that one can make close observations in the comfort of the home, classroom or laboratory, on such things as behaviour, life histories, growth, and so on. This is especially so with those species whose living space takes up a small area and whose food requirements are not too demanding.

Five examples are chosen here, from widely differing groups, in order to illustrate this interesting and rewarding form of study—a fish, insect, mammal, amphibian and a plant.

THE STICKLEBACK

This ubiquitous little fish is known to practically every child who has visited the pondside with jam-jar and net. It has also risen to fame in biological circles as a subject for detailed study into animal behaviour, due to its specialised breeding habits. The life story can be followed in the aquarium, if this is prepared in readiness for the breeding season between April and June. A suitable tank size is 2 feet by 1 foot tall, and is available at most aquarium dealers (Fig. 72).

The tank should be thoroughly cleaned and tested for leaks, then placed in its permanent position before it is filled. A good place is in medium lighting near a window where some sunlight can penetrate for part of the day. Aquarium sand, also available at the pet shop, or river gravel which is washed and sterilised with boiling water, is spread over the aquarium floor as anchorage for the plants. Finer material such as beach sand is not recommended because it tends to pack too tightly, and may harbour troublesome bacteria and lead to pollution. The sand should slope from about a 3-inch depth at the rear to 2 inches in front. This gives a better view of the contents and also facilitates removal of excessive debris which tends to drift towards the front. It can be removed by using a glass or rubber tube. Place a thumb over the tube, insert in the tank over the debris, remove the thumb so that it is sucked in, then replace the thumb and lift out.

A little of the waste material, called detritus,

may be left as a fertiliser for the plants. These can be bought or collected from a pond. They are fixed to the bottom by weighing down with small stones, so that they will root. Some floating plants can also be added to give surface cover. Some planning is necessary to give the aquarium scene a pleasing aspect. Tall-growing plants are grown along the back and sides, bushy growths more to the centre, and small plants in the foreground. A suggested plan and choice of common plants is given in Fig. 72. Note also the use of rockwork which gives a terraced effect as it rises towards the rear.

When this is done the water is added. To avoid disturbance to the plants this is slowly poured into a saucer laid on the bottom, which can afterwards be taken out. Rain water or clear pond water is advised, and if tap water is used then at least one jugful of pondwater should be added. This introduces the microscopic aquatic life found in all ponds, and forms the start of the biosere mentioned on page 59. This is usually noticeable in a few days when the water becomes cloudy as a bacterial colony builds up. The 'green water' stage is a healthy sign that free-swimming algae are multiplying, providing food for an increasing population of microscopic animals. If the green colour persists, the tank should be screened for a few days, or some *Daphnia* introduced.

In time clear water results which has an undefinable 'bloom' and is said to be in a mature condition. It is now time to introduce the fish.

A single healthy male with coloured throat which is devoid of visible signs of parasites is put in, and will probably start nest building in a day or so. If two specimens are introduced they may build their separate nests in opposite corners, so giving the rare opportunity of watching the fascinating behaviour in territorial fighting. Each respects the other's corner, treating the middle of the aquarium as a kind of no-man's land. There is much to observe at this stage, such as the actual nest-building, the threat positions and attacks of each attentive male towards an approaching rival. There is a rising intensity of red colour in the throat of

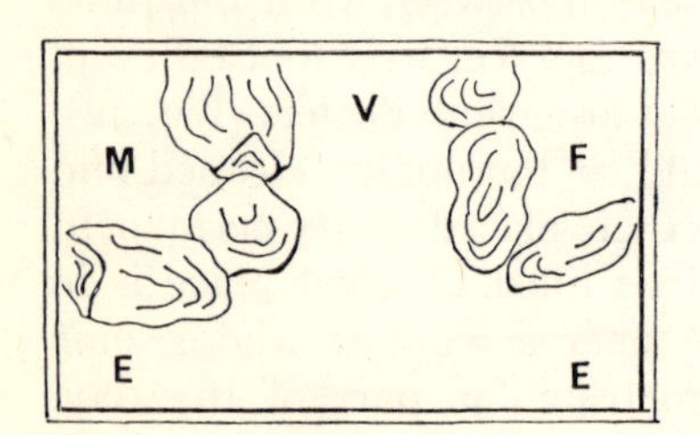

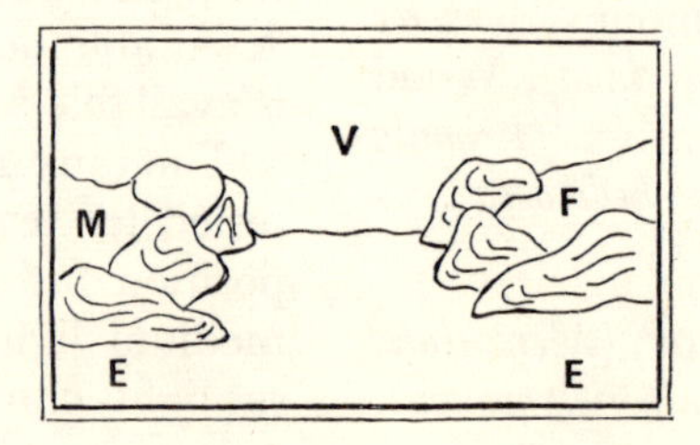

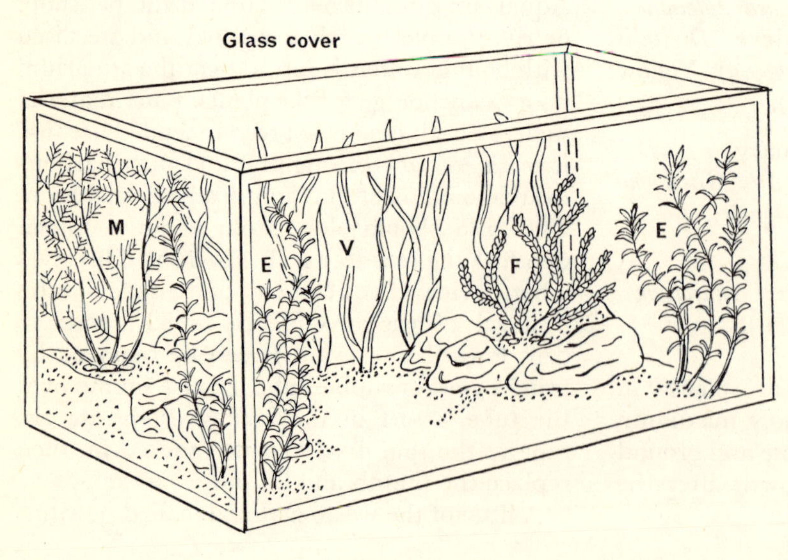

FIG. 72. Suggested layout of plants and rock material in an aquarium M, *Myriophyllum*. V, *Vallisneria*. F, *Fontinalis*. E, *Elodea*.

each victorious combatant after a bout, and a distinct paleness in the loser. In most cases the defender is the winner.

By now the nest building should be completed, and it is time to add a few females which are swollen with eggs. Then follows the captivating display of courtship, in which the zig-zagging males attempt to lure a female into the nest. Egg laying should follow, but is rarely seen. A sign that this has taken place is the change of heart of the male who now drives away the female, even to the point of bullying her, so it is best to remove her. The male is now in sole charge and can well manage on his own. Final success in this breeding operation is achieved when, one morning, a tiny cloud of babies is seen hovering over the nest. A harassed father guards them from all intruders, and will even attack a finger or stick placed near the nest. Any straying youngster is grabbed in the mouth and pushed back into the shoal. Finally, as father loses interest in his brood, the young fish begin to break up as their gregarious instinct fades.

During all these activities the parents show little interest in food, especially the male. Some daphnia and small earthworms can be dropped in now and then, but dried foods are best avoided as this may lead to pollution. An aquarist's best guide to a healthy aquarium is his nose. At the sign of any unpleasant smell, a warning that poisons are present, the cause of decay should be removed, and if necessary most of the water siphoned off and fresh added. In the confines of an aquarium the decay process can quickly snowball and kill all the occupants.

The Emperor Moth

Owing to their small size and elusive nature many insects and other invertebrates are only seen on the odd occasions when found by chance. This is often true of nocturnal species, and those living in remote places, where field observations have to be brief. This makes studies of habits and life histories difficult, and can only be followed from start to finish when specimens are kept in confinement. This is often possible if the species normally confines itself to a certain food-plant or micro-habitat which can be set up in a cage.

The example chosen here is a handsome moth which lives mostly where the heather grows on heaths and moorlands. The larva feeds on the leaves and blossom of the ling (*Calluna vulgaris*) during summer, pupates in the early autumn, and the adult emerges in spring. Eggs may be searched for in late spring, the caterpillars more easily during summer by looking for heather clumps which have been eaten, and pupae found tucked away in their cocoons during winter.

The cage to use should be light and airy, about 1 foot square by 18 inches tall, and made as a wooden frame with glass sides (Fig. 73). The roof, a frame of perforated zinc on a hinge, will allow for ventilation and act as a door. Narrow necked bottles filled with water are placed inside, underneath a false floor, to hold the sprays of heather. If this is not available, then plum or pear foliage can be used. The plant food is changed from time to time as it is eaten, until the pupal stage is reached. A wise precaution is to plug the bottle openings with cotton-wool, to prevent the caterpillars from crawling inside and drowning. A separate jar of water with perforated zinc top will provide humidity, and important consideration with an indoor cage, since the pupae may otherwise dry out and fail to develop.

With a little care it is possible to complete a life-cycle, then repeat it with a second brood. Adults will mate shortly after emergence, lay eggs, then probably die. The Emperor Moth has a number of interesting things to offer a naturalist. The caterpillars are fine examples of the camouflage guises adopted by insects. Their green colouring and bright orange-red dots blend very well with the green heather foliage and the flowers, especially during the final instar before pupation. This usually coincides with the September flowering of the food plant. Cocoon spinning in this species, which is a relative of the silk-worm moth, can be watched at close hand. The caterpillar weaves a shroud around itself, and finishes off the job by attaching a cone of stiff silk hairs at the head end, built to act as a one-way trap door. The emerging moth can push its way free, but enemies cannot enter. The male moth has a remarkable ability of finding its mate from a long way off. Its more feathery antennae are said to detect the scent of the female. If a fresh female is placed in a box and carried across the heather, any males in the vicinity will soon appear, and even cling to one's person in an eager search for the hidden female. The author once placed a newly emerged female in a small wire cage on the window sill, and released some marked males from the same brood about half a mile away. Later in the day one appeared at the window, entering the room to settle on the cage.

The Woodmouse

Keeping wild mammals is limited by their size, food requirements and tolerance to captivity. There seems no point in keeping an animal which may suffer in health through confinement, or which refuses to feed. Some of our smaller mammals, however, such as the ubiquitous woodmouse of woods and hedgerows, will thrive and breed in captivity if a suitable home is given. To help in this one should aim at providing as near as possible the natural surroundings to satisfy their needs. Some details of the food, habits, etc. of woodmice are given on page 163.

Depending upon the space available, a cage with a floor area of about 3 square feet is suitable for a pair of woodmice. A wire cage, better still, an aquarium which gives a clearer view, may be used. If wood is used, then the inside should be lined with metal sheeting to prevent gnawing. Height is important, and should be at least 1 foot, since these rodents are extremely agile and can easily jump to this height. They are also expert climbers.

To avoid any risk of escapes when feeding or cleaning out the cage, a doorway is fitted to the roof. The floor is covered with a generous layer of soil and leaf litter to resemble a woodland floor, and on which are laid some dead branches and pieces of flat bark. There is no need to use a nest box, as the mice will quickly tunnel below the bark and clear a space for a nest. Dry grass, leaves and moss can be provided for nest building.

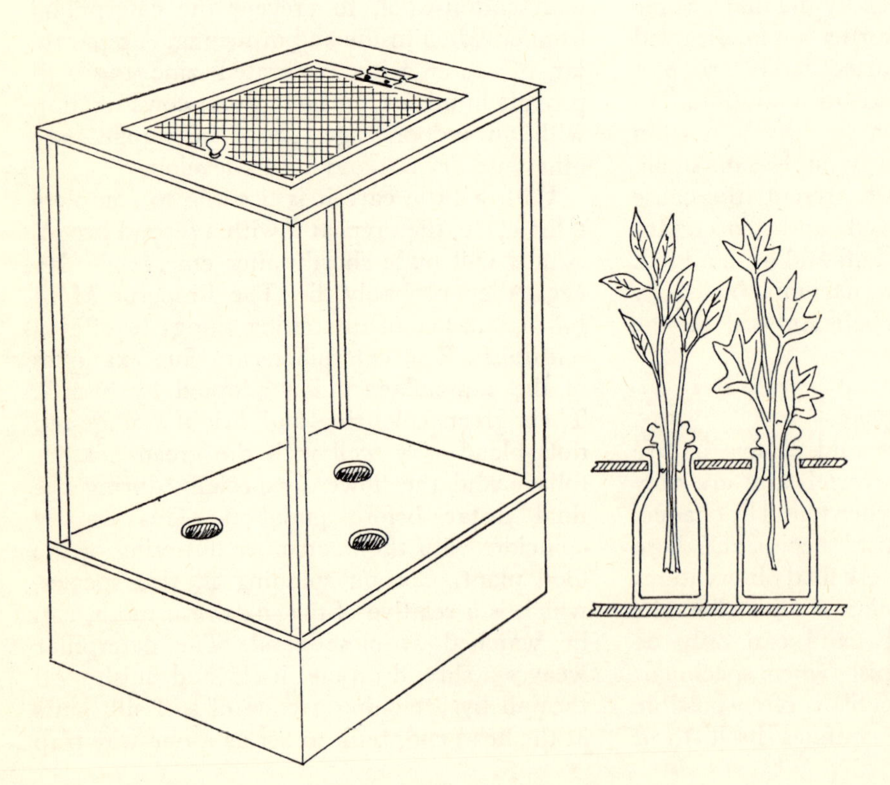

Fig. 73. An Insect Cage with false bottom for holding bottles in which food plants can be kept fresh.

Plenty of natural food is available from the countryside (see p. 163) to which is added an occasional handful of grain, nuts, some fruit and various small insects. Here, again, no feeding tray is necessary, since the mice will most probably make their own hidden store close by the nest.

Drinking water is important, but not in a bowl as this will soon be fouled. Use a drip-feed bottle, made simply from a container with a cork through which a glass tube is fixed. This is attached to the side of the cage. Finally, some branches may be added to allow for climbing.

With this type of home installed in the author's study it has been possible after dark to spend hours in quiet observation into the behaviour of these mice. One can watch their movements, social and feeding habits, mating behaviour, and the attention given to the young. On one occasion, in a half-exposed nest, a mother suckled and washed her babies in full view. All this is best watched in a torch beam covered with a red filter. Most mammals seem to ignore red, and go about their affairs as if oblivious to human presence.

Plants

Wild plants which are grown in confinement can vary from small herbs to young trees, the latter being sown as seeds. In some cases a complete life-cycle can be followed, and this is perhaps of greater interest and instruction in the case of the lower plants (e.g. liverworts, mosses and ferns) whose life story is more complicated and often overlooked. Such plants actually benefit from confined space, and should grow well if given the right soil and humidity. This can be achieved by growing the plant inside an aquarium which has a glass cover raised on corner supports, so as to cut down ventilation (Fig. 74).

Alternatively, a small specimen can be grown inside a jam-jar or similar glass container, or given a glass cover such as a bell-jar. The amount of condensation which settles on the glass serves as a guide to the amount of ventilation or watering which is needed. In all cases it is best to use rain water, or water which has been standing for a while.

With a good hand-lens the various organs can be examined at close quarters, such as the capsules and sporangia of the sporophyte generation, also the reproductive organs of the gametophyte. These cryptogamous plants, lowly and overlooked, are worthy of our closer attention into their intricate and often beautiful structures, and strange life-cycles. As early arrivals on bared ground they have an important role to play in the biosere which colonises and reclaims the land (see Project 5, p. 161).

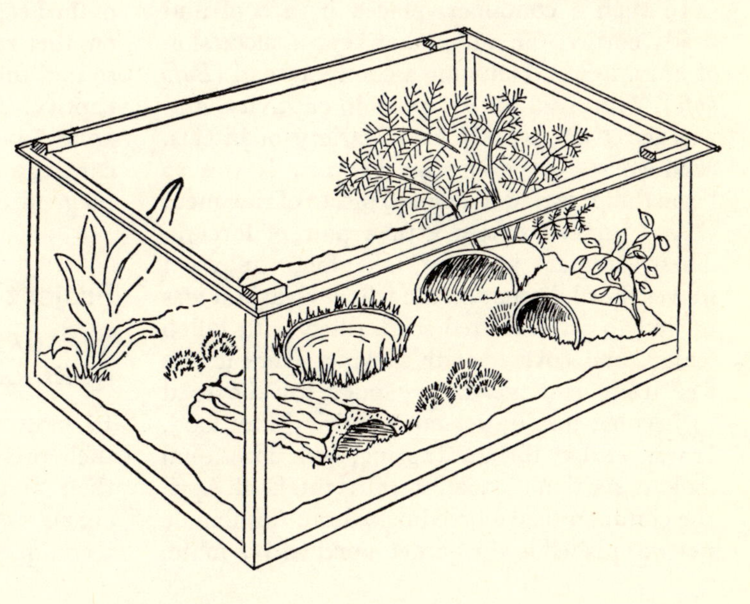

Fig. 74. An aquarium laid out as a vivarium for keeping amphibians such as a toad, also as a container for damp-loving plants such as mosses and ferns.

The Toad

The aquarium which is used in the above manner for keeping shade- and moisture-loving plants, lends itself very readily for the housing of many amphibians. In this form, known as a vivarium, the contents are assembled so as to reproduce as closely as possible the kind of habitat to which most amphibians (frogs, toad and salamanders) are accustomed. This is usually within the shelter of undergrowth and rocks, close to the ground, and in humid and gloomy places such as a ditch, woodland floor, cave, or in a marsh.

The floor of the aquarium is covered with a layer of loamy soil such as that found in most woodlands, or it can be made up from garden soil to which is added some sand, also peat or leaf mould. In this is planted a selection of mosses, ferns, liverworts, and any of the house plants which grow best in shade, and can be bought at florists. Hiding places for the inmates can be made out of pieces of curved bark laid between the plants, small caves of stones, or half flower pots placed on their sides. To cut a pot in half, make a scratch down one side, across the bottom, and up the other. Give it some gentle taps and the pot should split neatly in two. The only water necessary is in a shallow container sunk into the earth, and kept filled with clean water at all times.

In such a container, placed in a cool and shady corner, the author has kept a succession of amphibians, including a common toad (*Bufo bufo*). This lived for ten years in captivity, and was fed twice weekly on a variety of insects, earthworms, slugs and mealworms. It was so tame that it would take small pieces of raw meat waved under its nose with a pair of forceps. These animals will not take food unless it moves. For hibernation the whole vivarium was moved to an unheated shed, filled with fallen leaves, and covered with sacking. Here it was free from frost, yet cold enough for the toad to become torpid, yet not requiring any food. It was left like this until spring, with occasional looks to see that the toad was all right. Each April the common toad's breeding season, this female pet was placed in the garden pond with a mate, and allowed to spawn. This only seemed to occur if the toad had hibernated. If, as happened on one or two winters, she was left in the house in warm surroundings and fed continually, no spawning occurred the next spring, in spite of an obliging male attendant. This may possibly be due to some physiological need for a cold rest period, i.e. a winter sleep.

Discoveries of this kind, and many others which can be made through keeping a toad in confinement under close observation, make this kind of pastime a rewarding hobby in one's search for knowledge about animals. Among the Amphibia, the toad is particularly suitable since it readily adapts itself to close quarters, is placid in movement, quickly tamed and easy to feed.

The foregoing will have suggested the useful possibilities of keeping certain live specimens for personal study. They also serve as valuable aids to biological instruction and classroom demonstration, and do not take up much room. A small collection of cages and glass tanks containing a variety of animals and plants are a useful standby for teaching purposes. It is not always possible to get hold of the right material at short notice, or when required. 'A bird in the hand. . . !'

In conclusion a timely word of caution is perhaps not out of place. Animals and plants kept in confinement are at the complete mercy of the keeper. One should never attempt to take on this responsibility unless competent to do so and fully aware of the proper needs of each captive. There is a moral duty in this. A sick animal or ailing plant, the result of improper care, is a silent reproach to its owner, and does no good to anyone.

Project 14

THE GRID MAP, TRANSECT AND QUADRAT

In most surveys of plants some indication of their presence and position in the community, their abundance or rarity, can be shown on charts and maps by using well-tried plotting techniques. This was carried out in an area, some 30 yards square, along the border of Epping Forest where it joins the fields at Debden (see p. 151). It shows part of a field, a hedgerow and ditch, footpath, and part of Forest ground beneath the trees (Fig. 75).

To prepare a grid, four corner-pickets are placed in position and a base line chosen so that it is visible on all sides, in this case along the line AB, in the open ground between the Forest and field (Fig. 76). Along this, canes are put in at 6-foot intervals. To build up the grid, white tape or cord is run out from each cane, at right angles to the base line. To ensure an accurate right angle, measure off 9 feet from A and put in a temporary cane at C. Two lengths of tape are then used, a 12-foot length attached to A and a 15-foot length to C. Where the tapes intersect at D put in another cane. DAC will then make a right angle. Repeat this at the other end of the base line, at B, and put in a further cane at E. The shape of the grid is now fixed, and the perpendiculars can be run out. Across these the horizontals are then laid out, at 6-foot intervals and parallel to the base line. Should the base line happen to run across the middle of the survey area, as in this case, the squares will cover both sides, i.e. FGHI.

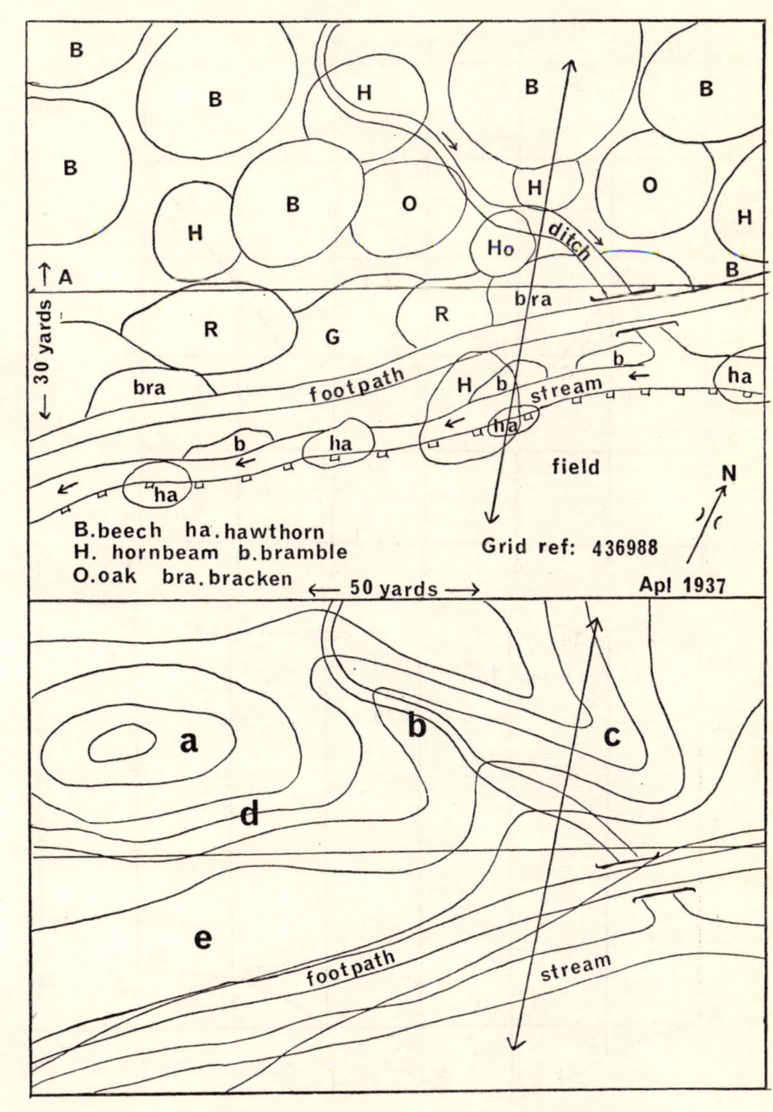

FIG. 75. Grid and contour maps for the Epping Forest survey area. See also Fig. 77. The arrowed line marks the position of the belt transect.

The tapes should be drawn tight, and pegged down where they cross. To avoid hopeless tangling when these are removed, all parallel tapes should be laid down above or below those at right angles. Where trees or heavy undergrowth interfere with the tapes, cane markers can be placed on the spots where the tapes would intersect. With such a grid in position the work can begin by mapping out the whole area. Each square is examined in turn and the contents sketched out in the field notebook. Number each square and mark in outline each patch of vegetation, as well as such features as a path, tree, ditch, bare soil, etc. Later, all the squares can be brought together on a single large sheet of graph paper so as to reproduce the whole area. If the work is done carefully, then each square should fit into the whole map like a jig-saw. Patches of the different plants can be outlined in Indian ink and coloured in. Isolated plants can be shown as symbols. Add a key to the map, together with scientific names, also the date, locality (grid-reference), local geology of the ground and any other useful details. Such a map need only give the general picture of the area, indicating where the different associations of plant occur, such as grass, rush, bracken, heather, trees, and

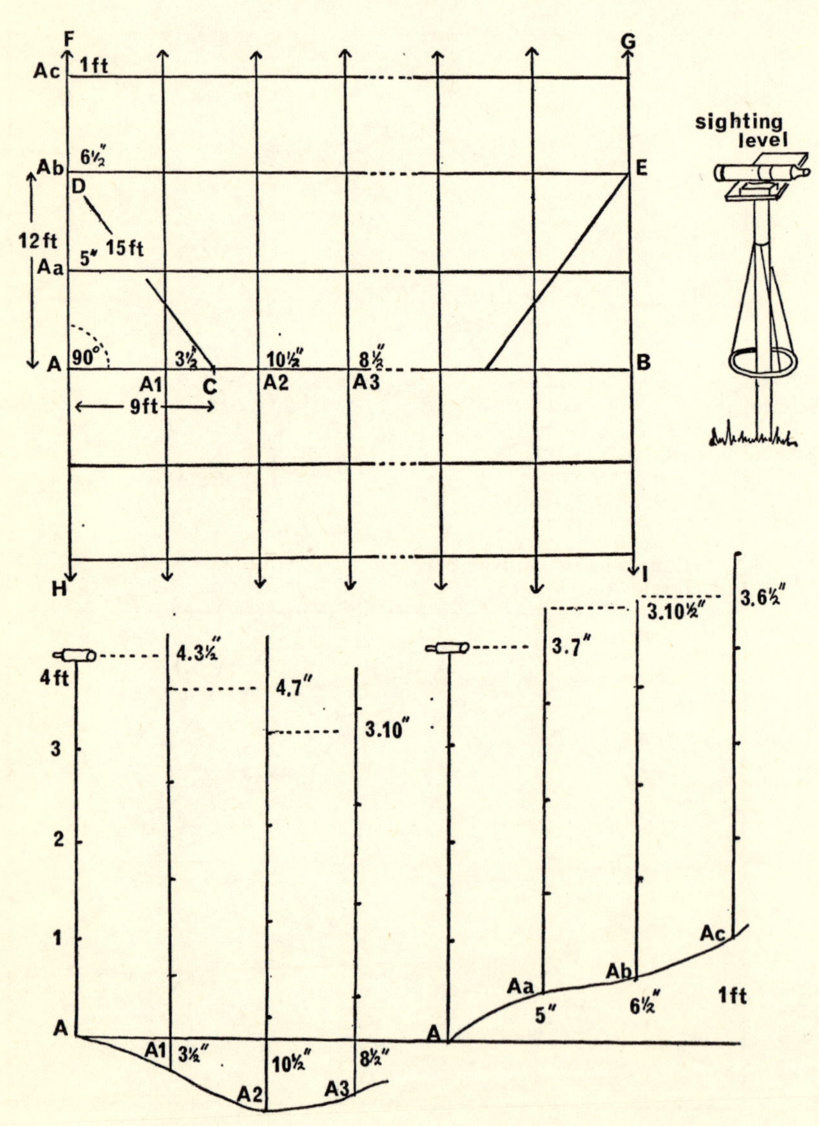

FIG. 76. Layout of grid for mapping out the survey area shown on Fig. 75 and, below, technique of levelling from which the contour map was produced. For details see text.

so on. More detailed information is obtained by taking out transects and quadrats in selected places (see on p. 190).

LEVELLING

In some areas even a slight alteration in the slope or height of the ground can make a difference to the vegetation. A sheltered hollow where water collects can support a different plant association to that which might be found on an exposed ridge, even in close proximity to one another. The contours of the survey area can be marked onto the grid map by starting from a zero point and working out the different heights at each grid intersection. This will not determine the actual heights in altitude, but only those relative to the area, which is all that really matters. From this the features of ditch and mound, path and embankment, etc., can be brought out and compared with the plants and animals which they support.

The apparatus required for this work is a sighting level. This consists of a small telescope through which, by means of a mirror, the image of an air bubble in a spirit level can also be seen. When this is bisected by cross-wires seen through the telescope, the level is horizontal. A strong wooden upright, e.g. a broom stick with a flat square base to prevent it from sinking into the ground, has a clamp to which the telescope can be fixed in the horizontal position by means of a fixing screw. The pole will be in the vertical position when the air bubble is in the cross-wires. To correct any lateral slope, attach a metal ring to the pole by means of three cords, so that it hangs around the pole. Alternatively, if one is available, a camera tripod with adjustable platform, may be used.

To obtain readings of relative height the telescope is sighted at a measuring pole which is marked in feet and inches. Three people work best at this exercise, one to hold the measuring pole, another to do the sighting, and a third to take down the readings. Assume that A is chosen as the zero point (Fig. 76). Remove the stake and place the leveller in position. A convenient height above ground for the telescope is 4 feet. Hold the measuring pole at the next grid mark A_1 along the base line. The reading is, say 4 feet $3\frac{1}{2}$ inches. Since A represents zero, i.e. o, there is a drop in ground level of $3\frac{1}{2}$ inches at B. This is marked in on a prepared grid ruled out on some graph paper. Next, move the leveller to A_1 and the measuring pole to A_2 for the next reading. This gives, say, 4 feet 7 inches, a further drop in ground level of $10\frac{1}{2}$ inches relative to A. A_3 reads 3 feet 10 inches, this time a rise in ground level but a relative drop of $8\frac{1}{2}$ inches below A. The figure will help to make this clear. Continue in this manner all along the base line to the end. Now it is the turn of the perpendiculars. Starting again from the zero point A, take a reading of Aa. This gives, say 3 feet 7 inches, a rise of 5 inches. Ab gives 3 feet $10\frac{1}{2}$ inches when sighted from Aa, a further rise of $6\frac{1}{2}$ inches. Ac reads 3 feet $6\frac{1}{2}$ inches, a rise of 1 foot. Repeat for all perpendiculars by starting from A_1, A_2, A_3, etc., until the whole grid area is plotted. With the assistance of a student friend the author completed this task at Debden in about four hours, a morning's work.

With the height marked in on the grid map the contour lines can now be worked out, bearing in mind that contour lines never cross and should be spaced between relative heights, as shown (Fig. 75). Useful tips to remember about contours can be seen on this map. Rising circular ground shows up as concentric contours (*a*). Arrow-shaped contours pointing uphill denote valleys, in this case the ditch (*b*). Arrow-shaped contours pointing downhill denote ridges or watersheds (*c*). Closely applied contours mean a steep slope (*d*). Those widely spaced indicate a flatter landscape (*e*).

Transects and quadrats

These exercises are carried out to determine the presence of various plant species. The transect is qualitative and will show the plant's relation to the community, its dominance or otherwise, whereas the quadrat is more quantitative.

The transect is intended to show the changes in vegetation which occur across the survey area. It marks the position of plant species, rather than their quantity, along a line which can be illustrated in vertical section—a line transect, or in plan—a belt transect (Fig. 77).

The line transect

A tape is laid across the survey area, or one of the existing grid lines could be used. A section is chosen which will bring out any contrasts in ground level and changes in vegetation. If the transect line can cross different soils, so much the better. It is drawn out on graph paper as a vertical section by referring to the contour map. Heights will have to be increased several times that of the horizontal scale, so as to bring out the dips and rises of mounds and hollows. Any obstacle crossed by the tape, such as a rock, tree stump, log, ditch, etc., should be marked in, since these may harbour their own patch of flora. Along this graph line note down at 6-inch intervals the plant species which occurs at that position. Choose the species which holds the dominant position, whether it is small or large, common or rare. Beware of selecting only the showy plants, or those which are perhaps better known to the recorder, and take particular note of those species which are typical of the habitat. Also, bear in mind the

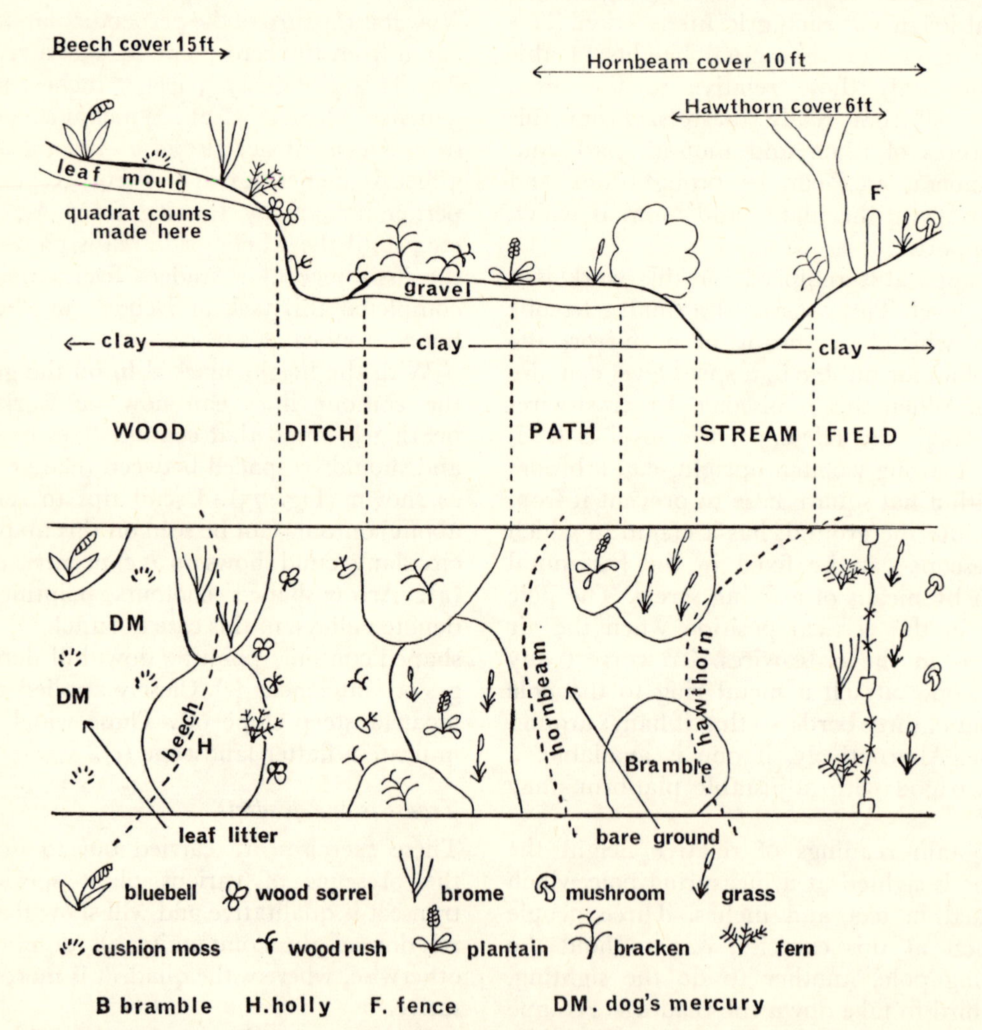

FIG. 77. Line and Belt Transects covering the area chosen for the Epping Forest Survey. (See also Fig. 75).

micro-habitats which may exist in the shelter of logs, in ditches, and so on.

The belt transect

Mark out a strip of ground across the survey area, to follow as closely as possible the line transect, by pegging out two parallel tapes, one yard apart. Mark this off at yard intervals. Work along the belt, yard by yard, and record the species which occur. These can be marked in the notebook by means of shading for each patch of similar vegetation, e.g. a carpet of grass, moss, bracken, etc., and using symbols for individual plants, as in the case of the main grid map. Such a belt reveals more than a line transect, since it shows the amount of actual space covered by each species. From this a *Cover Index* can be compiled, in which the percentage of each square occupied by each recorded species can be shown on a chart. The left-hand column gives the number of each square in the belt transect. In the top columns are written in the names of the species and below these, in the appropriate square which they occupy, the percentage area covered is written in. If graph paper is used in which each square contains a hundred small squares, it is a simple matter to count those which are shaded in to get the percentages.

It is important to remember that a belt transect map and a cover index chart will only show the amount of space covered by those species which are visible from above. Thus, a patch of bracken may appear to cover an entire square, but this would only be correct at a level of some 2-3 feet above ground where the fronds spread out. Below this is plenty of space between the frond stalks, and this may be occupied by a different plant such as a moss or grass. In order, therefore, to get a proper estimate of the cover for all species, especially in a habitat with different growth layers such as in a wood, a *Profile Transect* can be made. Using the same ground covered by the belt transect, the plants are now drawn in as viewed from the side (Fig. 78). A measuring tape is placed along the profile so that the drawing can be made to scale. Select a specimen at every 6 inches and

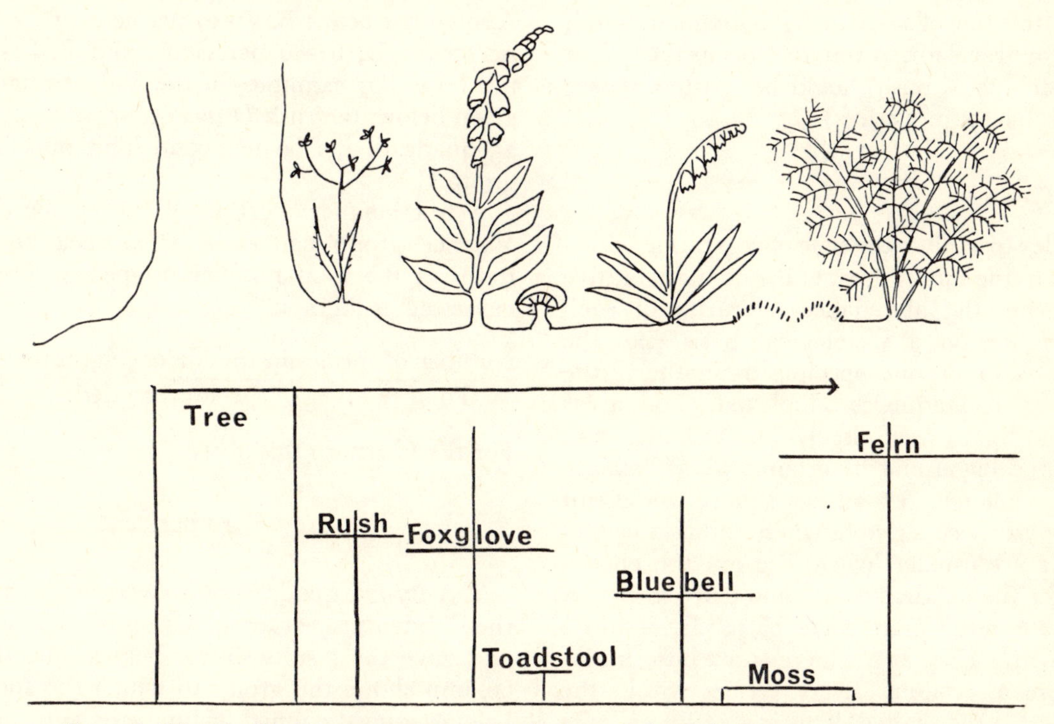

FIG. 78. The Profile Transect. A sectional view of the Belt Transect drawn in two ways, showing the amount of cover for different species of plant.

draw in its outline to scale after measuring its height and width. Bracken will have a broad top and slender base, moss will form a low and extended carpet, and bluebell will have a thick lower region of leaves but a sparser upper layer of separate flower stalks. Actual outline drawings can be made, or more simply a vertical line for height and a horizontal one at the appropriate level for width. Work in the field can be done on separate graph sheets, one for each yard of transect, then later compiled into an extended chart. If all plants are to be drawn in to scale from, say, the mosses to the larger herbs, then the chart may have to be continued two or three times across the same graph page. Trees and shrubs are usually too big to fit into the scale, but their bases, e.g. a tree-trunk width or that of a bush, can be marked along the bottom of the chart and the cover of the upper foliage above it. Ideally, this kind of transect should show the root structure of the various species, where layering also occurs (see Woodlands, p. 77). This is not always possible without disturbing the ground to expose the roots, but may already be exposed if the transect runs along the top of a ditch, embankment, sand, chalk or gravel pit. Where they occur the height of each tree or bush should be written in (see Appendix 17, p. 234).

The Quadrat

In order to analyse the type of vegetation which grows in the survey area, either on a qualitative basis (i.e. the abundance or rarity of each species) or on a sociological basis (i.e. the relationship of one species to another), the quadrat technique is employed. For areas covered by a closed association of plants, as in grassland or on marshy ground, a yard square is large enough. A 6-10 foot square would suit a bushy or wooded area where the numbers of species are usually lower. Peg out the chosen spot for the quadrat count, and drive in corner stakes. A yard square could be subdivided into four with string, and a larger area into similar convenient sections. Every species within the quadrat is then noted. This is a useful exercise in plant identification. With the aid of a flora and perhaps the assistance of a specialist friend this can be a helpful means of getting to know some of the more difficult plant groups, such as the grasses, sedges, rushes and mosses.

In order to obtain a balanced picture of the plants within the survey area, random sampling should be employed. For this a portable quadrat frame is needed. It is tossed over the shoulder so as to avoid any element of selection, and every plant species within the frame noted. This is repeated 100 times and the results noted on a chart. Species names are written in the first column, and a tick or cross placed in the subsequent columns for the appropriate quadrats in which they occur. The total number of occurrences is then entered in the last column, and expressed as a percentage.

Since this work may take some time, a lesser number of quadrat samples may be taken, such as 25, in which case the results need to be multiplied by four to give a percentage figure. This result is called a Valence Analysis of the species, and can be grouped into a frequency series within the following percentage ranges—A, 0-20 per cent.; B, 21-40 per cent.; C, 41-60 per cent.; D, 61-80 per cent., and E, 81-100 per cent. For example, in the Valence figures given below, the rush *Luzula* occurs in 15 of the 25 quadrats, i.e. 60 per cent. This puts it in group C.

A fractional proportion of species for each Valence group can now be worked out in terms of the total number of species. This is expressed as follows:

$$\frac{\text{Number of species in the Valence group}}{\text{Total number of species recorded}} \times 100$$

For the C group, this gives:

$$\frac{2}{20} \times 100 = 10 \text{ per cent.}$$

Twenty-five quadrat counts were made within the Debden project area by random sampling, and gave the results shown below. The third column shows the group to which the species belongs, and the fourth column the percentage of the total number of species:

Species	*Quadrat counts*	*Valence % i.e. × 4*	*Group*	*% of total species*
Cushion Moss				
Leucobryum	24	96	E	5
Dog's Mercury				
Mercurialis perennis	19	76	D	
Hairy Brome Grass				
Bromus ramosus	18	72	D	15
Lesser Celandine				
Ranunculus ficaria	16	64	D	
Hairy Woodrush				
Luzula pilosa	15	60	C	10
Wood Melick Grass				
Melica uniflora	14	56	C	
Bluebell				
Endymion non-scripta	10	40	B	
Wood Sorrel				
Oxalis acetosella	8	32	B	20
Dog Violet				
Viola riviniana	7	28	B	
Sweet Vernal Grass				
Anthoxanthum odoratum	8	32	B	
Holly				
Ilex aquifolium	4	16	A	
Yellow Pimpernel				
Lysimachia nemorum	2	8	A	
Bird's-nest Orchid				
Neottia nidus-avis	1	4	A	
Red Campion				
Melandrium dioicum	1	4	A	
Hedge Woundwort				
Stachys sylvatica	3	12	A	
Honeysuckle				50
Lonicera periclymenum	1	4	A	
Jack-by-the-Hedge				
Sisymbrium officinale	3	12	A	
Cuckoo Pint				
Arum maculatum	1	4	A	
Male Fern				
Dryopteris felix-mas	5	20	A	
Sanicle				
Sanicula europaea	4	16	A	

From these final figures a histogram can be compiled (Fig. 79). The horizontal axis is divided into the five groups, and the percentage number of each species blocked in along the vertical axis. The histogram gives a clear picture of the composition of the vegetation and its association, showing which is the common species and which are average or scarce. It has a typical J-shaped outline, in which the largest number of species occurs in the smaller percentage groups, i.e. the long arm of the J, whereas only a small number of species occurs in most of the trials. These can be regarded as the dominant and co-dominant species within the area. This assumption only applies to those plants at the same growth layer, in this case the ground flora, and would apply to places like open ground. Where trees occur, as in the Debden locality, these would have to be treated as the dominant plants, in spite of their smaller numbers.

In some surveys a permanent quadrat can be laid out with corner stakes driven into the ground, so that it may be visited at regular intervals. This is useful in following a change in flora where there is plant succession resulting from some interruption caused by a fire, felling of trees, ploughing, and so on, as in the case of the burnt ground studied in Project 5 on page 161.

Project 15

ADDERS

The adder, Britain's only venomous snake, is still shrouded in mystery and misunderstanding, as are most of its breed, even in these enlightened times. A sensible naturalist, however, does not usually subscribe to the loathing, hate or fear to which snakes are still subject. If only it could be more widely known that an animal does not attack a human merely because he is a human, then this fear bogey which so often prompts the use of a stick to smash the 'evil' serpent might be replaced by a more humane attitude. One might even come to admire a snake for its beauty of colour and grace of movement. Because of its remarkable build and unusual habits this is an animal worthy of closer appreciation and study.

Needless to say, a venomous animal or poisonous plant should be recognised before it is handled or eaten, and the adder is no exception, even though it is a shy and retiring serpent which never makes the first move. Bites seldom occur, and deaths are rare, and when this happens it is due to carelessness or ignorance on the part of the handler, as happened to me when a schoolboy. Over-eager to add a snake to my animal collection, I unwisely bent down one hot summer's day to pick up a handsome adder in the bare hand, and without regard to its identity. I nearly died from the resulting bite, in spite of hospital treatment. Yet, far from being deterred by this painful experience, I made a firm decision to learn all that I could about reptiles.

The adder (*Vipera berus*) is widely distributed through mainland Britain, being absent from Ireland, and occurs through most of Europe north of the Alps. It is likely to turn up in places such as woodland glades, hillsides, quarries, cliffs, moors and heaths, in fact, in those places where it is left undisturbed. Even so, it can exist in proximity to human habitation, even where people gather. The author has known a glade in Epping Forest, close to the main road where an adder colony has existed for some thirty years. At the approach of any ramblers, horse riders, or picnic parties, the adders quietly slip away into the thick growth of thorn bushes which dot this glade. They come out to bask in sunny patches, often on old grass and lichen-covered mole hillocks, but always with a safe retreat near by. In this way they often remain all day without detection. They may come out as early as six on a sunny summer

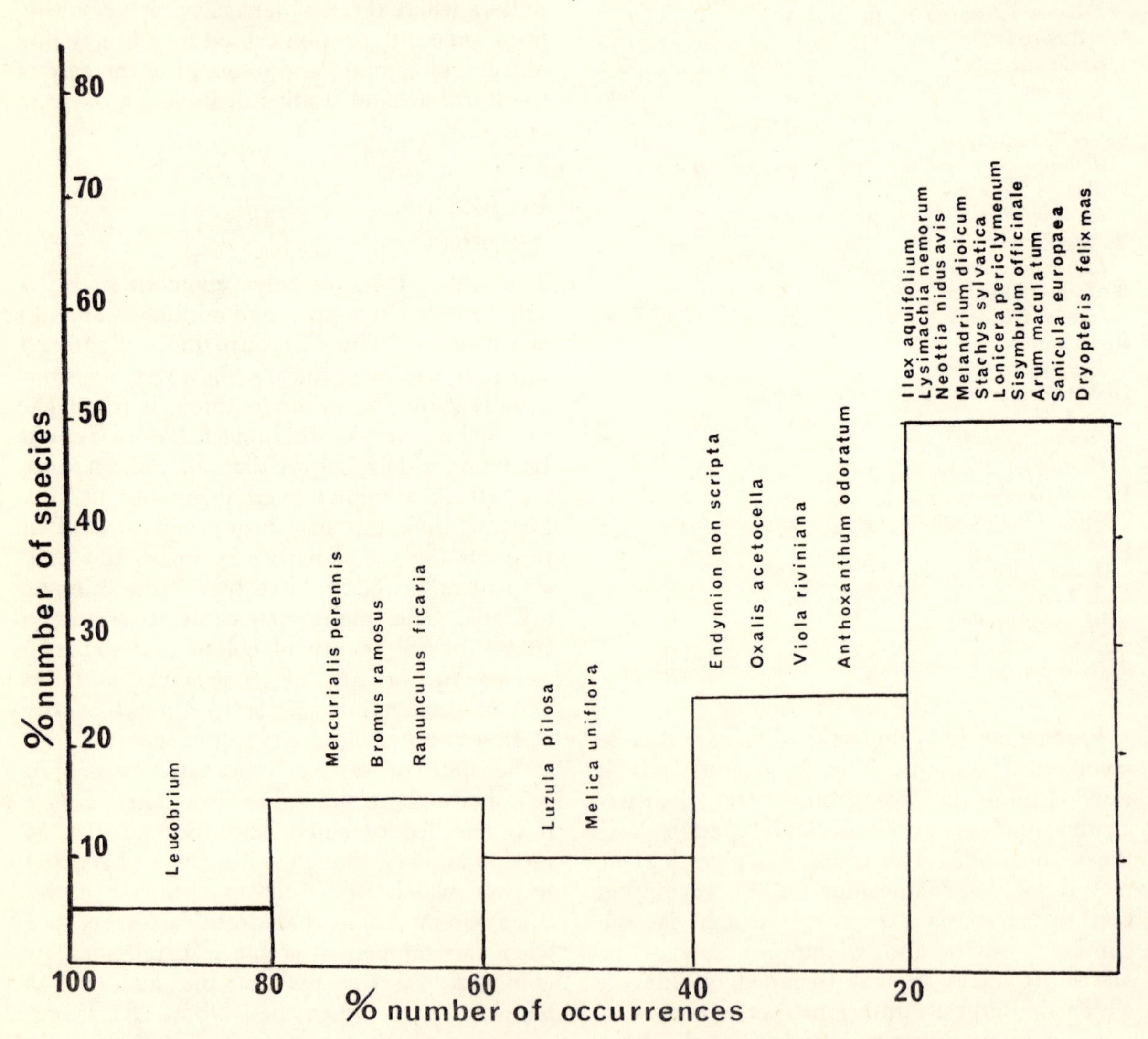

FIG. 79. A histogram based on quadrat sampling for the Epping Forest project (See text).

morning, also in any winter month during a warm and sunny period when snow may be lying about. This snake is a northerly species and can exist even within the Arctic Circle. It will retreat during the hot mid-day, but can be abroad on summer nights.

When basking, the adder is usually reluctant to move, and will remain on the defensive when approached. This is when bites may occur, especially to children and dogs. Being a live-bearing snake (ovo-viviparous), the female carries her embryos to the point of hatching, and is, in effect, a kind of living incubator. She comes out in sunny spells so as to provide the unborn young with warmth, and retires on dull days. Her heat sensitivity is acute, and is a safeguard against the approach of frost.

Adders under the author's observation have been marked from time to time by removing a sub-caudal scale and noting the markings and colour in the field notebook. In this way the habits and growth rates of individuals have been followed. The longest time recorded for any one specimen was for a female which turned up regularly in roughly the same area over a period of five years.

Adders tend to use the same sunning patches each day, each animal resenting the appearance of another, except during the courtship months, in May and July. It is then that the males make aggressive displays, rearing up one against another in the so-called 'adder dance' which was once thought to be a form of courtship. The author has only witnessed this twice, once in the New Forest, and again in his home reptiliary. From a hidden observation seat it was possible to watch the movement and behaviour of some captive adders which appeared to behave normally, since their roomy enclosure was well laid out with plant cover and rocks to resemble a piece of hillside. Movement, swimming, drinking, and a few times the stalking of prey ending with the lightning strike, could be witnessed.

Visits to the Epping Forest site meant early arrival before the adders were about, and sometimes long intervals of waiting before anything happened. This adder watching can then become even more frustrating than badger watching! Occasionally on warm summer nights the author and a friend would lie out inside the adder glade on a groundsheet, and with torches handy. Once, after a small camp-fire had been built, a circle of some six adders had formed up around us, each curled up in the fire's glow.

The highest count for this colony which lived in a glade of some 50 yards square, was 33 in one year, 10 females and 23 males. In late summer the broods of young were frequently seen shortly after birth, sometimes with the exhausted mother near by (though she never appeared to swallow any!). Two melanic specimens turned up during the survey, and the largest specimen, a female, measured 28 inches. Once a fight between adder and hedgehog was observed.

There is something exciting as well as rewarding in watching and uncovering the secrets of an animal with such notoriety, and learning the real truth about this beautiful serpent.

Project 16

MODELS OF SOIL PROFILES

Soil profiles which are measured and recorded as part of a field project can usefully be reproduced in miniature, by taking samples of the material and building these into a monolith. The result should look like one of the illustrations in Fig. 80.

On the site of the project an exposure may already be present (e.g. the top edge of a gravel or chalk pit, a cliff face, road cutting, etc.) or it may have to be dug with a spade. On private land the owner's permission should first be sought.

For collecting samples of the soil, also the bedrock if this is accessible, some stout, numbered canvas bags will be required for transport. Make a careful sketch or photograph of the chosen profile, which can be marked off with two parallel white tapes, about 2 feet apart, fixed vertically on the face of the section. Measure and note the thickness of each horizon, from topsoil down to bedrock, or as far down as one can go.

Samples of each horizon are collected in the bags. If possible these should be on the dry side. If the bedrock is very hard, such as granite, some pieces can be hammered loose, or pieces picked up in the vicinity. If surface litter is present, collect this too, also some specimens of the surface vegetation. The idea is to re-assemble all this material on to a length of wood by glueing, so as to make a model on a reduced scale.

Rough, unplaned wood strips, about 3 inches wide and ½ inch thick, should be obtained in readiness, and cut to suitable length. At a reduced scale of one-quarter, a 6-foot profile would need a wood strip 18 inches long. The material is stuck on to the board with the use of carpenter's glue, starting from the top, in the following manner:

1. Spread on some heated glue at one end of the board for the top horizon of plants and, if present, leaf litter. Gently press on some dried plant specimens (e.g. grass, rush, moss, heather, etc.) which will give some indication of the habitat and soil condition. Do this so that the shoot portions jut out beyond the board and the roots rest on the glue. Complete plants should, if possible, be used, in order to give the model realism and to show how deeply these penetrate into the soil.

2. Sprinkle the top soil between the roots, and to the required depth, including any other

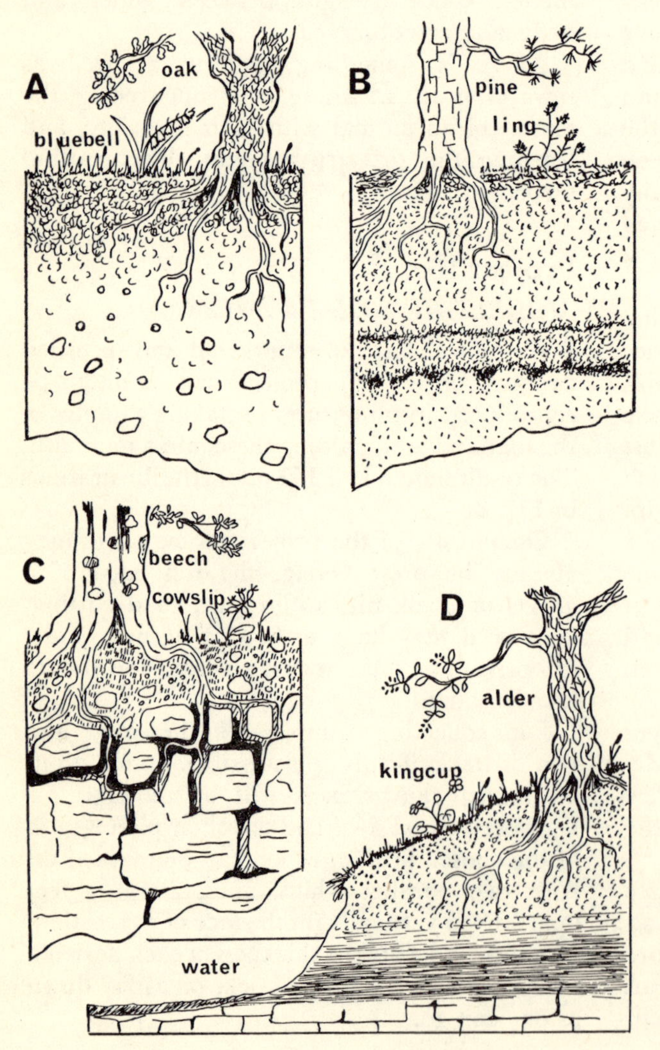

Fig. 80. Soil profiles showing different humus layers and bedrock; together with typical plants for each area. A Brown forest earth; B, Podsol; C, Rendzina and D, Glei.

ingredients which may occur, such as leaf litter, shells, stones, etc. Press all this gently with the fingers, filling in any gaps. Excess material can then be removed by carefully turning over the board when the glue begins to set. Try to keep the roots of the plants visible through this soil layer.

3. Spread on further glue for the next horizon, which may be a subsoil, and see that it runs into the one above so that there is no gap. If there is any grading of soils smear the two portions together with the fingers.

4. Continue in this fashion with any further horizons until bedrock is reached.

5. The bedrock may be soft or hard and can be applied in the following manner: (*a*) Soft rock (e.g. chalk, sand or loose sandstone). Crush the material until fine grained and sprinkle on to the glue, smooth out with the fingers and press flat with a strip of wood. (*b*) Hard rock (e.g. granite, limestone, millstone grit). If available from the site, choose a number of small flattish pieces which will cover the board section and can be fitted together in jig-saw fashion. If one large flat piece can be obtained and broken up with a hammer, this will fit more snugly together.

Peat can be stuck on in thin strips, but should be overlapped on the model as this tends to shrink on drying.

6. Prepare a label for the monolith, to include the following details: locality with grid reference, local geology and type of landscape, description of profile and horizons, prevailing climate and date.

7. When the model is completed and dry, cover with a transparent varnish.

A job like this should only take a few minutes and should be carried out smoothly and in order from top to bottom, adding more glue as one reaches the next horizon. Their thickness should be marked to the proper reduced scale along the edge of the board before glueing commences.

Project 17

SOIL TESTING

The type of soil which covers the survey area—its chemical and physical nature, water-holding capacity, sticky point and thickness, should be examined as part of the exercise, since the nature of soil has an important influence on the plants which can grow there. Apart from making soil profiles (Project 16, p. 193), samples of soil will be required for separate tests. Clear the ground of surface vegetation and loose surface humus, and collect material no deeper than 12 inches. Samples from different spots are put into separate tins marked with the grid position.

The natural water content of soil can be calculated in the following manner. A small quantity of the sample, say, 10 gm. is weighed before and after it has been slowly cooked in the oven. The loss in weight, i.e. the water content, is calculated as a percentage of the oven-dry weight of soil. If two samples are taken from each spot, one about 24 hours after rain and the other during a drought period, the results obtained will give some idea of the water-holding capacity.

A further figure to record would be the saturation capacity. This is the amount of water which completely fills the air spaces in the soil when it is saturated. For this a metal tube, about 2 inches long and 1½ inches across, is required. A ring of perforated zinc is soldered across one end. On this is placed a circle of filter paper cut to size. Thoroughly dry and pulverise a sample of soil, as finely as possible, and fill the tube to the rim. Hold this in water without immersing the open top end and wait until the water seeps through the soil to the top. Remove the tube, allow surplus drops of water to fall away, dry the tube as quickly as possible, then weigh. Oven-dry the tube and soil, then reweigh. To determine the water content of the saturated soil, the following gramme weights are required:

a. Weight of tube and filter paper.

b. Weight of tube, wet filter paper and saturated soil.

c. Weight of tube, dry filter paper and dry soil.

d. Weight of saturated filter paper.
e. Weight of tube.

This can be expressed as a percentage:

$$\frac{b-(d+e)}{c-a}\times 100, \quad \text{i.e.} \; \frac{\text{water in soil}}{\text{oven dry soil}}\times 100$$

The result gives the amount of water taken up by 100 gm. of soil.

The sticky point of soil is a useful indication of its physical property. It is the point at which the soil particles begin to pack together. The colloidal contents are then saturated, and all air has been expelled. Spread soil of the sample in its dry condition on a piece of glass, spray lightly with water and mix with the back of a spoon. Scrape off a portion and knead it between thumb and fingers until it begins to roll into a ball and will cut cleanly. Weigh the ball, then weigh again after thorough oven-drying. Express the two figures as a percentage as is shown above for the saturation capacity. It has been found that for most soils there is a relation between these two conditions, a constant difference of 23 per cent. Thus a sticky point of, say, 45 per cent. should give a saturation point of 68 per cent.

Organic contents of soil cannot be readily worked out in the field and are somewhat involved in the laboratory. What can normally be done is to calculate the amount of humus which oxidises. This is usually the most important soil constituent used in plant nutrition.

Weigh a gramme of finely sieved, oven-dry soil from the sample. Place in a beaker and add 10 cc. of distilled water and 10 cc. of hydrogen peroxide (1: 20 vols.). Stir the mixture over gentle heat until the oxygen bubbles cease. Add a further 10 cc. of hydrogen peroxide and repeat the heating. Filter off into a crucible through an asbestos layer. Thoroughly wash the filtered residue with hot water, adding this to the filtrate. Now dry the residue in oven heat and weigh. Pour the filtrate into a measuring cylinder, pour off a measured amount into an evaporating dish and allow to dry. Determine its weight by subtracting the weight of the dish. Ignite this residue, allow to cool, then reweigh to determine the weight of the mineral matter left over.

The figures so obtained should be as follows:

a. Weight of the dry soil, i.e. 1 gramme.
b. Weight of residue in the crucible.
c. Weight of the mineral matter from the filtrate.

If O represents the organic matter which has been oxidised, this can be expressed as $O = a-(b+c)$. By these means the following figures were obtained for the clay soil and beechwood soil in the survey area described in Project 14, p. 184.

	Saturation capacity	*Sticky point*	*% organic content*
Clay	56·4	31·5	2·0
Beechwood soil	65·5	40·5	5·0

Project 18

WATER TESTING

The physical and chemical nature of a pond or lake alters with the seasons, more so in shallow water, and this affects the life which it contains. As part of a pond or lake survey it is useful to take periodic readings and measurements during visits, such as those listed here:

Temperature

This tends to drop with depth, so that a series of readings along a vertical axis may be required, especially in the deeper waters of a lake or reservoir. A maximum and minimum thermometer is placed in a weighted container, and attached to a cord in which knots are tied at 1-foot intervals. The thermometer reading is noted before it is lowered to the required depth. Leave for a minute or two, then haul up. The new reading will give the water temperature at that depth. Reset with a magnet and repeat for the next depth. Figures obtained can be plotted on a graph. Temperature gradients are much greater in deeper waters than in shallow ponds (see p. 110). In the latter case water mixing takes place more readily by convection, and only two readings are usually needed, one at the surface and the other on the bottom. Working across a pond in a straight line and taking readings at, say, every 10 yards, the depth could also be measured with a weighted string or measuring stick. This would be required if a transect of a pond was to be recorded. During winter when ice has formed, holes can be bored through which a thermometer is lowered. In this situation the water at the pond bottom should be at a slightly higher temperature than that just below the ice. This peculiar physical property of water is of great importance to the winter inhabitants (see p. 117).

Transparency

The penetration of light through water varies from pond to pond. Turbulence, suspended matter, overhanging bushes and trees, floating aquatics and flushes of algae, all these can cut off the light from above at certain times, thereby affecting the submerged life. The limit of visibility can be recorded by fixing a white disc which hangs horizontally from a graduated cord. This is lowered in the water till it disappears from sight, then raised until it comes back into view. The average of the two depth readings is noted. For more accurate results, use a surface viewer—a box or cylinder with a glass bottom. Pushed in the water, this gives a clearer view as it eliminates reflection and any ruffle on the surface due to wind. Fine measurements in shallow water can be taken by fixing a stiff length of polished wire to a rod calibrated in millimetres. At the other end, about 3 feet away is fixed a ring. The wire is viewed through this ring as the rod is lowered, and the depth noted where it becomes invisible to the eye.

Nature of the bottom

Information about the material on a pond or lake bottom may be important, as it could affect the benthic life which exists there. Samples may be obtained by fixing a weighted tin to a cord, and using it as a scoop. For a more carefully selected sample of the material actually on the surface of the pond floor, use a short length of piping filled with some tallow. As it strikes the bottom, the topmost material will stick to the tallow, and can be brought to the surface.

Chemical contents

Water analysis is usually carried out on samples taken back to the laboratory. Accurately measured quantities of water, and certain reagents are required, and this may be a little difficult to provide in the field. Reagents are available in tablet form supplied by some chemists, such as the 'Soloid' tablets made by Burroughs and Wellcome, and these could be used at the pondside for testing such contents as ammonia, nitrates, oxygen and carbon dioxide, organic matter, hardness, etc. Methods for testing may be found in various textbooks, such as McLean and Cook's *Practical Field Ecology*. The chart obtained for the Baldwin Pond survey was supplied to the author by a student friend who was at the time employed by an analytic chemist (Fig. 47).

Project 19

PARASITES

In the course of the Epping Forest pond survey (Baldwin's Pond) the author on a number of occasions collected and bred sticklebacks in the aquarium (see Projects 11 and 13). At one stage a stickleback had completed its nest, so a female was introduced. It died shortly after. A second female was then put in, with the same result. Both had been severely attacked by the male. The dead fish were closely examined. Although both were very swollen and lacking in any red pigmentation, they turned out to be males. Whereas the nesting male had apparently recognised his opponents as rivals, the author had been taken in completely.

On post-mortem it was discovered that both fish were heavily infested with a tape-worm *Schistocephalus gasterostei*. The average length of these was 3 cm., and one 3-inch fish contained nine parasites! It is a wonder that it could have survived. As it was the condition, sluggishness and lack of colour must have been due to parasitism.

In this tape-worm three hosts are involved, forming part of a food chain. The adult worm is found in the intestine of fish-eating water birds, such as divers, grebes and ducks. The egg released into the water with the bird excreta hatches into a swimming ciliated *coracidium.* This is swallowed by a cyclops crustacean, where it forms into a hooked individual, called a *procercoid.* The cyclops is eaten by a stickleback and the parasite turns into a *plerocercoid.* These were found in the above fish. When an infected stickleback is eaten by a water bird the adult parasite stage is reached. The life-history of this tape-worm follows a similar pattern to that found in humans, namely *Diphyllobothrium latum.*

A second accidental discovery of an internal parasite occurred during the project on the fallow deer carcase (p. 174). When the mutilated body and internal organs were examined casually, the condition of the liver was noted. On closer inspection it was found to contain a number of flat, worm-like parasites still showing signs of movement. These were later identified as the notorious liver fluke found in sheep, namely *Fasciola hepatica.* Cases of this fluke in fallow deer have since been reported from the New Forest. After confirmation of the species a search was made for the *redia* and *cercaria* stages found in the intermediate host, a water snail (*Limnaea truncatula*), but without success.

At the time of discovery of the carcase (1938) a number of dead fawns had been reported by the keepers, who said that this might be due to 'scouring from wet grass' during an unusually wet summer. One often reads in old records and documents about the deaths which occurred among park fallow deer. They 'died of the murrein'. This word, written variously as murreyn, murrain, moryn, etc., is derived from Anglo-Latin, and possibly stems from mort—death. The word was used indiscriminately for deaths among humans, cattle, sheep, deer, and even bees. It could have referred to some parasitic infection.

COLLECTING EQUIPMENT AND FIELD GEAR

The following list of gear and containers is a selection of what can be usefully employed during field outings and on surveys, either at a water area or on dry land.

Freshwater Studies

1. The Pond Net. This is a general all-purpose net for catching visible pond animals in open water (e.g. beetles, newts, fish), or for sweeping through water plants for hidden animals (e.g. larvae of dragonfly and beetle, snails, water spider). A net of strong and fine-mesh netting, with $\frac{1}{8}$-$\frac{1}{4}$ inch sized holes, about 1 foot across and tapering to a point, is fitted to a triangular frame of strong wire attached to a stout cane or wooden handle. For convenience in transport the frame can be made of metal strips hinged at the outer corners, having a screw thread at the attachable end which fits into a socket on the handle (Fig. 82e). Such a net can then be collapsed when not in use. A further refinement, in order to protect the netting from wear and tear, is a double wire frame connected by rings. The inner frame holds the net and is protected by the outer frame which joins the handle. The frame of a fisherman's landing net is ideal and only requires a net of suitable mesh for use as a pond net.

2. Plankton Net. Used for collecting samples of the small, drifting pond-life such as water fleas, young insect larvae, diatoms, etc. A strong circular frame, about 9 inches across, holds a net of fine-mesh nylon or silk through which the water can pass but which will hold the catch. These are then trapped in a small glass or polythene tube fitted through a hole at the pointed end of the net and held in position with a rubber band, or tied on (Fig. 81). A tube with a rim should be used to ensure a tight fit. The plankton net is swept slowly through the water to and fro a number of times, in places where the tiny animals and plants are concentrated, usually in the sunny spots. If the tube has an open end which can be fitted with a cork stopper then the contents can be released into a collecting bottle.

3. Drag Net. A bag of stout small-mesh netting about 2 feet across and tapering to a point, is fitted to a strong triangular metal frame. Cords, one to each corner, are attached to a ring which can be joined to a long throwing-cord. The net should be fairly weighty so that it can be swung out some distance over the water when attached to the throwing-cord (do not leave go of the other end!). The net will sink slowly, and owing to its shape one side will rest on the bottom. It is then slowly hauled in so that it scrapes along the bottom and picks up benthic life, such as mussels, snails, caddis larvae, etc. This device is particularly useful during winter studies when much of the pond life and resting plants have settled on the pond bottom. One annoyance can be the hazards of submerged objects such as sunken logs and branches, not to mention discarded bedsteads, planks and other junk which snag the net. There is little one can do to overcome this, apart from carefully choosing the spots to be dragged.

4. The Grapple, (Fig. 81). This alternative drag is less likely to catch on to immovable snags. Three or four lengths of metal rod, about ½ inch thick, are firmly fixed together at one end. The free ends are pushed through a metal

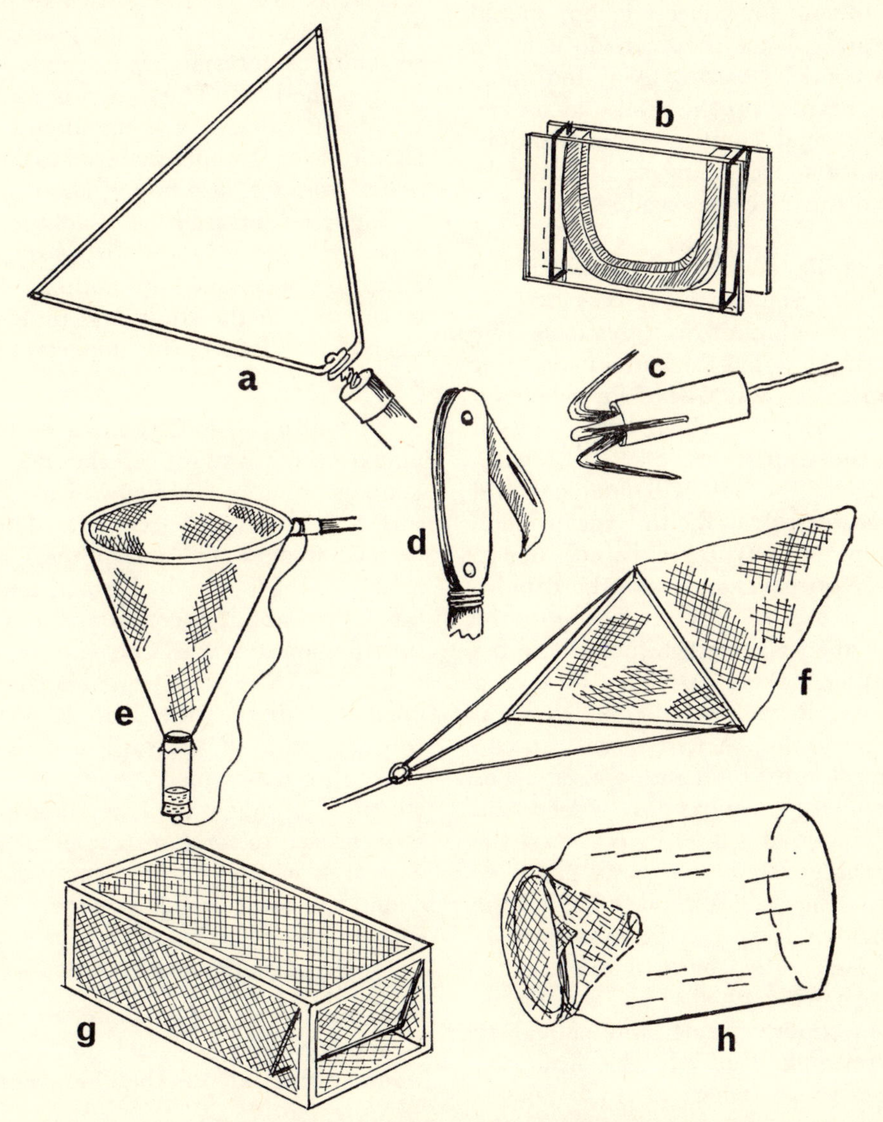

Fig. 81. Apparatus used in freshwater collecting. a, pond net; b, 'Aquarium' viewer; c, grapple; d, knife cutter; e, plankton net; f, drag net; g, crayfish trap; h, minnow trap.

ring and bent back on themselves so as to stand out at an angle. Cord is fitted to this so that when dragged in it will catch on to water plants and other objects. Apart from the plant specimens obtained, these will contain a quantity of trapped animals. The grapple could also be used for retrieving anything which may have fallen in accidentally. The author actually saved a friend who fell out of a boat one day, and could not swim.

NOTE. The Drag net and Grapple are both valuable equipment for collecting, but should be used sparingly, since they can do a lot of damage to a pond by tearing away too much plant life and stirring up the mud and sunken leaves. Like a good camper a conscientious ecologist will leave everything as he found it, with as little disturbance as possible.

5. The Baited Trap. Any animal which can be attracted with some kind of bait may be trapped in a cage containing suitable food. The trap is so constructed that the animal can enter but not escape. A simple design for this is on the lobster-pot principle which has a conical entrance. If the trap is made of open wire-mesh the scent of the bait will float out and lure the creature inside. By this means such animals as minnows, crayfish, snails, eels, newts, and a number of scavengers, may be caught. Some decaying animal flesh is put into the trap, which is then placed in a spot where the creatures may be lurking. Attached by a cord to the bankside, it can then be pulled in at intervals for inspection. A wire-mesh trap can have the conical entry built into one end. For crayfish a trap-door entry has proved successful (Fig. 81). The animal pushes its way past the door, which is hinged at the top to open inwards so that it can swing back into place behind the captive. A minnow trap can be made simply by using a glass bottle (Fig. 81). A cone of perforated zinc is pushed into the bottle opening. Use a square of zinc large enough to overlap the opening. Cut into this from one corner to the centre, where a small hole is made. Fold the zinc into a cone, fit into the bottle and fold over the corners, which can be fixed to the bottle rim with some wire. Minnows are attracted by bright objects, and some silver paper inside the bottle should do the trick. As soon as one minnow has entered, many others should quickly follow, and in a short while the collector has far more than he needs.

For best results minnow traps and other baited cages should have the opening pointing downstream in moving water, since animals tend to face the current.

6. Towing Net. For use in open water where a boat is necessary (e.g. a lake or reservoir) a net of the type used for plankton work in sea-water is necessary. It is simply towed along behind the boat. The design is the same as for the hand-worked net mentioned in 2, but should be of stronger material to withstand the water pressure, also larger, about 3 feet across. A big concentration of plankton can sometimes be collected near to the surface, and at other times practically nothing. This useful work helps in the study and understanding of the ebb and flow of this important life in open waters.

7. The Floating Cage. In addition to the plankton life caught in the net above, the animals, mainly insects, which hatch at the surface can also be caught. These include various midges and mosquitoes. The trap is made out of light ½-inch wood and built as a box frame, about 2 feet square, and covered on the sides and top with fine wire-netting (muslin will do in fine weather). The bottom is left open and fitted with floats so that the contraption rests on the surface without any air gaps through which emerging insects can escape. The cage is anchored in a suitable open spot, or tied to some surface plant if available. This trap will help in the study of the various water creatures which emerge as adults, the species which occur, their concentration, and the times of emergence.

8. Some accessories for pond study. A number of smaller gadgets which are useful aids to water studies include the following:

a. Strainer. This can be obtained from a store where kitchen utensils are sold, such as a tea-strainer or perforated spoon. This is fixed to a cane or wooden handle and used for collecting surface life near the water's edge. It is also useful for straining off excess water when sorting out the collection.

b. Mud sampler. A round tin such as a cocoa or other beverage tin, is fixed to a stick with the opening facing in the direction of the handler. This can then be pulled along the pond bottom in order to fill it with samples of the mud, debris, and so on.

c. Knife cutter (Fig. 81). For obtaining a plant specimen which is out of reach (e.g. a flowering stalk of reed-mace, lily flower, a tree branch overhanging the water) a suitable pruning knife obtainable at a gardening shop can be used by attaching it to a long cane. The blade should be fixed in the half-open position so that it can be angled on to the stalk, which is then cut free.

d. Bottle clamp. For samples of water out of normal reach a clamp of the type used in chemical experiments, and fixed to a stick, can be used for holding a collecting bottle. For samples of water at lower depths a bottle which has a hinged lid and is spring-loaded can be used. The bottle is lowered empty (its weight will take it down if it is attached to a piece of iron or lead) and at the required depth the lid can be opened with a cord, and then allowed to close when the bottle is filled.

As specimens are brought in they will need to be examined and sorted, and unwanted material returned to the water. For this the following equipment will prove useful:

A. Sorting Dish. A small white polythene or white enamel pie-dish into which specimens can be tipped is very handy for examining and selecting those which are needed, especially the smaller ones. A hand lens will be useful for this. With a spoon or pipette the animals can then be picked out and placed into their travelling containers.

B. 'Aquarium Viewer'. A transparent glass or polythene specimen container which has flat sides, as used in museums and laboratories, and obtainable from dealers in biological equipment, is most useful for looking closely at specimens, by holding the container up to the light. A temporary viewer can be made on the spot with two pieces of clear glass. These are separated by a length of solid rubber tubing (a strip cut from a rubber tyre would do) which is shaped into a U. The two glasses are held tightly against this with stout elastic bands or spring clips. This container should then hold water which does not leak out (Fig. 81).

C. Collecting Containers. For transport of the catch a number of containers of various sizes will be needed. If possible these should have screw caps, such as preserving jars and those in which beverages are sold. Ideal containers for small specimens are the polythene or glass bottles in which medical supplies are sold. These can sometimes be obtained, free of charge, from one's local doctor or chemist. For the large specimens (e.g. newts, fish plants) the bait cans used by fishermen may be necessary. With a selection of containers the specimens can be sorted out to avoid overcrowding and to keep vulnerable animals away from the predators. Even during a short journey home a carnivorous beetle can play havoc with the collection.

Land Studies

Specimens to be collected, sampled and identified from a land surface will be found at various levels, from the trees to below ground, and will require various techniques for capture. The following are some examples—

A. On the Trees and Bushes. An open umbrella, hooked to a branch, or a white cloth spread over the ground below, will collect a fair sample of the animal life hidden among the foliage. Only the firmer branches should be tapped or shaken, so as not to damage foliage. A special beating-tray held by a short handle can be made out of some dark cloth which is fitted to a light wooden frame, about 3 feet square. If this is made of two wooden slats bolted together at the centre, it can be collapsed for easy transport. The cloth is removable, and fits on to the frame by having corner pockets which go on to the ends of the slats (Fig. 82).

B. On Tree Trunks. Material removed from a tree trunk, such as moss, lichen and loose bark, also decaying wood, can be broken up and teased out on a white cloth for hidden animals (see also Berlèse Funnel and Aspirator). Fruit can be opened up, or kept in a box until the animal inside emerges (e.g. nuts, acorns, also galls).

C. Herbage and Long Grass. A sweep net pushed through the undergrowth should dislodge and catch many specimens. A stout triangular frame, about 2 feet across and hinged at the corners, can be screwed to a short handle, as for the pond net. Use some stout, finely woven material for the net, such as linen or calico, which should be rounded at the base, about as deep as wide. This net will have to stand up to hard wear. For detailed examination, individual plants should be examined, bearing in mind that some are food plants for certain species, also that some of the feeders are diurnal, others nocturnal.

D. On Roots. Examine infected material, for which a sharp knife and small trowel may be useful.

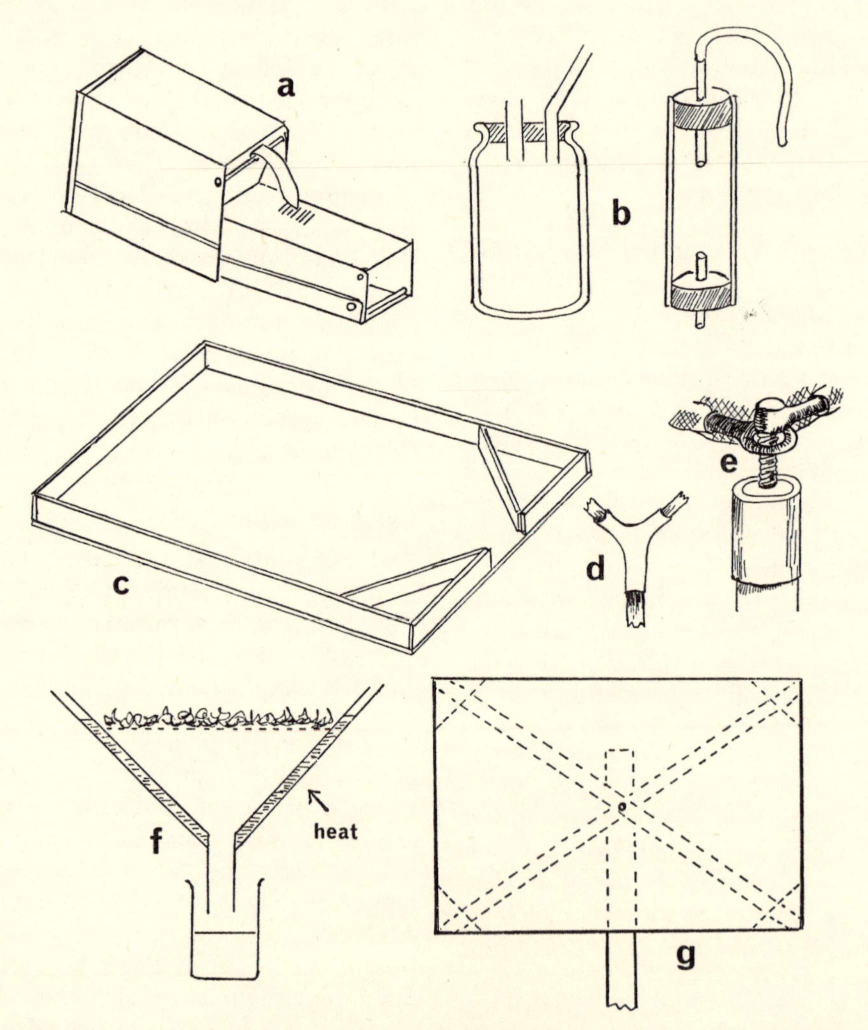

Fig. 82. Apparatus used in land studies. a, Longworth trap; b, two types of aspirator; c, sorting tray; d, Y-piece joint for aerial kite net; e, screw attachment for sweep net; f, Berlèse funnel; g, beating tray from underside.

E. In the Air (e.g. dragonflies, flies, hymenoptera and lepidoptera). A butterfly or kite net is essential for day work. This is lightly built out of a circular ring of stiff wire or split cane, about 18 inches across, which can be fitted into a tubular Y-piece and joined to a short handle (Fig. 82). Muslin or mosquito netting is used for the bag, and should be deep enough to fold back on itself so as to trap the captive. Dark material helps to show up a specimen in torchlight after dark. For night work a bright light shone on to a suspended white cloth (torch, car headlamps, etc.) will produce some results. This can be improved by sugaring. Mix some brown sugar or molasses with stale beer, adding a little glucose for stiffening, and smear on to the suspended cloth or a board.

NOTE. Sugaring is forbidden in some areas, and should never be applied where it may disfigure a tree or fence.

F. In the Soil. There are a number of methods of bringing soil animals to the surface, apart from digging them up, as well as different baits and traps. For general samples the flotation method can be used. Sprinkle some broken-up soil into water and stir in some salt and paraffin, when most of the animals will float to the surface and can be picked out. They will probably have been killed. To eliminate any froth, add a drop or two of iso-butyl alcohol. Worms may be collected by clearing a patch of ground and pouring on a mild solution of potassium permanganate (about 2 gm. to a gallon of water which should then turn pink). Sunken jars or tins baited with food will catch some animals, including small mammals (but see on). These traps should be covered with a roof of glass or slate raised on supports, to keep out rain. Bran, meal, bread or some cereal is suitable bait, and may also attract slugs and beetles. Damp sacking or a piece of wet bark is good for woodlice, sliced potato for millepedes and sugar for ants. Smoke or water can be used for holes and cracks to drive out the inmates.

G. In Leaf Litter. Since most leaf-litter fauna is on the small side it is better to take home samples of the leaf-mould, etc., also moss, lichen, or whatever needs examining. Use stout canvas bags for this, carefully labelling each as a reminder of the locality of each sample. Do not rely on memory. Later, each sample can be tipped onto a sorting tray (Fig. 82). Over this is suspended a strong light to illuminate the litter. The bright light and the drying out of the sorted litter will reveal the hidden animals, which can be picked out with forceps or an aspirator (see on). For delicate specimens a temporary pair of forceps can be made out of two thin slivers of wood fixed to a cork with an elastic band. This will do less damage to specimens than a steel pair. Specimens needing further identification are put into bottles placed in readiness. A permanent sorting tray can be built on a spare table by fixing strips of wood along the edges to prevent overspill. As the sorted material is finished with it is pushed through the gap into a waste container. A large amount of litter can be examined in this way by a team of sorters.

The Berlèse funnel

Named after its inventor, this apparatus consists of a small sieve for holding the litter, which is placed inside a funnel surrounded by a water jacket. This can be heated electrically or by using a spirit lamp or bunsen burner. The hot water warms the air inside the funnel which then rises into the litter and drives out the animals (Fig. 82). They fall through the funnel tube into a bottle underneath. Only small quantities of litter can be treated this way, but very few animals escape. Berlèse funnels can be purchased from dealers in biological equipment, or made by hand. A strong light is suspended over a funnel such as may be used for pouring petrol. It has a wire mesh inside on which the litter is placed, and the combined heat and light from bulb should drive out most of the animals.

The aspirator

This handy equipment is used for picking out small animals from leaf litter, moss, tree bark, walls, etc., which are otherwise difficult to capture. It consists of a cork-stoppered bottle with two lengths of glass tubing through the cork. A length of rubber tubing is attached to one as a mouthpiece and the inner end covered with muslin. The second tube lets in air. Two designs are shown (Fig. 82). Specimens are drawn into the tube by sucking the mouthpiece.

The Longworth trap

Manufactured by the Longworth Scientific Instrument Co. Ltd., Abingdon, Berks. (Fig. 82), this excellent small mammal trap is now widely accepted as the standard model for trapping in the field. Made of a rustless light alloy, it is in two parts. The box compartment forms a temporary prison for the animal, and is filled with some loose bedding such as dried grass, moss, etc., and suitably baited with corn, bread or raw meat (for shrews). Into the open end is fitted the trap. In order to reach the food the animal passes through this over a delicate trip-wire which releases the door. This falls at the lightest touch and locks behind the captive. These traps are best set in the evening, and must be examined first thing next day. To avoid losing traps in heavy undergrowth, mark each spot with a piece of bright cloth or ribbon tied to an exposed branch near by.

NOTE. The foregoing list of equipment and containers may convey the impression that a naturalist out on a field project looks somewhat like a walking pantechnicon, rather like a camping novice who takes everything with him 'bar the kitchen sink'. In actual practice, for most outings only a fraction need be taken, and the author keeps in readiness a basic kit which can be taken out at a moment's notice. This consists of a collapsible pond net and butterfly net, a grapple with cord, a penknife, some spare string, small trowel, polythene bags, a bait can, and a number of small stoppered-bottles. All this packs comfortably into a rucksack for journeys on foot, and only the can and bottles need be carried in the hand. These fit into compartments built into a box with a strap. Binoculars and camera are also included.

No doubt the reader will be able to think up his own ideas for equipment, which can often be handmade, and with experience will learn to ration himself to the bare essentials. A final piece of advice is to remember that a good craftsman or labourer, be he carpenter, gardener or naturalist, takes care of his tools. After each field trip clean out and dry all containers and nets as soon as they are finished with, and see that the captives are properly housed before taking that bath or well-earned meal!

Some general equipment and accessories useful in field work

1. A sharp, serviceable penknife, of the kind supplied to campers and boy scouts.

2. A hand chisel for splitting off old bark and rotting wood.

3. A collecting bottle. Useful for collecting lively, small specimens such as some beetles, spiders and centipedes. The cork stopper has a glass tube pushed through it, with a smaller cork to fit. Once the specimen is inside there is little chance of escape.

4. A small sieve for sifting material in the field.

5. A camel-hair brush and fine-pointed forceps for removing small specimens. For delicate specimens use two strips of thin shaving-wood fixed to each side of a cork with an elastic band, so that they operate like a pair of tweezers.

6. Specimen tubes with, preferably, screw caps, and of polythene. These are sometimes available as surplus at the chemist, and can be packed into a suitable box.

7. Collecting tins with perforated lids, for transporting medium specimens, such as insects and their larvae, molluscs, crustaceans, etc.

8. A hand lens. A three-in-one set of lenses, with magnifications of different strength, is probably a best buy. When in use it should be hung around the neck by means of a cord or tape, in order not to lose it, which could so easily happen in the open,

9. A small trowel for digging, and for lifting plants with roots attached.

10. Binoculars. These are always handy, not just for animal study but for spotting plants in difficult places, such as on cliffs and over water. Since it is the lenses which matter, look for a good make, even a second-hand one. A recommended size is 8 × 40.

11. A camera. No longer a luxury these days, a camera can now be purchased, at reasonable prices, in many makes. If possible, choose a 35 mm. reflex camera. This ensures a full view of the subject when focusing, an important factor in close-up work. A flash attachment is useful when working in shady corners, especially when photographing plants. Fast speeds may be necessary for animal shots. The author uses an Exa 2 Reflex which has stops from *f* 3·5 to 16, speeds up to 1/250 sec. and has a built-in pre-set diaphragm stop. This means that the viewer sees a clear picture through wide open lens up to the point of releasing the shutter, which then closes the diaphragm to the required stop. For clear colour slides used in lectures the author prefers Kodachrome 2. Kodachrome X is used for fast-moving subjects.

Epilogue

The future of the British countryside and its wildlife

Much of the charm of Britain's landscape lies in its diversity. This is due to its geology, climate, and 2,000 years of human settlement. In the south the half-wild patches of heathland, woods and downs are largely dominated by farmland, in a setting of tamed rural scenery. In the north and west the more rugged hills of grass and heather end in cliffs and mountain crags which dwarf the rural scene, and the view is wilder and more primitive.

Taken as a whole there is more varied landscape and wildlife in these islands than anywhere else on earth within so small a space. Since man is now the biggest ecological factor at work, his behaviour must be considered when assessing the influence he has on his habitat. By his supremacy over other living things he assumes the land as his by right and inheritance, to do with as he pleases. The slow but inevitable process of adapting it to his needs is a measure of his culture and progress, and this must be accepted, however painful the result.

It is within the present century that more changes and destruction of natural habitats have taken place than at any other time in Britain's long history. It is true that the destruction began as far back as the Neolithic days when the first farmers began to clear the woodlands for living space, and introduced farm crops and animals into Britain. It is also true, on the other hand, that for centuries large tracts of land were safeguarded from disafforestation by the Crown, for the purpose of following the sport of venerie which was introduced by the Normans and controlled by British law (see p. 74). Much as many of us might frown on the activities of the Norman huntsmen, we might also approve their motives in preserving the land, for in this they were excellent conservationists. In hunting circles it is necessary to preserve the quarry in order to continue this sport. The irony of this is in protecting wildlife in order to kill it.

Traces of these old hunting forests are still with us, a witness to the pride of ownership of land, and the preservation of its natural beauty. This is perhaps inherent in man, and applies as much to private land, whether a garden or an estate. Many landlords in the past stubbornly resisted the temptation to profit by leasing or selling their land.

Within the past 100 years, however, new threats to the land have arisen. Increased taxation, death duties, lack of maintenance, and the increased pressure on more land through a rising population, have so weakened the security, that private estates, in some cases even Crown land, have been broken up. The needs for housing, agriculture, forestry, industry and national defence are strong and demanding.

Some would say that these human needs are essential to our welfare, and must take priority. Open spaces, however attractive and enjoyable, must give way to human demands, particularly so as we are an industrial nation. People must be properly housed, fed and educated, with full employment and an adequate wage packet. If the rural scene can in any way help in this, they say, then it should be 'developed' to meet the demand. This should even apply to farming, the most rural of industries, and which requires modernising in order to keep up the food production. Even here there is a threat to animal and plant habitats.

There is opposition to this utilitarian view by other people, and there are many who argue that man cannot 'live by bread alone' in a materialistic world, and that mental and spiritual refreshment are as important to him as a full stomach and secure job. The open countryside can be a powerful antidote to the stress and strain of urban living where human density is now reaching a dangerous level due to overcrowding. This is one reason why many town dwellers seek out quiet surroundings for spare-time activities, and as holiday centres. It has been suggested that those who prefer the eternal noise and 'high lights' of towns may be in a state of mental sickness brought on by an unnatural environment. A love of nature and open air, from which we all stem, is latent in all humans, yet how many these days have never had the chance of realising it?

The amenity value of a rural countryside and its natural beauty and fresh air is of inestimable value in maintaining a healthy balance in our artificial lives. People whose interests lie mainly in politics and economy, often those in authority, and who can do so much good or harm to the land, should be made aware of this great asset. Better still, a nation as a whole, and especially its children, should be made aware of the importance of preserving as much as possible of the countryside which has so much to offer—in amenity, scientific research and education. A people who can understand and accept this will act as a powerful brake on the spread of industry and urbanisation.

Is it possible, then, to resolve this conflict between a materialistic and economic demand on our land, as opposed to the aesthetic, scientific and even sentimental regard for its preservation? In other words, can land be put to human use but at the same time retain a more or less natural aspect and beauty so as to preserve the habitat for its wildlife? Can we live with nature?

Some examples of the problems involved, and the ways in which they are being tackled so as to hold a balance between man's demands and the safety of land and wildlife are given here:

A. *Poisons*. Modern methods of control of undesirable plants and animals, the so-called 'weeds' and 'pests', have resulted in some danger to wildlife generally. Persistent chemicals such as the organochlorine pesticides (e.g. dieldrin) are said to have threatened the lives of some of the 'apex' animals at the end of food chains (see p. 64). Water birds and predators such as the birds-of-prey, fox and stoat have been found dead from such poisons. In some areas there is a danger of poison entering rivers or the sea. Indeed, contaminated eggs of penguins have been discovered as far away as the Antarctic. The careless use of sprays may ruin a complete hedgerow and eliminate the food plants of some useful insects such as bees, and attractive ones like butterflies. A proper sales control of clearly marked products and how to use them, especially those sold to gardeners and farmers, should help to minimise this danger. In 1968 a conference of manufacturers, ministry officials and naturalists met in London to discuss and approve a Code of Conduct on the proper manufacture, distribution and use of pesticides by farmers, councils, gardeners and other users.

B. *Demand for water*. The increasing demand for water, both for industrial and domestic use, means a steady drain on rivers and reservoirs which may reach a dangerous level in drought periods. Further storage may be necessary to supply the needs, and new reservoirs created.

This may threaten areas of biological and scientific importance, as in the case of the Teesdale reservoir project, where a unique flora of Ice Age relicts exists. Alternative sites should be considered before such an irreparable step is taken. The fate of this unique valley now appears certain (1968).

C. *Waste disposal.* The problem of disposal of human waste increases with the population. For health reasons, if only this, such disposal should be properly controlled. Solid waste normally ends up as foundation material, or in some disused gravel or chalk pit, or quarry, and water-borne waste, via pipes and drains, at the sewage works. In the former case a colony of interesting, even rare, plants may exist until the pit is filled in. Some of these places might be preserved as miniature reserves. A tolerant local council or landlord, if approached, might turn out to be a useful ally. A problem in sewage works has been the increasing use of detergents which have upset the biological control by those bacteria which can render the waste harmless. A further hazard is the careless discharge of improperly treated factory waste into rivers.

D. *Recreation.* For health reasons a visit to the countryside by town dwellers is to be encouraged. However, wherever humans gather, even in play, disturbance may be caused. It applies as much to well-behaved people as to vandals. This may occur at beauty spots, seaside resorts, camping and caravan sites, and so on. There is also the increased use of motor cars and horses. Each is now penetrating into quiet places. The rising popularity of sailing and boating means that many inland waters, such as lakes, rivers, canals, even the larger drowned pits and quarries, are in demand for this pastime.

All this is bound to cause disturbance to wildlife, and the Central Council of Physical Recreation is planning a Code of Conduct to give some protection to nature by setting aside restricted areas where these pastimes are carried out.

E. *Shooting, hunting and collecting.* Hunting is deep in the fibre of man, whether it is expressed by chasing a fox or catching wildlife with a camera. This should not be condemned out of hand, but rather used as a means of control over animal populations, as well as control over those who carry it out. Indeed, too many bird-watchers can probably do more harm than one deer stalker. Controlled shoots of animals which become too numerous, as in the case of red deer and grey seal, are sometimes necessary. With game birds and mammals which are regularly hunted, a close season is observed, as in the case of freshwater fishes. Apart from this, the need for occasional thinning out of numbers is important, since too many animals of one species may upset a habitat and so affect other species. This is conservation working both to the advantage of the huntsman and naturalist, in a so-called 'shotgun wedding'. A good example of this is the voluntary agreement of wildfowlers not to shoot in severe winters, and also to stay away from some marshland areas which can act as reservoirs of wildlife, so as to keep up their numbers.

F. *Pests.* When does an animal or plant species become a pest? This is usually when it interferes with man's welfare and interests by attacking his person or possessions, by spreading disease or by becoming a nuisance. In most cases this is due to an increase in populations. These need to be controlled, and this is where A and E can be put to effective use. The grey squirrel and rabbit, both alien invaders, have become pests, not because of their species so much as because of their high numbers and tolerance to new environments. Strict control is needed. In parts of Scotland and in Europe the native red squirrel is shot for the same reason. Control measures, apart from shooting and poisoning, include trapping, and in all this necessary work a need for humane killing is desired by most people. The law now governs this to some extent by banning the use of certain poisons and traps, and by a Firearms Act which limits the use of guns.

At a Conference held in March 1965, many bodies concerned with the countryside and wildlife (farmers, huntsmen, foresters, naturalists, etc.) unanimously agreed on setting up a

Code of Behaviour in our approach to wildlife, in particular towards the mammal predators (fox, badger, otter, stoat, weasel, polecat, wild cat, pine marten and feral mink). It is encouraging to find that the dogmatic enmity towards these animals in past years is fast disappearing (see Bibliography). This also applies to our birds-of-prey, now recognised as worth saving by the Bird Protection Act of 1954.

G. *The countryside in 1970*. An important Conference under this title, held in 1963 with the Duke of Edinburgh as chairman, met to discuss measures to be taken in order to safeguard the countryside in the future. The formation of various study groups was organised in order to look into the many problems now facing wildlife, such as those mentioned above. The policy behind most of these measures and recommendations may be summed up in one word—conservation.

Conservation has been described as the 'maintenance of the flow of living energy', involving such things as food chains, the correct balance between species, and the right kind of habitat in which this can be carried out. To understand fully what this implies it is necessary to realise that no animal or plant is ever entirely on its own. It is adapted to a particular environment and influenced by all the factors and other living things with which it comes into contact.

In a virgin land, rarely met in Britain, nature is in control, and always strives towards the climax community for a particular area, whether it be a grassland, heath, shingle or wood. But where land is in a semi-natural state, as in most of our rural areas, or highly artificial as in towns and other built-up areas, then man takes over. Due to his presence and activities the normal climax is often upset and the biosere interrupted.

Actually this may be welcomed in some cases, since the result is not only beneficial to man but also to certain animals and plants. One example is the increased numbers of alien conifers planted by the Forestry Commission. This may have eliminated a number of species more at home on heathland where many plantations now exist, but it has undoubtedly helped in the spread of red squirrels and roe deer. Or again, on the southern downlands where trees once stood, a woodland climax has been kept down by the grazing of centuries of sheep and rabbit. The result is a richness of grassland flora, including orchids, which might never have happened if the trees had remained. Also, the pleasures of a cool summer breeze under open sky, and the feel of springy turf beneath the feet, may never have come our way.

This book has been mainly concerned with the British countryside, its origin and the wildlife which it supports, and how this may be studied and better known through the pursuit of Ecology, or what is sometimes termed scientific Natural History.

There is still much to do and learn, with many places and habitats to explore. Much is now in a semi-natural state, which is all the more reason why we should get to understand it better, in order to appreciate what it can offer—in material use, in scientific discovery, in education and in aesthetic pleasures. Land will only give back what we put into it, and will repay in full measure the love and care we are prepared to offer it.

Behind all this is an even greater threat to the land, and even to man himself. In the long run this may override all the measures and compromises which we make in trying to preserve what countryside and wildlife is still left to us. Today a steadily rising birthrate which goes on unchecked, and a longer lifespan due to improved medical care, as well as an increased laxity in moral behaviour, is creating a human population explosion. If this continues at its present rate, then the future of our wildlife looks bleak. The time has perhaps arrived when the conservation measures which man is carrying out on wildlife should be tried on himself. There is now a growing realisation that man may one day have to face up to his greatest enemy, the most dangerous animal on earth—himself.

Appendices

1. Mosses as soil indicators
2. A selection of plants and animals in Britain
   - A. Plants in an oak wood
   - B. Woodland animals (birds and mammals)
   - C. Chalk plants
   - D. Limestone plants
   - E. Trees and herbs in a beech wood
   - F. Rare plants of chalk and limestone
   - G. Plants on acid soil (heath and moor)
   - H. Plants in acid bogs and dune slacks
   - I. Plants in coastal habitats (sea cliff, sand-dune, salt marsh and shingle)
   - J. Plants on mountains
   - K. Mountain animals
   - L. Plants of pond, river and marsh
   - M. Plants of waste ground
   - N. Garden weeds
   - O. Hedgerow and roadside plants
   - P. Plants of grassland
   - Q. Animals of buildings
3. A famous quartet
4. A Devon lane
5. Road casualties
6. Wildlife in a garden sanctuary
7. Biology of sewage farms and waterworks
8. The Country Code
9. Fungi in Epping Forest
10. Plaster casts
11. Food niches in Epping Forest
12. Life on the oak tree
13. Micro-habitats—log, carcase and cowpat
14. Aquatic life in Baldwin's Pond, Epping Forest
15. pH
16. The herbarium
17. Measuring angles and heights
18. Main habitat regions in Britain
19. British freshwater fishes
20. British reptiles and amphibians
21. Animals in the post
22. Cleaning a skeleton
23. The Grid reference

APPENDIX 1: Mosses as soil indicators

The following list of mosses is placed in order of habitat, ranging from a good decomposed humus soil, passing through podsol to a poor, peaty soil:

Good humus

Mnium undulatum. Good hardwood in shady moist situations.
Mnium hornum. As above, slightly more acid soil.
Eurhyncium striatum. On basic or acid soil.
Catherinea undulata. As for *Mniums*, but drier.
Thuidium tamariscinum. Mainly W. Scotland under spruce.

Medium humus

Brachythecium purum. Poorer hardwoods.
Hylocomium spp. N. and E. Scotland in wetter parts along woodland borders.
Hypnum spp. Conifer woods, on dry, acid soil with heather.
Dicranum majus. See *Hypnum*.

Poor humus

Polytrichum commune. Wet, acid soil in open places.
Polytrichum formosum. Acid soil in dry shady woods.
Leucobryum glaucum. Highlands, also leached soil in beechwoods of S.E. England, e.g. Epping Forest.
Sphagnum spp. Wet, acid places.
Rhacomitrium lanuginosum. Mountain rocks with Deer grass (*Scirpus caespitosus*) and lichens, on poor, acid peat. Of little use for trees.

These notes are taken from the Forestry Commission Booklet No 1, entitled *Woodland Mosses* (H.M. Stationery Office, 1947, 2s.).

APPENDIX 2: A selection of plants and animals in Britain

2A. *Some plants associated with pedunculate oak-woods* (*Quercus robur*)

Hawthorn, *Crataegus monogyna*.
Crab apple, *Malus sylvestris*.
Wild cherry, *Prunus avium*.
Bird cherry, *Prunus padus*.
Hornbeam, *Carpinus betulus*, mainly S.E. England.
Ash, *Fraxinus exelsior*, may be co-dominant.
Maple, *Acer campestre*.
Holly, *Ilex aquifolium*.
Hazel, *Corylus avellana*.
Guelder rose, *Viburnum opulus*.
Wild service, *Sorbus torminalis*.
Dog rose, *Rosa canina*.
Bramble, *Rubus fruticosus*.
Honeysuckle, *Lonicera periclymenum*.
Ivy, *Hedera helix*.
Bluebell, *Endymion non-scripta*.
Wood anemone, *Anemone nemorosa*.
Wood sorrel, *Oxalis acetosella*.
Lesser celandine, *Ranunculus ficaria*.
Primrose, *Primula vulgaris*.
Foxglove, *Digitalis purpurea* mainly acid soils.
Lesser periwinkle, *Vinca minor*, local.
Common lungwort, *Pulmonaria officinale*.
Oxlip, *Primula elatior*, eastern counties on clay.
Yellow archangel, *Galeobdolon luteum*.
Common valerian, *Valeriana officinalis*.
Gladiolus, *Gladiolus illyricus*, rare New Forest.
Wood cranesbill, *Geranium sylvaticum*.
Herb robert, *Geranium robertianum*.
Red campion, *Melandrium dioicum*.
Greater stitchwort, *Stellaria holostea*.
Wild daffodil, *Narcissus pseudo-narcissus*, local, especially in S.W.
Common cow-wheat, *Melampyrum pratense*.
Dog violet, *Viola riviniana*.
Wood dog violet, *Viola reichenbachiana*, commonest in south.
Moschatel, *Adoxa moschatellina*.
Cuckoo pint, *Arum maculatum*, less common in north.
Ground ivy, *Glechoma hederacea*.
Bugle, *Ajuga reptans*.
Enchanter's nightshade, *Circaea lutetiana*.
Wood goldilocks, *Ranunculus auricomus*, local.
Wild strawberry, *Fragaria vesca*.
Pignut, Earthnut, *Conopodium majus*.
Ground elder, *Aegopodium podagraria*.
Wild angelica, *Angelica sylvestris*.
Wood spurge, *Euphorbia amygdaloides*.
Wood forget-me-not, *Myosotis sylvatica*.
Common figwort, *Scrophularia nodosa*.
Creeping jenny, *Lysimachia nummularia*.
Yellow pimpernel, *Lysimachia nemorum*.
Early purple orchid, *Orchis mascula*.
Twayblade, *Listera ovata*.
Common helleborine, *Epipactis helleborine*, local.
Sweet vernal grass, *Anthroxanthum odoratum*.
Wood meadow grass, *Poa nemoralis*.
Hairy brome, *Bromus ramosus*.
Wood melick, *Melica uniflora*.
Slender false brome, *Brachypodium sylvaticum*.
Great fescue, *Festuca gigantea*.
Drooping sedge, *Carex pendula*.
Hairy woodrush, *Luzula pilosa*.
Great woodrush, *Luzula sylvatica*.
Male fern, *Dryopteris felix-mas*.
Lady fern, *Athyrium felix-femina*.

2B. *A List of woodland birds and mammals*

Woodpigeon.
Stockdove.
Turtle dove (s).
Green woodpecker.
Great spotted woodpecker.
Lesser spotted woodpecker.
Nuthatch.
Tree creeper.
Song thrush.
Blackbird.
Mistle thrush.
Redwing (w).
Carrion crow.
Jay.
Jackdaw.
Magpie.
Starling.
Tree sparrow.
Chaffinch.
Bullfinch.
Hawfinch.
Brambling (w).
Woodcock.
Tawny owl.
Kestrel.
Buzzard.
Sparrow hawk.
Pheasant (i).
Wren.
Common Redstart.
Great tit.
Blue tit.
Longtailed tit.
Nightingale (s).
Woodlark.
Cuckoo (s).
Willow warbler (s).
Wood warbler (s).
Chiff chaff (s).
Blackcap (s).
Redbacked shrike (s).
Fallow deer (i).
Sika (i).
Muntjac (i).
Roe.
Badger.
Fox.
Grey squirrel (i).
Red squirrel.
Wood mouse.
Yellow necked mouse.
Dormouse.
Stoat.
Rabbit (i).
Common shrew.
Lesser shrew.

(s) summer and (w) winter visitor; (i) introduced.

2C. *Some plants associated with calcareous soils*

Species found mainly on chalk in East and South-East England:

Horseshoe vetch, *Hippocrepis comosa.*
Kidney vetch, *Anthyllis vulneraris.*
Bird's-foot trefoil, *Lotus corniculatus.*
Salad burnet, *Poterium sanguisorba.*
Dropwort, *Filipendula hexapetala.*
Yellowwort, *Blackstonia perfoliata.*
Common rockrose, *Helianthemum nummularium.*
Hairy violet, *Viola hirta.*
Dove's scabious, *Scabiosa columbaria.*
Carline thistle, *Carlina vulgaris.*
Stemless thistle, *Cirsium acaulis.*
Marjoram, *Origanum vulgare.*
Deadly nightshade, *Atropa belladona.*
Chalk eyebright, *Euphrasia pseudo-kernei.*
Sweet briar, *Rosa rubiginosa.*
Hoary plantain, *Plantago media.*
Squinancywort, *Asperula cynanchica.*
Ploughman's spikenard, *Inula conyza.*
Clustered bellflower, *Campanula glomerata.*
Harebell, *Campanula rotundifolia.*
Burnet saxifrage, *Pimpinella saxifraga.*
Wild thyme, *Thymus drucei.*
Wild basil, *Clinopodium vulgare.*
Sheep's fescue, *Festuca ovina.*
Quaking grass, *Briza media.*
Upright brome, *Bromus erectus.*
Green hellebore, *Helleborus viridis.*
Cowslip, *Primula veris.*
Spring sedge, *Carex caryophyllea.*
Fragrant orchid, *Gymadinia composa.*
Man orchid, *Aceras anthropophorus.*
Pyramidal orchid, *Anacamptis pyramidalis.*
Musk orchid, *Herminium monorchis.*
Fly orchid, *Orchis insectifera.*
Autumn lady's tresses, *Spiranthes spiralis.*

2D. *Some plants associated with limestone soils*

Species found mainly on limestone in the West and North of England:

Dark flowered helleborine, *Epipactis atropurpurea.*
Vernal squill, *Scilla verna.*
Autumnal squill, *Scilla autumnalis.*
Gladdon, *Iris foetidissima.*
Spring sandwort, *Arenaria verna.*
Globe flower, *Trollius europaeus.*
Lily-of-the-valley, *Convallaria majalis.*

Bloody cranesbill, *Geranium sanguineum.*
Pellitory-of-the-wall, *Parietaria officinalis.*
Bird's-eye primrose, *Primula farinosa.*
Blue gromwell, *Lithospermum officinale.*
Burnet rose, *Rosa spinosissima.*
Red valerian, *Centranthus ruber.*
Wall pennywort, *Umbilicus pendulinus.*
Stonecrop, *Sedum* spp.
Caper spurge, *Euphorbia lathyrus.*
Butterfly orchid, *Platanthera chlorantha.*
Hart's tongue, *Phyllitis scolopendrium.*
Rusty-back, *Ceterarch officinarum.*
Wall-rue, *Asplenium ruta-muralis.*
Limestone polypody, *Gymnocarpium robertianum.*

2E. *Trees, shrubs and herbs associated with chalk scrub and beechwood*

Beech, *Fagus sylvatica*, typically on chalk.
Ash, *Fraxinus excelsior*, typically on limestone.
Yew, *Taxus baccata.*
Box, *Buxus sempervirens*, N. and S. Downs, Chilterns, Cotswolds.
Juniper, *Juniperus communis.*
Spindle, *Euonymus europaeus.*
Hawthorn, *Crataegus monogyna.*
Wayfaring tree, *Viburnum lantana.*
Dogwood, *Cornus sanguinea.*
Clematis, *Clematis vitalba.*
Field maple, *Acer campestre.*
Privet, *Ligustrum vulgare.*
Dog's mercury, *Mercurialis perennis.*
Bugle, *Ajuga reptans.*
Cuckoo pint, *Arum maculatum.*
Yellow archangel, *Galeobdolon luteum.*
Wood spurge, *Euphorbia amygdaloides.*
Spurge laurel, *Daphne laureola.*
White beam, *Sorbus aria.*
Woodruff, *Galium odoratum.*
Wood sanicle, *Sanicula europaea.*
Yellow bird's-nest, *Monotropa hypopitys.*
Bird's-nest orchid, *Neottia nidus-avis.*
Broad-leaved helleborine, *Cephalanthera damasonium.*
Narrow-leaved helleborine, *Cephalanthera longifolia.*

2F. *Some rare and local chalk and limestone plants*

Pasque flower, *Pulsatilla vulgaris*, chalk turf, Chilterns up to Yorkshire.
Hoary rockrose, *Helianthemum carnum*, limestone cliffs, Gower, N. Wales, Lancashire and Galway Bay.
Rare cuckoo pint, *Arum neglectum*, chalk, south coast.
Hair-leaved goldilocks, *Linosyris vulgaris*, limestone cliffs, W. England.
Blue gromwell, *Buglossoides purpuro-caerulea*, limestone cliffs and woods, S.W. England and S. Wales.
Spring gentian, *Gentiana verna*, limestone turf, Teesdale.
Autumn gentian, *Gentiana autumnalis*, chalk turf, Chilterns.
Spiked speedwell, *Veronica spicata*, Breckland.
Western spiked speedwell, *Veronica hybrida*, limestone, Avon Gorge and Wales.
Honewort, *Trinia glauca*, limestone, Berry Head, Avon Gorge and Mendips.
Round-leaved garlic, *Allium sphaerocephalon*, Avon Gorge.
Round-headed rampion, *Phyteuma tenerum*, chalk, N. and S. Downs.
Yellow whitlow grass, *Draba azoides*, limestone cliffs and ruins, Gower.
Cut-leaved germander, *Teucrium botrys*, chalk and limestone, Kent to Cotswolds.
Rock pepperwort, *Hornungia petraea*, limestone and walls, Mendips, Gower, Craven and Pennines.
Clove-scented broomrape, *Orobranche caryophyllacea*, on Chalk Bedstraw (*Galium mollugo*) W. Kent.
Portland spurge, *Euphorbia portlandica*, limestone cliffs, Dorset and Devon.
Spotted cat's-ear, *Hypochaeris maculata*, chalk, Eastern counties.
Downy woundwort, *Stachys germanica*, limestone, N.W. England.
Crested cow-wheat, *Melampyrum cristatum*, chalk, S. England.
Baneberry, *Actaea spicata*, limestone rocks and ash woods, N. England.
Martagon lily, *Lilium martagon*, Surrey and Wye Valley.
Oxlip, *Primula elatior*, Eastern counties.
Cheddar pink, *Dianthus gratianopolitanus*, Cheddar Gorge.
Dark-flowered helleborine, *Epipactis atropurpurea*, limestone, N. England.
Lady orchid, *Orchis purpurea*, Kent.
Lizard orchid, *Himantoglossum horcinium*, chalk and limestone, S. England. Erratic in growth.
Monkey orchid, *Orchis simia*, Chilterns.
Late spider orchid, *Orchis fuciflora*, S.E. Kent.
Military orchid, *Orchis militaris*, Chilterns. Very rare.
Red helleborine, *Cephalanthera rubra*, Cotswolds.
Maidenhair fern, *Adiantum capillus-veneris*, moist limestone rocks and cliffs, S.W. and N.W. England, Wales and W. Ireland.

The following calcicoles are strong indicators of chalk soil:

Chalk milkwort, *Polygala calcarea.*
Dark mullein, *Verbascum nigrum.*
Upright brome, *Bromus erectus.*
Clematis, *Clematis vitalba.*

2G. *Some plants associated with acid soils on heath and moorland*

Gorse, *Ulex europaeus*.
Lesser gorse, Dwarf furze, *Ulex minor*.
Petty whin, *Genista anglica*, local.
Broom, *Sacothamnus scoparius*.
Bilberry, Whortleberry, *Vaccinium myrtillus*.
Cowberry, *Vaccinium vitis-idaea*, north and west Britain.
Crowberry, *Emeptrum nigrum*, north and west Britain.
Ling, *Calluna vulgaris*.
Bell heather, *Erica cinerea*.
Cross-leaved heath, *Erica tetralix*.
Tormentil, *Potentilla erecta*.
Heath dog violet, *Viola canina*.
Heath speedwell, *Veronica officinalis*.
Betony, *Betonica officinalis*.
Common cudweed, *Filago germanica*.
Heath bedstraw, *Galium saxatile*.
Lesser stitchwort, *Stellaria graminea*.
Leafy hawkweed, *Aphyllopoda* spp.
Sheep's sorrel, *Rumex acetosella*.
Greater broomrape, *Orobanche rapum-genistae*, on broom or gorse.
Common dodder, *Cuscuta epithymum*, on gorse or ling.
Wild thyme, *Thymus drucei*, also on chalk and in dunes.
Heath spotted orchid, *Dactylorchis maculata*.
Deer grass, *Scirpus caespitosus*.
Tufted hair grass, *Deschampsia caespitosa*.
Wavy hair grass, *Deschampsia flexuosa*.
Purple moor grass, *Molinia caerulea*.
Star sedge, *Carex echinata*.
Moor sedge, *Carex binervis*.
Oval sedge, *Carex ovalis*.
Heath rush, *Juncus squarrosus*.
Fine bent grass, *Agrostis tenius*.
Hard fern, *Blechnum spicatum*.
Bracken, *Pteridium aquilinum*.
Foxglove, *Digitalis purpurea*.

2H. *Some plants associated with heath and moorland bogs and dune slacks*

Spiked sedge, *Carex spicata*.
Star sedge, *Carex echinata*.
Marsh violet, *Viola palustris*, mainly in north.
Marsh cinquefoil, *Potentilla palustris*, uncommon in south.
Common sundew, *Drosera rotundifolia*.
Long-leaved sundew, *Drosera longifolia*.
Greater sundew, *Drosera anglica*, mainly in north.
Marsh pennywort, *Hydrocotyle vulgaris*.
Cranberry, *Vaccinium oxycoccus*.
Grass of Parnassus, *Parnassia palustris*, north and west, local.
Cotton grass, *Eriophorum angustifolium*.
Bog asphodel, *Narthesium ossifragum*.
Marsh bedstraw, *Galium palustre*.
Marsh gentian, *Gentiana pneumonanthe*.
Early marsh orchid, *Dactylorchis incarnata*.
Heath spotted orchid, *Dactylorchis maculata*.
Marsh helleborine, *Epipactis palustris*.
Lesser twayblade, *Listera cordata*, mainly in north.
Bog myrtle, *Myrica gale*.
Lousewort, *Pedicularis sylvatica*.
Sharp flowered rush, *Juncus acutiflorus*.
White beak sedge, *Rhynochospora alba*.
Creeping willow, *Salix repens*.
Royal fern, *Osmunda regalis*.
Bog moss, *Sphagnum* spp.

2I. *Some plants associated with coastal habitats*

Sea Cliffs

Sea pink, thrift, *Armeria maritima*.
Sea campion, *Lychnis maritima*.
Wall pennywort, *Umbilicus rupestris*.
Rock sea lavender, *Limonium binervosum*.
Red valerian, *Centranthus rubus*.
Scurvey grass, *Cochlearia officinalis*.
Tree mallow, *Lavatera arborea*, mainly S.W.
White stonecrop, *Sedum album*.
Wall pepper, *Sedum acre*.
Spring squill, *Scilla verna*, S.W. and Scotland.
Autumn squill, *Scilla autumnalis*, local in S.
Sea wormwood, *Artemisia maritima*.
Kaffir Fig, *Mesembryanthemum* (*carpobrotus*) *edulis*, S.W.
Maidenhair fern, *Adiantum capillus-veneris*, S.W., N.W., Wales and W. Ireland.

Sand-dunes

a. Drift line on foreshore

Saltwort, *Salsola kali*.
Sea beet, *Beta vulgaris*.
Sea rocket, *Cakile maritima*.

b. Front or 'white' dune.

Marram grass, *Ammophila arenaria*.
Sand sedge, *Carex arenaria*.
Sand couch grass, *Agropyron junceiforme*.
Common sea kale, *Crambe maritima*.
Sea holly, *Eryngium maritimum*.
Sea bindweed, *Calystegia soldanella*.
Bird's-foot trefoil, *Lotus corniculatus*.
Ragwort, *Senecio jacobaea*, also wet pastures.
Carline thistle, *Carlina vulgaris*.

Great sea stock, *Matthiola sinuata*, v. local, N. Devon and S. Ireland.

c. Fixed or 'grey' dune

Red fescue grass, *Festuca rubra.*
Mouse-ear hawkweed, *Hieracium pilosella.*
Burnet rose, *Rosa pimpinellifolia.*
Spear thistle, *Cirsium vulgare.*
Sea sandwort, *Honkenya peploides.*
Evening primrose, *Oenothera erythrosepala.*
Lesser hawkbit, *Leontodon taraxocoides.*
Common centaury, *Centaurium erythraea.*
Field gentian, *Gentianella campestris.*
Felwort, *Gentiana amarella*, also chalk turf.
Sea pansy, *Viola curtisii*, N. and W. Ireland, also Breckland.
Viper's bugloss, *Echium vulgare*, also chalk.
Sea stork's-bill, *Erodium maritimum.*
Rest harrow, *Ononis repens*, also chalk.
English stonecrop, *Sedum anglicum.*
Hound's-tongue, *Cynoglossum officinale*, also chalk.
Sea spurge, *Euphorbia paralias*, Norfolk, S. and W.
Pyramid orchid, *Anacamptis pyramidalis*, also chalk.
Privet, *Ligustrum vulgare*, also chalk.
Wild thyme, *Thymus drucei*, also chalk, heaths, rare in S.E.
Sea buckthorn, *Hippophae rhamnoides.*
Hawthorn, *Crataegus monogyna*, common to many soils.
Moss and Lichen, spp.

Salt Marsh

a. Bare marsh.

Bladderwrack, *Fucus spiralis* and *vesiculosus.*
Grass wrack, or Eel grass, *Zostera marina.*
Sea-weed grass, *Enteromorpha intestinalis.*
Sea lettuce, *Monostronia grevillei* and *Ulva lactuca.*

b. Low marsh

Marsh samphire, Glass-wort, *Salicornia europaea.*
Grass wrack, *Zostera marina.*
Cord or rice grass, *Spartina townsendii.*
Reflexed meadow grass, *Poa distans.*

c. Main marsh

Annual sea blight, *Sueda maritima.*
Sea arrow grass, *Triglochin maritima.*
Sea lavender, *Limonium vulgare.*

d. High marsh

Thrift, sea pink, *Armeria maritima.*
Sea plantain, *Plantago maritima.*
Sea rush, *Juncus maritimus.*
Sea spurrey, *Spergularia marina.*
Sea wormwood, *Artemisia maritima.*
Sea couch grass, *Agropyron pungens.*
Sea Milkwort, *Glaux maritima.*
Sea club rush, *Juncus maritimus.*
Marsh mallow, *Althaea officinalis.*
Sea mugwort, *Artemisia maritima.*
Common mallow, *Malva sylvestris.*

e. Salt creeks

Sea purslane, *Halimione portulacoides.*
Silver goosegrass, *Galium aparine.*

Shingle

Common seakale, *Crambe maritima.*
Sea pea, *Lathyrus japonicus*, local S. and E. England.
Yellow horned poppy, *Glaucium flavium.*
Shrubby sea blight, *Sueda fruticosa*, Wash to Chesil Bank.
Bittersweet, *Solanum dulcamara.*
Sea holly, *Eryngium maritimum.*

2J. *Some plants associated with mountains*

Sub-montane region (up to 2,000 feet), three zones:

a. Woodland. Durmast oak wood (*Quercus petraea*), a fairly rich but limited ground flora:

Bluebell, *Endymion non-scripta.*
Bracken, *Pteris aquilinum.*
Wood sorrel, *Oxalis acetosella.*
Whortleberry, *Vaccinium myrtillus.*
Foxglove, *Digitalis purpurea.*
Honeysuckle, *Lonicera periclynum.*
Soft grass, *Holcus mollis.*
Wood sage, *Teucrium scorodonia.*
Bramble, *Rubus fruticosus.*
Mountain lady fern, *Athyrium alpestre*, local, Highlands.
Beech fern, *Thelypteris phegopteris.*
Brittle bladder fern, *Thelypteris robertiana.*
Limestone polypody, *Cystopteris montana.*
Hart's tongue, *Phyllitis scolopendrium.*
Hard shield fern, *Polystichium aculeatum*, rare, Scotland.

Pine-wood (*Pinus sylvestris*). Preserved in remnants of the Caledonian Forest, or cultivated in Forestry plantations. Poor in flowering plants, but rich in mosses and fungi (see Appendix 1). Typical grass is the Tussock Grass (*Deschampsia caespitosa*).

Common birch (*Betula pubescens*). A montane species occurring in natural woods, or as coloniser in felled areas. Associated with ling and bracken.

Subordinate trees are hazel (*Corylus avellana*) with oak, ash (*Fraxinus exelsior*) on limestone, juniper (*Juniperus communis*), and alder (*Alnus glutinosa*) with willows (*Salix*) along streams and lakesides.

b. Lower grassland (sheep country)

On the upper drier and richer slopes—bent grass (*Agrostis stolonifera*) and alpine meadow grass (*Poa alpina*).

On lower, wetter soil—mat grass (*Nardus stricta*).

On valley alluvial soils the cultivated grasses—rye (*Lolium*) and meadow grass (*Poa annua*).

c. Moorland (grouse moors and deer forests)

Drier places—heather moors with ling (*Calluna*) and berries; sedge moors with deer grass (*Scirpus caespitosa*) and wavy hair grass (*Deschampsia flexuosa*) on the poorer soils.

Wet places—a cotton-grass moor with cotton grass (*Eryophorum*), bent grass (*Agrostis*), purple moor grass (*Molinia*) and tufted hair grass (*Deschampsia caespitosa*). The boggy places, as on the lowland moors, include:

Bog myrtle, *Myrica gale.*
Bog asphodel, *Narthecium ossifragum.*
Moor rush, *Juncus squarrosus.*
Sundew, *Drosera* spp.
Grass of Parnassus, *Parnassia palustris.*
Tormentil, *Potentilla erecta.*
Bog moss, *Sphagnum* spp.

Montane (above 2,000 feet), four zones:

a. Summit Heath (leached acid areas, usually on hard base-deficient rock). A poor heath community containing ling (*Calluna*), berry plants, mat grass (*Nardus*), dwarf willow (*Salix herbacea*), moor rush (*Juncus squarrosus*), and various lichens.

b. Summit grassland (on richer but more insecure base rock subject to 'frost heaving'). A moss-heath community with hair grass (*Deschampsia*), sheep's fescue grass (*Festuca ovina*), and the following:

Tormentil, *Potentilla erecta.*
Mountain bedstraw, *Galium montanum.*
Thyme, *Thymus serpyllum.*
Eyebright, *Euphrasia officinalis.*
Alpine Lady's mantle, *Alchemilla alpina.*
Alpine meadow rue, *Thalictrum alpinum.*
Fir clubmoss, *Urostachys selago,* N. and W. also Ireland.
Stag's-horn clubmoss, *Lycopodium clavatum.*

c. Colonising zone (a succession of plant seres as the ground settles):

1. Bryophyte 'flushes' of mosses and lichens on bare rock and stones which fluctuates with the rains.
2. A herb and grass growth between stones protected from winds, including woolly hair moss (*Rhacometrium*), hair grass (*Deschampsia*), mountain sedge (*Carex montana*), moss campion (*Silene acaulis*) and reindeer moss, a lichen (*Cladonia rangiferina*).
3. A summit grassland containing sheep's fescue (*Festuca*), club-mosses, berry plants and ferns.
4. A summit heath, as in *a*, should the area turn acid due to leaching (for an analysis of different heath and moorland types, see page 88).

d. A mountain flush (an alpine community on well-drained places with a regular water supply and washed-in soil, found mainly on the richer calcareous rocks). The species include:

Mountain avens, *Dryas octopetala.*
Mountain bedstraw, *Galium boreale.*
Alpine cinquefoil, *Potentilla crantzi.*
Mountain saxifrage, *Saxifraga aizoides.*
Purple saxifrage, *S. oppositifolia.*
Mossy saxifrage, *S. hypnoides.*
White saxifrage, *S. stellaris.*
Highland saxifrage, *S. rivularis,* Cairngorms.
Spring sandwort, *Minuartia verna.*
Roseroot, *Sedum rosea.*
Moss campion, *Silene acaulis.*
Sea campion, *S. maritima.* Also on coastal cliffs.
Thrift, *Armeria maritima.* Also on coastal cliffs.
Mountain sorrel, *Oxygia digyna.*
Globe flower, *Trollius europaeus.*
Butterwort, *Pinguicula vulgaris.*
Mountain cranesbill, *Geranium sylva.*
Mountain melick, *Melica nutans.*
Mountain pansy, *Viola lutea.*
Azalea, *Louiseleura procumbens.*
Viviparous sheep's fescue, *Festuca vivipara.*
Alpine bistort, *Polygonum viviparum.*

2K. *Mountain animals*

BIRDLIFE. *High peaks and rocks* (for roosting and nesting): Golden eagle, raven, peregrine, chough, rock pigeon (mainly coastal).

Moorland and grassland: Ptarmigan, red grouse, curlew, golden plover, dotterel, snow bunting (nests on summits), wheatear, ring ousel and whinchat (summer), blackcock (moorland fringes).

Woodland: Buzzard, kite (C. Wales only), kestrel (also coastal cliffs), capercailzie ,goldcrest, firecrest, crested tit, redpoll, hooded crow, crossbill, long-eared owl, redstart (summer).

MAMMALS. *High ground in summer:* Red deer, mountain hare, wildcat, otter, making for valleys and coast in winter.

Woodland: Wildcat (in heavy undergrowth), roe deer (young plantations), pine marten, badger and fox (also in cairns), red squirrel, polecat (central mountain valleys in Wales), woodmouse, field vole (in thick grass).

Rocky coasts and islands: Grey seal, otter (almost marine on some Scottish islands). Visited by fox, rat and various scavenging birds.

REPTILES AND AMPHIBIANS: Adder (moorland and rocks), viviparous lizard (moorland and grassland), palmate newt (largely montane in pools and tarns), common frog (often at high altitude where it is a late breeder).

In mountain lakes: Otter, water vole, water shrew, seal and dolphin (with access to the sea), osprey (a welcome returning species), cormorant, greenshank, kingfisher, heron, black-headed gull (summer breeder), sedge warbler, divers, and various waders and duck (some as winter visitors, others to breed.) Fish—mainly pike and lake trout (*Salmo trutta*), also rare and localised Ice Age relicts such as the char and vendace.

2L. *Some plants associated with ponds, drowned earth diggings and river backwaters*

Marsh zone (including wet places in ditches, hollows and river banks:

Great water grass, *Glycera maxima.*
Flote grass, *Glycera fluitans.*
Marsh marigold, *Caltha palustris.*
Water forget-me-not, *Myosotis scirpioides.*
Water figwort, *Scrophularia nodosa*, mainly in south.
Water mint, *Mentha aquatica.*
Gipsywort, *Lycopus europaeus*, thinning northwards.
Marsh bedstraw, *Galium palustre.*
Amphibious bistort, *Polygonum amphibium.*
Common spiked rush, *Eleocharis palustris.*
Toad rush, *Juncus bufonius.*
Monkey flower, *Mimulus guttatus*, local.
Water avens, *Geum rivale*, absent in S.E.
Ragged robin, *Lychnis flos-cuculi*, common in fens.
Meadow sweet, *Filipendula ulmaria.*
Marsh St. John's-wort, *Hypericum elodes.*
Purple loosestrife, *Lythrium salicaria*, also in fens.
Yellow loosestrife, *Lysimachia salicaria*, also in fens.
Common comfrey, *Symphytum officinale.*
Water speedwell, *Veronica anagallis-aquatica.*
Brooklime, *Veronica beccabunga.*
Butter-burr, *Petasites hybridus.*
Hemp agrimony, *Eupatorium cannabinum.*
Marsh cinquefoil, *Potentilla palustris.*
Glaucus sedge, *Carex flacca*, common on chalk fens.
Jointed rush, *Juncus articulatus*, common on chalk fens.
Southern marsh orchid, *Dactylorchis praetermissa*, also dunes and fens in south.

Swamp zone

Reed, *Phragmites communis*, local, also in brackish water.
Bulrush, *Scirpus palustris.*
Pond sedge, *Carex acutiformis.*
Sweet flag, *Acorus calamus*, local, mainly south.
Greater spearwort, *Ranunculus lingua*, local.
Lesser spearwort, *Ranunculus flammula*, common in north and west.
Mare's-tail, *Hippurus vulgaris*, local.
Great water dock, *Rumex hydrolapathum.*
Bog-bean, buck-bean, *Menyanthes trifoliata*, local.
Arrowhead, *Sagittaria sagittifolia*, local.
Common water plantain, *Alisma plantago-aquatica.*
Flowering rush, *Butomus umbellatus*, local, mainly in south.
Branched bur-reed, *Sparganum erectum.*
Great reedmace, *Typha latifolia.*
Sedge, *Cladium mariscus*, local, especially E. Anglia.
Trifid bur marigold, *Bidens tripartita.*

Rooting aquatics

Water crowfoot, *Ranunculus aquatilis.*
River crowfoot, *Ranunculus fluitans*, local in fast streams.
Ivy-leaved crowfoot, *Ranunculus hederaceus*, usually on mud.
Water lily, *Nymphaea alba.*
Yellow water lily, Brandybottle, *Nuphar lutea.*
Water starwort, *Callitriche palustris*, often in puddles.
Fringed water lily, *Nymphoides peltatum*, local.
Water violet, *Hottonia palustris*, local, rare in north.
Broad pondweed, *Potamogeton natans.*
Perfoliate pondweed, *Potamogeton perfoliatus*, local.
Curly pondweed, *Potamogeton crispus.*
Opposite pondweed, *Groenlandia densa*, often in fast water.

Floating aquatics

Frog-bit, *Hydrochaeris morsus-ranae*, local.
Water soldier, *Stratiotes aloides*, mainly E. Anglia.
Greater duckweed, *Lemna polyrhiza*, local.
Common duckweed, *Lemna minor.*
Ivy duckweed, *Lemna trisulca.*
Fat duckweed, *Lemna gibba*, local, mainly in south.
Least duckweed, *Wolffia arrhiza*, very local, does not flower.

Submerged aquatics

Hornwort, *Ceratophyllum demersum*, local, flowers underwater.
Small bladderwort, *Utricularia minor*, local in acid pools, England.
Greater bladderwort, *Utricularia vulgaris*, widespread but uncommon.
Tape grass, *Vallisneria spiralis*, naturalised in canals, especially in Yorkshire and Lancashire.
Whorled water milfoil, *Myriophyllum verticellatum*, local.
Spiked water milfoil, *Myriophyllum spicatum*, local.
Stoneworts, *Chara* and *Nitella*, widespread, mainly in clear, often brackish water.
American waterweed, Canadian pondweed, *Elodea canadensis.*

Some waterside trees

Crack willow, *Salix fragilis*, less common in north.
White willow, *Salix alba.*
Osier, *Salix viminalis.*
Pussy willow, *Salix capraea.*
Alder, *Alnus glutinosa.*

2M. *Some plants of waste ground, former bombed sites, etc.*

Annual meadow grass, *Poa annua.*
Cock's-foot, *Dactylis glomerata.*
Soft brome, *Bromus mollis.*
Barren brome, *Bromus sterilis.*
Common couch, *Agropyron repens.*
Wall barley, *Hordeum murinum.*

False oat, *Arrhenatherum elatius.*
White campion, *Melandrium album.*
Common mouse-ear, *Cerastium holosteoides.*
Common chickweed, *Stellarea media.*
Henbane, *Hyoscyamus nigrum.*
Fat hen, *Chenopodium album.*
Black nightshade, *Solanum nigrum.*
Thorn apple, *Datura stramonium.*
Smooth sow thistle, *Sonchus olearaceus.*
Beaked hawksbeard, *Crepis versicaria.*
Smooth hawksbread, *Crepis capillaris.*
Dandelion, *Taraxacum officinale.*
Pineapple weed, *Martricaria martricarioides.*
Spear thistle, *Cirsium vulgare.*
Scentless mayweed, *Tripleniospermum maritimum.*
Creeping thistle, *Cirsium arvensis.*
Nipplewort, *Lapsana communis.*
Ribwort plantain, *Plantago lanceolata.*
Rat's-tail plantain, *Plantago major.*
Groundsel, *Senecio vulgaris.*
Coltsfoot, *Tussilago farfara.*
Teasel, *Dipsacus fullonum.*
Scentless mayweed, *Tripleurospermum maritimum.*
Red deadnettle, *Lamium purpureum.*
Common toadflax, *Linaria vulgaris.*
Common sorrel, *Rumex acetosa.*
Persicaria, Redleg, *Polygonum persicaria.*
Fool's parsley, *Althusa cynapium.*
Common bird's-foot, *Ornithopus perpusillus.*
White mustard, *Sinapis alba.*
Charlock, *Sinapis arvensis.*
Rosebay, Fireweed, *Chamaenerion angustifolium.*
Canadian fleabane, *Conyza canadensis.*
Black bindweed, *Polygonum convolvulus.*
Field bindweed, *Convolvulus arvensis.*
Great bindweed, *Calystegia sepium.*
Creeping buttercup, *Ranunculus repens.*
Burdock, *Arctium minus.*
Shepherd's purse, *Capsella bursa-pastoris.*
Common forget-me-not, *Myosotis arvensis.*
Oxford ragwort, *Senecio squalidus.*
Hedge mustard, *Sisymbrium officinalis.*
Dove's-foot cranesbill, *Geranium molle.*
Cut-leaved cranesbill, *Geranium dissectum.*
Scarlet pimpernel, *Anagallis arvensis.*
Corn spurrey, *Spergula arvensis.*
Elder, *Sambucus nigra.*

2N. *Some garden invaders treated as weeds*

White clover, *Trifolium repens.*
Red clover, *Trifolium pratense.*
Yarrow or milfoil, *Achillea millefolium.*
Daisy, *Bellis perennis.*
Ribwort plantain, *Plantago lanceolata.*
Ratstail plantain, *Plantago major.*
Hoary plantain, *Plantago media,* mainly on chalk.
Scarlet pimpernel, *Anagallis arvensis,* especially on chalk.
Common couch grass, *Agropyron repens.*
Common chickweed, *Stellaria media.*
Dandelion, *Taraxacum officinale.*
Black nightshade, *Solanum nigrum.*
Groundsel, *Senecio vulgaris.*
Nipplewort, *Lapsana communis.*
Smooth hawksbeard, *Crepis capillaris.*
Common sorrel, *Rumex acetosa.*
Common cat's-ear, *Hypochoeris radicata.*
Persicaria or redleg, *Polygonum persicaria.*
Rosebay, *Chamaenerion* (*Epilobium*) *angustifolium,* especially burnt ground.
Shepherd's purse, *Capsella bursa-pastoris.*
Great bindweed, *Calystegia sepium.*
Field bindweed, *Convolvulus arvensis.*
Creeping buttercup, *Ranunculus repens.*

2O. *Plants associated with hedgerows and road verges*

Soapwort, *Saponaria officinalis.*
Black bryony, *Tamus communis.*
Bittersweet, *Solanum dulcamara,* often on shingle.
Welted thistle, *Carduus crispus.*
Hedge bedstraw, *Galium mollugo.*
Lords and Ladies, Cuckoo-pint, *Arum maculatum.*
Hedge woundwort, *Stachys sylvatica.*
White deadnettle, *Lamium album.*
Common mullein, *Verbascum thapsus.*
Rough chervil, *Chaerophyllum temulentum.*
Cow parsley, Wild Chervil, *Anthriscus sylvestris.*
Hedge parsley, *Torilis japonica.*
Hemlock, *Conium maculatum.*
Hogweed, *Heracleum sphondylium.*
Garlic mustard, Jack-by-the-hedge, *Alliaria petiolata.*
Tufted vetch, *Vicia cracca.*
Meadow pea, *Lathyrus pratensis.*
Bramble, *Rubus fruticosus.*
Herb robert, *Geranium robertianum.*
Creeping cinquefoil, *Potentilla reptans.*
Herb bennet, *Geum urbanum.*
Agrimony, *Agrimonia eupatoria.*
Field rose, *Rosa arvensis.*
Dog rose, *Rosa canina,* rare in Scotland.
Creeping buttercup, *Ranunculus repens.*
Greater celandine, *Chelidonium majus.*
Fumitory, *Fumaria officinalis.*
Sweet violet, *Viola odorata.*

Mugwort, *Artemisia vulgaris.*
Goosegrass, *Galium aparine.*
Toothwort, *Lathraea squamaria.*
Stinging nettle, *Urtica dioica.*
Elm, *Ulmus procera.*
Maple, *Acer campestre.*
Hawthorn, *Crataegus monogyna.*
Blackthorn, *Prunus spinosa.*
Hazel, *Corylus avellana.*
Wall rue, *Asplenium ruta-muraria*, walls and rocks.
Hart's tongue, *Phyllitis scolopendrium*, walls and rocks.
Common spleenwort, *Asplenium trichomanes*, walls and rocks.

NOTE. Hedgerows are artificial extensions to woodlands and contain the kinds of plants usually found in sheltered places between bushes and trees. There is great variation in species due to the type of soil (i.e. acid, alkaline or neutral), the degree of moisture (i.e. a dry or wet soil, the presence or absence of a ditch, etc.), and the influence of man (i.e. sprays, cutting, neglect, coppicing, etc.).

2P. *Plants associated with grassland and grassy places*

Drier grassland

White bent, *Agrostis stolonifera.*
Timothy grass, *Phleum pratense.*
Meadow foxtail, *Alopecurus pratensis.*
Perennial rye grass, *Lolium perenne.*
Smooth meadow grass, *Poa pratensis.*
Red clover, *Trifolium pratense.*
White clover, *Trifolium repens.*
Common vetch, *Vicia sativa.*
Yellow trefoil, *Trifolium dubium.*
Hop trefoil, *Trifolium campestre.*
Bird's-foot trefoil, *Lotus uliginosus.*
Corn poppy, *Papaver rhaeas*, more common in south.
Long-horned poppy, *Papaver dubium*, more common in north.
Meadow cranesbill, *Geranium pratense*, local.
Cut-leaved cranesbill, *Geranium dissectum.*
Bladder campion, *Silene vulgaris.*
Daisy, *Bellis perennis.*
Greater hawkbit, *Leontodon hispidus.*
Autumn hawkbit, *Leontodon autumnalis.*
Goat's-beard, Jack-go-to-bed-at-noon, *Tragopogon pratensis.*
Mouse-ear hawkweed, *Hieracium pilosella.*
Ox-eye daisy, *Chrysanthemum leucanthemum.*
Tansy, *Chrysanthemum vulgare.*
Greater knapweed, *Centaurea scabiosa.*
Black knapweed, Hardhead, *Centaurae nigra.*
Common ragwort, *Senecio jacobaea*, also dunes.
Yarrow, Milfoil, *Achillea millefolium.*
Common cat's-ear, *Hypochoeris radicata.*
Field scabious, *Knautia arvensis.*
Harebell (Bluebell in Scotland), *Campanula rotundifolia*, largely on chalk.
Hoary plantain, *Plantago media.*
Common broomrape, *Orobanche minor*, on clover.
Selfheal, *Prunella vulgaris.*
Yellow rattle, *Rhinanthus crista-galli.*
Bulbous buttercup, *Ranunculus bulbosus.*
Good Friday grass, *Luzula campestris.*
Moonwort, *Botrychium lunaria*, local, also among rocks and in dunes.
Adder's tongue, *Ophioglossum vulgatum*, local, often overlooked.

Damper grassland

Hairy sedge, *Carex hirta.*
Meadow fescue, *Festuca pratensis.*
Rough meadow grass, *Poa trivialis.*
Meadow buttercup, *Ranunculus acris.*
Lady's smock, Cuckoo flower, *Cardamine pratensis.*
Silverweed, *Potentilla anserina.*
Fritillary, *Fritillaria meleagris*, Thames valley.
Meadow saffron, *Colchicum autumnale*, Cotswolds, Wiltshire and E. England.
Soft rush, *Juncus effusus.*
Common fleabane, *Pulicaria dysenterica.*
Common skullcap, *Scutellaria galericulata.*
Devil's-bit scabious, *Succisa oratensis.*
Marsh thistle, *Circium palustris.*
Marsh St. John's-wort, *Hypericum elodes.*
Common spotted orchid, *Dactylorchis fuchsii.*
Southern marsh orchid, *Dactylorchis praetermissa*, south.
Northern marsh orchid, *Dactylorchis purpurella*, north.

2Q. *Animal invaders of buildings*

The following is a list of animals likely to enter buildings for shelter, for breeding purposes, or in search of food:

Mammals

House mouse, *Mus musculus.*
Common or brown rat, *Rattus norvegicus.*
Ship or black rat, *Rattus rattus.* Now a rarity in Britain.
Edible dormouse, *Glis glis.* Local, in the Chilterns.
Bats, mainly the Pipistrelle and Noctule.

Birds

House sparrow.
Feral rock dove or London pigeon.
Starling.
Jackdaw.
House martin.
Swallow. Mainly barns, sheds and outhouses.
Swift. Tall buildings.
Black redstart. Ruins.

Amphibians

Common toad, *Bufo bufo*. Enters basements and cellars to hibernate.

Invertebrates

House fly, *Musca domestica*.
Lesser house fly, *Fannia canicularis*.
Autumn fly, *Musca autumnalis*.
Bluebottle, *Calliphora erythrocephala*.
Silver fish, *Lepisma saccharina*.
Fire brat, *Thermobia domestica*.
Furniture beetle, *Anobium punctatum*.
Death-watch beetle, *Xestobium rufovillosum*.
Powder post beetle, *Lyctus* spp.
Fruit fly, *Drosophila melanogaster*.
Cluster fly, *Pollenia rudis*.
Stable fly, *Stomoxys calcitrans*.
Common cockroach, *Blatta orientalis*.
German cockroach, *Blattella germanica*.
American cockroach, *Periplaneta americana*. Mainly in bakeries and hot-houses, reptile houses, etc.
Australian cockroach, *P. australasiaea*. Mainly in bakeries and hot-houses, reptile houses, etc.
Clothes moth, 3 species, *Tineola bisselliella*, *Tinea pellionella* and *Trichophaga tapetzella*.
House moth, 2 species, *Borkhausenia pseudospretella* and *Endrosis lactella*.
House ants—Pharaoh's ant, *Monomorium pharaosis* and Argentine ant, *Iridomyrmex humilis* (in bakeries).
Garden ant—Small black ant, *Lasius niger*.
House cricket, *Gryllulus domesticus*.
Common earwig, *Forficula auricularia*.
House spider, *Tegenaria domestica*.
Booklice—various species of Psocids.

The following beetles are associated with stored foods and other commodities in buildings:
Curculionidae (Weevils) stored food.
Nitidulidae, Sap feeding beetles, stored food.
Cleridae. Cured meats, ham, bacon and cheese.
Dermestidae. Furs and woollens.
Anobiidae. Furniture and tobacco.
Tenebrionidae. Mealworms, etc., stored food, meal, flour, grain. *Tenebrio molitor*—the yellow mealworm. *Tribolium castaneaum*—the red-rust flour beetle.

The following parasites are associated with humans:
Bed bug, *Cimex lectularius*.
Flea, *Pulex irritans*.
Head louse, *Pediculus humanus*.
Pubic louse, *Phthiris pubis*.

Lighted rooms with open doors or windows will attract a large selection of night-flying insects, especially moths, during summer months. Also, many garden insects are introduced with cut flowers, or on clothing.

APPENDIX 3: A famous quartet

The Oxford ragwort (*Senecio squalidus*). The story of this successful invader of built-up areas is typical of the many plants which were recorded during the London bombed sites survey. Its original home is believed to be on the larval soils of the central Mediterranean area, and it corresponds closely to the species found on Mount Etna. It was first cultivated here during the seventeenth century at the Oxford Botanic Gardens. By 1800 it had escaped on to old walls in the town, and in 1879 was discovered growing on the cinders of the railway tracks. Each passing train carried successive generations of parachute fruits further along the lines, so that it reached places as far away as Cornwall, the Midlands and London. At the outbreak of war in 1939 it was well inside the capital's suburbs, and beginning to grow on the air-raid shelters and other well-drained spots. By 1942, after the Fire Blitz, it entered the bombed sites. Soon afterwards it was found to have crossed with another invader, the Viscid groundsel (*S. viscosus*), a local species found in coastal areas of Britain, and probably brought to London with sand used for the defences. This hybrid, called *Senecio* × *londiniensis* was discovered in 1943 by the London botanist J. E. Lousley.

Today it appears to be extinct in London, but the Oxford ragwort still flourishes on poor, dry soil in many places, especially along railway tracks. Another invader, the rosebay *Chamaenerion* (= *Epilobium*) *angustifolium*, a willow-herb attracted to burnt ground and old walls, spreads by means of countless parachute seeds liberated from elongated pods. One estimate is 80,000 seeds from a single plant. The creeping underground shoots enable it to spread over a considerable area. It became known as the Bomb-weed to Londoners, and is called the Fire-weed in Canada, where it flourishes after a prairie fire.

The Canadian fleabane *Conyza* (= *Erigeron*) *canadensis* is a native of North America which arrived in Europe during the seventeenth century, and was seen in London after the Great Fire in 1690 by the great John Ray. It is now a common sight on waste ground in S.E. England.

The London rocket *Sisymbrium irio*, a crucifer with small yellow flowers, also appeared after the 1666 blaze in London, and was anticipated by botanists after the 1940 Blitz. However, it proved to be a rarity, and many of the records actually refer to closely related crucifers. It is not easy to identify, and still remains one of Britain's rarest plants.

APPENDIX 4: A Devon lane

During a week's stay at a village near Totnes in South Devon, in the spring of 1941, the following plants and animals were recorded for a 50-yard stretch of lane. It was flanked in part by a water ditch, a thick hedgerow, and had a grass verge and steep bank. At one end it emerged from a small wood and at the other entered the village. Invertebrates were not included in the list, unless seen in the open.

In the wood: Oak, wild crab, hazel, guelder rose—wood anemone, bluebell, wild arum, ivy, wood sorrel, wood dog violet, early purple orchid—woodmouse, badger (occupied sett), fox (heard barking), common dormouse (picked up dead)—adder—tawny owl, great spotted woodpecker, nuthatch, sparrow hawk, woodcock.

Hedgerow: Oak, hazel, hawthorn, field maple, black bryony, bittersweet—wild arum, foxglove—bank vole (seen), rabbit, dormouse (nest)—chaffinch, blackbird, hedge sparrow—honey bee, queen hornet—peacock, brimstone, small tortoiseshell.

Grass bank, verge and ditch: Creeping buttercup, nettle, cow parsley, primrose, thistle, bramble, honeysuckle, lesser celandine, marsh marigold, water avens, yellow flag, brooklime, water starwort—common frog, grass-snake, slow-worm.

Sunken lane: seen on early morning and evening walks—carrion crow, jay, magpie, common rat, stoat, hedgehog, fox, partridge.

Fields: part arable and part meadow grass—Lapwing, corn bunting, common hare, partridge—milfoil, daisy, common ragwort, red and white clover.

Village: House martin (nests under cottage eaves), swallow (nests on barn rafters), song thrush, blackbird, barn owl (seen flying), pipistrelle bats. On a recent visit to the same village in 1966, that is, twenty-five years later, the author was saddened to discover the changes in this and many other Devon lanes. The ditch had been filled in and the hedges severely trimmed to allow room for a rising tourist traffic. The verge showed evidence of spray treatment (most of the primroses were missing). The village had extended along the lane, and most of the neighbouring fields were now under the plough. Only the wood seemed to be undisturbed. Fox, badger, dormouse and tawny owl were still in evidence. The wood had been neglected for years and was up for sale.

APPENDIX 5: Road casualties

The following list of road casualties among animal 'jay-walkers' was submitted by a student as a piece of individual study which he undertook during a London University extra-mural course on the Wildlife of Epping Forest run by the author during 1956-7. This student was in the habit of cycling to work at 6 a.m. across a portion of open Forest land flanked by woodland, then returning home for breakfast at 9 a.m.

The road casualties seen on the outward journey, mostly victims of vehicles, run down in darkness, were recorded over a period of one year, and included the following:

Fallow deer. 'One specimen in a ditch to which it had crawled after a car collision at night, and reported by the driver at the police station.' December.

Grey squirrel. 'Six animals all apparently run over in daylight during different months.'

Common rat. 'One animal run over.' March.

Badger. 'Reported dead by the roadside but not seen.'

Hedgehog. 'Ten run over, mainly in late summer prior to hibernation.'

Woodmouse. 'Three animals flattened by car wheels,' June, October and January.

Common shrew. 'A fair number throughout year. Most specimens appeared uninjured.'

Common hare. 'Two dead, a third found injured.'

Rabbit. 'Four dead or dying from myxomatosis.'

Grass snake. 'Three squashed in road.' Summer.

Adder. 'One dead by roadside, not squashed but with back broken from a blow.'

Common toad. 'Numerous flattened bodies mostly along one stretch of roadway, in March-April.'

Birds. 'Various casualties, including wood-pigeon, chaffinch, blackbird, tree pipit and tawny owl.'

An analysis of this report (the author's comments) may suggest some of the reasons for these deaths:

The fallow deer, one of the few remaining in the Forest, is but one of the road casualties among our larger mammals resulting from road traffic. This is now a major problem on many of our open moors and commons, i.e. unfenced, and applies equally to the wild ponies, sheep and cattle. Deer and ponies in particular are sudden and unpredictable in their movements, and especial care should be taken when motoring across their territory. We owe much to these animals for the freedom of our open spaces.

Badger and hedgehog are both inoffensive night prowlers, and inclined to blunder about in carefree fashion, relying much on their sense of smell rather than on sight and hearing, and seem indifferent to road dangers.

Woodmice normally keep to overhead cover, but may wander across roads in search of new territory.

The rat will forage along a highway at night, in search of carrion. This is a case of 'the biter bitten'.

There has been an increase of hares in the Forest after

the thinning out of rabbits from disease. A hare tends to run along a road, rather than across, whereas a rabbit bolts for cover.

Grass snakes are frequent summer casualties along roads bordered by suitable habitats, and will sometimes come out in early morning to bask on the warm road surface. Grass snakes are known to occur in this area.

The adder died from a blow with a stick, the usual fate meted out to this shy and inoffensive reptile. Adders never bite until molested. The toads were almost certainly on migration, crossing the road to reach their breeding pond. This occurs mainly at night.

The various birds were killed by a blow and not run over, due to misjudging the speed of the vehicles. The owl is interesting, since it was run over, near the corpse of the hare. Could it have been feeding on the carcase? It happened in winter-time when food might have been scarce.

On the outward journey the student noticed a number of scavengers feeding on the night's casualties. These included magpie, carrion crow, jackdaw, rat, grey squirrel, fox and badger (once). On the homeward journey these diners had finished their work, and the road was more or less 'cleaned up and ready for the next morning's breakfast.'

APPENDIX 6: A list of wildlife in a London garden sanctuary

The following animals and plants have been established in the author's woodland garden, in existence since 1954:

Dominant trees—old apple and pear trees with a canopy layer between 20 and 40 feet above ground.

Shrub layer—young oak, hornbeam, hazel, sycamore, guelder rose, ash, maintained at 5-10 feet.

Ramblers and climbers—honeysuckle, bittersweet, clematis, dog rose and bramble.

Ground flora—a herb layer between 6 inches and 3 feet—bluebell, primrose, oxlip, wood anemone, wood sorrel, lesser celandine, wood sanicle, ivy, ground ivy, bugle, wood garlic, woodruff, wood violet, hart's tongue fern, male fern, lady fern, hound's tongue, dog's mercury, enchanter's nightshade, wood geranium, wild gladiolus, ground elder, butcher's broom, herb robert, early purple orchid, twayblade, sweet vernal grass, wood melick, hairy woodrush, thistle, nettle, willow herb, Oxford ragwort.

Bryophyte layer—below 6 inches—liverworts and mosses: *Pellia*, *Marchantia*, *Funaria*, growing mainly on stones and logs; fungi—little success with introductions, but the following have appeared—*Coprinus*, *Pholiota squarrosa*, an earthstar (*Geaster bryantii*), *Marasmius* (see also log project 10).

Invertebrates—see list for log project (10), also the following insects: Lepidoptera—large and small cabbage white*, peacock, small tortoiseshell*, comma, red admiral*, clouded yellow, privet* and elephant hawk moths, yellow underwing, puss moth, garden tiger moth*, angle shades, silver Y, magpie moth*. Coleoptera—Nut weevil on hazel (*Curculio nucum*)* cockchafer (*Melontha vulgaris*)*, ground beetle (*Carabus monilis*), click beetle (*Corymbites lineatus*) also its wireworm larva, seven-spot ladybird (*Coccinella septempunctata*) Diptera—Cranefly (*Tipula oleracea*), drone fly (*Eustatis tenex*)* also larvae in pond, i.e. rat-tailed maggot, gnat (*Culex pipiens*)*, pear midge (*Contarinia pyricora*)*. Hymenoptera—common wasp (*Vespa vulgaris*)*, hive bee (*Apis mellifica*), bumble bee (*Bombus terrestris*)*, leaf-cutting bee (*Megachile centuncularis*)*, black ant (*Acanthomyops nigra*)*.

Birds—mistle thrush*, song thrush*, blackbird*, robin*, hedgesparrow, chaffinch*, greenfinch*, goldfinch, woodpigeon*, collared dove (since 1966)*, magpie, jay, tawny owl, carrion crow, black-headed gull, house sparrow*, house martin*, tree sparrow, starling*, willow warbler, chiff-chaff, jackdaw, blue-tit*, great tit*, spotted flycatcher, wren.

Mammals—badger†, fox†, hedgehog*, woodmouse*, dormouse†, rabbit†, weasel†, common shrew, bank vole, grey and red squirrels†.

Reptiles—adder†, grass snake†*, smooth snake†, viviparous lizard, sand lizard†, slowworm*, green lizard†, wall lizard, European and Spanish terrapins, Spur-thighed, Hermann's and Horsfeld's tortoises.

Amphibians—common frog*, common toad*, edible frog, European tree-frog, midwife toad, European salamander*, crested*, smooth* and palmate newts*.

In addition to the above a good selection of water life, both plant and animals, also marsh plants live in and around the small pond.

* Has bred in the sanctuary.
† Kept in cages or tanks.

APPENDIX 7: Biology of sewage farms and waterworks

Sewage farms. The fauna of a sewage farm is composed mainly of those animals normally associated with mud at the pondside or river backwater. The concentrations of bacteria and fungi in the percolating filter beds supply food to the hosts of protozoans, such as the stalked ciliates *Vorticella* and *Epistylis*, which in turn are devoured by worms and fly larvae, of which the following are typical:

Oligochaete worms—*Lumbricus rubellus* and *Pachydrilus* (= *Lumbricillus*) *lineatus*: two species of fly—*Psychoda alternata* (moth-fly) and *Anisopus fenestralis* (window-fly); springtails (*Collembola*); and midges—*Spaniotoma* and *Metriocnemus* spp.

Some 14 species make up the sewage farm population, but even this restricted number may be limited by the strength of the sewage and the routine of operation. Periodic dousing alternating with exposure can kill off the filter bed fauna. Even so, up to 40 million flies per acre per day may hatch in summer weather. Whereas the midges normally keep to the farm, the feebler flying *Psychoda* and *Anisopus* get carried away by wind, to settle in dark places. This may happen to neighbouring dwellings, to the annoyance of the occupants. There is little risk of disease from this, and the truly domestic species, such as house- and blow-fly which settle on human food, do not breed in the filters.

Waterworks. The pipe fauna. In 1886 Professor K. Kraepelin made an investigation of the animals which were living in the pipes of Hamburg's water supply, partly in the hope of finding examples of blind and pigmentless crustaceans which occur in dark, underground waters, or to discover any possible modifications which may have evolved among any species living in this unique habitat. The whole area was trapped by fitting wire-gauze frames to the street mains, and catching the animals as they poured out with each gush of water. Although no blind, pale or in any way modified specimens turned up, the result was astonishing. Below the Hamburg streets there lived a huge population representing some 50 different genera, of which the following are some examples:

Protozoa—the fixed *Vorticella* and free-swimming *Paramaecium* and *Stentor;* Sponges—*Spongilla lacustris* and *Ephydatia fluviatilis*, both freshwater; Hydrozoa—*Hydra fusca* (but not *Hydra viridis* because of the darkness) and *Cordylophora lacustris;* Annelids—*Limnodrilus, Chaetogaster* and *Naias;* Leeches—*Herpobdella* and *Glossiphonia;* Nematodes—Anguillula and Echinorhynchus (in eels); Tremadodes—*Planaria;* Molluscs—*Unio, Physa, Ancylus, Acroloxus, Bythinia, Limnaea, Planorbis, Pisidium, Dreissensia.* Of these *Dreissensia polymorpha*, the Zebra mussel, was extremely common, and practically choked some of the pipes; Polyzoa—*Cristatella, Paludicella, Plumatella* these 'moss-animals', as the workmen called them, were the most abundant of the pipe fauna. They bud off by means of statoblasts which can be carried on birds' feet; Crustaceans—mainly *Gammarus pulex* and *Asellus aquaticus*. Due to the strong current there were very few small free-swimming forms such as water-fleas; Insects—only a very few aquatic larvae; Fishes—enormous number of eels, up to 1 foot long. This was perhaps the most surprising discovery.

There was no sign of specialisation in these pipe animals, because of a short history, since pipes had only been started some 30 years before. Also, the population was continually added to from the river Elbe. In this special and secure habitat, a dark and narrow tunnel subject to sudden gushes of a one-way water flow, the successful inmates were those who could cling tightly to the pipe walls, or were small enough to cling onto or hide between their larger neighbours. Also, they were adapted to feeding by filter methods or by browsing on the microscopic algae (i.e. diatoms) and organic debris.

This 'animal Eldorado', as Professor Kraepelin called it, came to a sudden halt with the introduction of sand filtration in 1894. Tiny organisms were eliminated, and the pipe fauna died of starvation.

Plants in waterworks. Because of the cleaning process in sand filters, and the concrete sides and deep waters of storage reservoirs, few flowering plants manage to exist. The flora consists mainly of bacteria and algae such as diatoms, green and blue-green algae. These are very useful as oxygenators, and in making the film on the filter-bed floor. The filamentous greens such as *Spirogyra, Tribonema* and *Cladophora*, the so-called 'carpet' weeds, and in some cases the diatoms, make up this valuable biological filter. On the other hand these plants may become a nuisance after sudden flushes of growth and subsequent death, by causing unpleasant smells, bad taste and discolouration. The blue-greens such as *Rivularia, Oscillatoria* and *Anabaena* can turn a reservoir a dirty brown, due to a concentration of sulphur and phosphorus. The watermen used to say that the water was going 'through its period of sickness' (see p. 139).

Iron Bacteria. Another problem is the furring of the pipes due to the presence of a common bacterium *Crenothrix*. This forms a brownish pile on stones, water plants, static water animals, and so on, in stagnant pools, lakes, wells and reservoirs. The pile, about ¼ inch thick, is composed of little tufts of filaments, hence the name which means 'thread of springs', in German *Brunnenfaden*. It is peculiar to subsoil waters, and often latent in springs and wells. Sometimes it can turn water blood-red, and there are frequent references to this in the Bible. Ferrous or 'iron' salts in solution in soil water are oxydised by the bacteria into insoluble ferric salts which are deposited in the filament sheath. The redness tends to appear during hot and dry spells.

Today the use of chemicals helps to allay these nuisances. Permanganate of potash helps to eliminate the bad taste and copper sulphate acts as an algicide. Other chemicals act as coagulants, binding the algae into clumps which then sink and can be more easily removed. Ferrous salts are removed by aeration to encourage oxidation, so that ferric salts can be filtered off. All this, coupled with the efficient sand filtration and the delicate film of algae, was initiated by James Simpson, the consultant engineer to the Chelsea Water Company, who began experiments in sand filtration in 1826. It is to him that we owe the reliable and steady flow of pure water from our taps today.

APPENDIX 8: The Country Code

The Country Code booklet, obtainable from Her Majesty's Stationery Office or through any bookseller, price 6d., urges all who enter the countryside to observe the following:

Guard against all risk of fire.
Fasten all gates.
Keep dogs under proper control.
Keep to the paths across farmland.
Avoid damaging fences, hedges and walls.
Leave no litter.
Safeguard water supplies.
Protect wild life, wild plants and trees.
Go carefully on country roads.
Respect the life of the countryside.

The *Code* goes on to say: 'The countryside is largely what men have made of it through many centuries of toil and care. The work of conservation and improvement demands and creates in the countryman a spirit and a way of life which has great value for the nation. All who understand this spirit will preserve the beauty and order of the countryside, which we of the present hold in trust for the future generations.'

APPENDIX 9: A list of fungi found in Epping Forest

As part of a week-end nature course arranged by the author at Debden House, the Further Education Centre of the former Borough of East Ham Education Committee, a Fungus Foray was held in the Forest in October 1963. This turned out to be a good season for fruiting fungi during a mild autumn followed by an indifferent and wet summer. With the assistance of a local mycologist, Mrs Stuart Boardman, 14 students during a 2-hour foray collected and identified 76 species:

Class BASIDIOMYCETES. Fungi with spores (usually in 4's) growing from a *basidium*.
Order Hymenomycetes. Basidia produced externally on gills, spines, etc.
Family Agaricaceae. Fungi with gills, or toadstool shape, which includes the mushrooms.

Amanita. Toadstools with white gills, spots on the cap, a ring and volva present, and including the most poisonous species.
- *A. phalloides* (Death cap). White all over, with a greenish tinge on cap. Not very common. Deadly poisonous.
- *A. citrina* (False death cap). Whitish to pale lemon-yellow. Poisonous but not deadly, smelling of new potato. Very common in 1963.
- *A. muscaria* (Fly agaric). Easily recognised by its red cap with white spots. Under birch. Poisonous.
- *A. rubescens* (Blusher). Brownish. Flesh turns pink when bruised. Edible.
- *A. pantherina* (Panther cap). Rather similar to Blusher but does not bruise. Suspicious. May be confused with *A. spissa*.

Amanitopsis. Toadstools with a volva but no ring.
- *A. fulva* (Tawny grisette). Reddish brown, tall and slender with fork-like marks around edge of cap. Edible.
- *A. vaginata* (Grisette). Similar but greyish and larger. Edible.

Armillaria. Tree toadstools with a membranous ring.
- *A. mellea* (Honey fungus or Forester's curse). A parasitic toadstool which attacks living trees. Yellow to brown cap, scaly when young, creamy gills, stalk white above ring, yellow below.
- *A. mucida* (Beech tuft). Pure white, slimy and translucent. On dead beech.

Lepiota. Parasol mushrooms growing in rough pasture, edible.
- *L. procera* (Parasol mushroom). Large toadstool up to 10 inches with a cap 8 inches across, covered with brownish scales. Found in a clearing.

Agaricus. True mushrooms. Whitish with pink to purple gills according to age. A ring but no volva.
- *A. campestris* (Field mushroom). Debden fields.

Laccaria. Smallish Forest toadstools with wide spaced gills joining the stalk.
- *L. laccata*. Reddish brown to bright red.
- *L. amethystina*. As above, deep violet, very common this season.

Collybia. Toadstools with a cartilaginous stalk and incurved edge to cap when young.

C. maculata (Coco-dust toadstool). Whitish cap with wavy shape and rusty brown patches.

C. radicata (Rooting shank). Long slender stem penetrating soil and tapering to a point. Yellowish brown and usually slimy.

Marasmius. Small toadstools, rough and leathery, often forming 'fairy rings', e.g. the well-known Champignon of lawns.

M. personatus (Wood woolly-foot). Deep brown cap, yellow-brown gills and stalk with a woolly coat.

M. ramealis and *M. erythropus* also found.

Mycena. Small toadstools with conical cap, slender stem, often in lawns and on tree-stumps.

M. galopus. On tree stump.

Pleurotus. Toadstools on tree-stumps, dead trunks and branches, the stalk to one side and the gills running into it.

P. ostreatus (Oyster mushroom). Bluish grey, smooth cap with little or no stalk. Gills white, edible.

Hygrophorus. Average sized toadstools with widespread waxy gills.

H. lactus. Only species found.

Cantharellus. A cup-shaped cap with gills running down stalk.

C. rubaeformis (False chanterelle). Its famous cousin the true Chanterelle (*C. cibarius*) is a delicacy found mostly in pine woods, and is uncommon in Epping Forest where conifers are almost absent.

Lactarius. The 'milky toadstools'. When broken, the flesh oozes a milky fluid. A large genus.

L. turpis. Olive brown, funnel shaped.

L. rufus. Brownish red with a hot, peppery taste.

L. subdulcis. Cinnamon colour. Milk at first tasteless, then bitter.

L. quietus. Dull, reddish brown, a mild taste.

L. camphoratus. Brick-red, strongly smelling.

Russula. Another large group, with no ring or volva. Flesh is firm and breaks with an audible snap. The white gills 'rustle' when stroked. Some suspicious, none deadly.

R. emetica (Sickener). Bright scarlet, fading to pink after rain. Said to cause vomiting.

R. fragilis. Similar but usually smaller, with a toothed edge to cap.

R. ochroceuca. Bright ochre yellow.

R. nigritans. Large, sooty black. When old may have a small toadstool growing on the cap (*Nyctalis*).

R. fellea. Straw coloured, bitter taste.

R. cyanoxantha. Yellowish cap with purple rim. Edible.

R. rosacea. Rose coloured. Suspicious.

R. mairei.

Pholiota. Toadstools with brown spores.

P. squarrosa. In clusters at base of stumps and dead trees. Yellowish with scales on cap. Stem smooth above ring, scaly below.

Cortinarius. A large genus, some with slimy caps and difficult to identify. All grow under trees.

C. elatior. Brownish, bell-shaped cap with furrows. Gills brown. Very common.

C. sanguineus. Blood red, including gills.

Stropharia. Gills joining stem, and ring membranous.

S. aeruginosa (Verdigris toadstool). Cap slimy, a bright bluish green with white rim. Suspicious.

Hypholoma. Gills wavy or sinuate, stem fibrous, no ring. Mainly on wood.

H. fasciculare (Sulphur-tuft). Cap lemon yellow with darker centre. In clusters on tree stumps. Common.

Psarhyrella spadicata.

Family Boletaceae. Fungi shaped like toadstools whose spores grow inside tubes under the cap.

Boletus. A large group of stoutly built toadstools with thick stems. Cap underneath spongy with large numbers of tubes.

B. edulis (Penny bun, *Cèpe* or *Steinpilz*). The famous edible toadstool resembling a bun, toast-brown on cap with darker centre and whitish rim. Not common in Forest.

B. scaber (Rough stemmed Boletus). Cap greyish, stem rough covered with warts. Common under birch.

B. versipellis. As above but with orange-tawny cap. Under birch.

B. badis. Brownish, turning blue-green when bruised. Usually under pine.

B. chrysenteron.

Family Polyporaceae. Bracket fungi. These grow out of the side of dead or dying trees and logs, flat or bracket-like in shape, and parallel to the ground. The flesh is leathery or corky, and the underside covered with small pores.

Polyporus. Woody or corky brackets with pores more or less rounded often peculiar to a specific tree.

P. betulinus (Birch bracket or Razor-strop fungus). Whitish or pale yellow, on dead birch, very common. Once used for stropping razors, as tinder, and for lining insect cabinets.

P. frondosus. Brackets arising from a common stem, greyish or tan colour, on tree stumps.

P. lucidus. Blood-red to chestnut, with a greyish powdery film. Takes on a rich polish when rubbed. Specimen from Debden garden.

P. radicatus, *P. adiposus* and *P. stipticus* also found.

Fistulina. Cap fleshy and juicy, with pink spores.

F. hepatica (Beef-steak). Resembling a lump of raw steak, even dripping 'blood', mainly on oak. Edible when young.

Fomes. Similar to *Polyporus* but with pores arranged in layers. Very woody, once used as tinder.

F. ulmaria (Dryad's saddle). Common on base of old elms. May grow to 1 yard across. Colour whitish to buff. Specimen from Debden garden.

F. applanatus. Cinnamon to grey with white margin.

Polystictus. Thin, corky brackets growing in tiers in fallen trees, logs and stumps.

P. versicolor. Very common. Has a soft velvety surface attractively marked in concentric rings of yellow, grey, brown and black.

P. radiatus. As above, but deep brown to dull orange.

Trametes. Similar to *Polyporus* but usually larger and thicker, and with large pores elongated radially.

T. gibbosa. Whitish to grey, on stumps and posts.

Daedelea. Like *Trametes* but thicker and firmer, with sinuous pores.

D. quercina (Oak bracket). Buff brown, uneven surface, pores elongated and wavy. Mainly oak stumps.

Family Hydnaceae. Toadstools which produce their spores on spines or projecting teeth under the cap. With or without a stalk.

Hydnum. Stalked kinds with spines under cap.

Hydnum repandum (Wood hedgehog). Pinkish or buff cap with short spines under cap running partly down stalk.

Family Theleporaceae. Sporophore of varying shapes, with or without stalk, cup or flesh shaped, or merely flattened. Spores develop on a smooth surface.

Thelepora terrestris. Flat, leathery and peel-like fungus, growing on ground usually between moss. Pinkish white to black with age. Often encrusts fallen branches.

Stereum hirsutum. As above, but yellowish to grey, with a 'hairy' upper surface. Resembles a cluster of small brackets growing on dead wood or on ground.

Family Clavariaceae. Called Fairy Clubs, Coral or Stag's-horn fungi. Small sporophores club-shaped, often branched, erect, from which spores grow. Mainly on ground.

Clavaria. A fleshy erect fruit body, simple, branched or clubbed.

C. cristata. Whitish, branched tufts growing on ground.

C. cinerea. Similar but greyish, very common.

Sparassis. Fleshy, erect, much branched in plate-like layers. On ground, often on roots.

S. crispa. Whitish, many branched, resembling a tiny cauliflower fruit. Common in some years, usually on bare ground along paths.

Calocera. Erect, jelly like, but hardens on drying.

C. cornea. Yellow-orange, awl-shaped branches, usually on buried wood.

Order Gasteromycetes. Basidiomycete fungi whose pores grow internally inside a *peridium*, and escape through a rupture in the wall.

Family Phalloidaceae (Stinkhorns). From an egg-shaped fruit body (peridium) a stalk or receptacle expands upwards to carry out a gelatinous, bell-shaped cap containing the spores buried in a slimy coat. Flies are attracted by the smell and assist in spore dispersal.

Phallus impudicus (Common stinkhorn). A tall whitish and robust fungus when expanded, up to about 9 inches. Spore surface of cap covered with dark olive-green slime which is removed by flies. A characteristic, unpleasant smell.

Mutinus caninus (Dog stinkhorn). Much smaller and more slender, usually in damp spots. Egg stage oval, receptacle pinkish, and spore surface orange-red.

Family Lycoperdaceae (Puff-balls). Rounded or oval fungi having a double wall which ruptures at apex and puffs out spores when disturbed, or with changes in atmosphere.

Lycoperdon perlatum (Common puff-ball). Snow-white then brown, at first covered with fragile spines which later wear off. In grass. Edible.

L. pyriforme. Rather similar but grows on wood. At first white covered in fine granules, turning brown.

L. giganteum (Giant puff-ball). The largest fungus in Britain, up to a foot or more across. Not found, but has occurred in the fields around Debden. Edible.

Family Sclerodermaceae (Earth-balls). Not to be confused with puff-balls, which they may resemble. There is a single wall which ruptures irregularly.

Scleroderma aurantum (Common earth-ball). Very common on ground, usually attached to roots. Rounded or bun-shaped, ochre-yellow.

Class ASCOMYCETES. A second major division of the Fungi. Spores are produced internally, in special chambers called *asci* (each containing 8 spores). Many species grow underground, such as truffles.

Order Discomycetes. Cup-shaped fruiting body, at first closed, then opens. May have a stalk. Includes edible forms such as the Morel.

Helvella crispa. Saddle-shaped wavy cap (spongy in Morel) on top of a stalk. Cap white, tinted yellow, stalk thick, white and grooved. Found in grass, normally not eaten.

Peziza. Called the Cup-fungi or Pixie Cups. The sporophore starts as a hollow thin, rubbery ball, opening into a cup, usually on bare soil. Some species are brightly coloured.

P. aurantia (Orange-peel fungus). Bright orange cup resembling orange peel.

P. sanosa and *P. badia*. Both dark coloured, also found on bare, burnt ground.

The third major group of Fungi, the PHYCOMYCETES, is largely made up of microscopic species, commonly known as moulds (saprophytes) and mildews (parasites), and was not included in this foray. All the above specimens were discovered in a beechwood area within a mile radius of Debden House.

APPENDIX 10: Plaster casts

The following equipment is needed for taking plaster casts of animal footprints:

(*a*) A good-quality plaster-of-Paris, as sold in hardware stores. Some makes harden faster than others. Dental plaster which is of fine quality, but more expensive, is not necessary. Plaster is usually sold in bags, and should be transferred to a tin with a tight fitting lid to keep it dry and to prevent spilling.

(*b*) Some strips of pliable cardboard, about 2 inches wide and 1 foot long. These can be rolled together for convenience. More durable strips can be made from plastic or tin sheeting.

(*c*) Paper clips, a pair of tweezers and a tin of vaseline.

(*d*) A bottle for containing clean water, and which has an attached stopper or screw top. It can so easily be mislaid.

(*e*) A mixing bowl, preferably unbreakable, and a spoon.

(*f*) Some sheets of newspaper and a cleaning cloth.

All this can be packed into a carrier such as a haversack or sachel. The author uses an old respirator case which has compartments.

Tracks of mammals and some ground and water birds can be found in soft ground or snow. Places to search are along the borders of ditches, ponds, rivers, etc., even around puddles and along muddy paths and rides. Marshland, estuaries and the seashore make good hunting ground. The procedure for casting is as follows:

1. Examine the track for clarity and freshness, and decide whether it is worth taking. So many tracks are spoilt by age, unevenness, or imperfection. A little 'touching up' might be allowed if the track is otherwise good and clear. Remove any leaves, twigs, etc., with the tweezers. Light material such as dust can be blown clear.

2. Shape a strip of cardboard into a circle or square, hold together with a paper clip, so as to cover the track area with a little space to spare. Smear vaseline on the inside to stop the hardened plaster from sticking, place the ring over the track, and gently press into the soil.

3. Pour some water into the mixing bowl and sprinkle on plaster until it reaches the surface. Stir with the spoon into a creamy mixture. The right amounts and thickness of mixture will come with practice.

4. With a steady continuous motion pour in the plaster with the aid of the spoon against one side of the cardboard to that it flows across the track. Plaster poured straight on to the track impression may trap air bubbles in the finer parts such as the claw impressions, and so spoil the finished result. Allow for about a half-inch thickness of plaster. A useful dodge where there are fine imprints such as claw marks, is to pour in a weak mixture, then sprinkle on some dry plaster to soak up the water, and top up with a normal mixture.

5. Gently tap the cardboard ring or 'puddle' the plaster with the spoon as this helps to drive out any air bubbles. Allow to set (about 10-20 minutes) and test the hardness by tapping with the finger nail.

6. Remove the cast. If it is firmly fixed to the soil, e.g. in clay, then carefully lever out with the aid of a pointed stick or penknife pushed underneath.

7. Remove any adhering lumps of soil, wrap in newspaper, and complete the cleaning later under a tap by using an old nail or tooth-brush

8. Wash and clean out the bowl immediately after each operation.

9. On the back of the cast scratch the name of the animal, the date and locality.

This operation will produce a negative cast of the track. To make a positive from this is not always easy if plaster is used, because of undercutting. Claws and hoof-tips tend to break off when the two casts are separated. The author uses a latex compound which is supplied by Macadam and Co, Rubber Merchants, of 5 Lloyds Avenue, London, E.C.3. The negative cast is placed on a board and a low wall of plaster is built around the outside. Into this is poured the latex liquid. This will slowly set into a rubber when exposed to air, and when a sufficient thickness has set the surplus can be poured back into tin. Such a positive cast, an exact copy of the original track, is flexible and will not break. It can be treated with oil colours to give it a very realistic imitation of the original.

Sometimes a number of tracks can be taken in one cast, so as to show the sequence of footprints, i.e. a trail. To strengthen a large piece of plaster drop in some pieces of perforated zinc or wire netting while the plaster is setting, in order to reinforce it.

Snow, unfortunately, is no good for plaster work, as this generates heat when mixed with water, so that the snow starts to melt before it is set. On the other hand snow is excellent for working out trails and methods of locomotion, where scale drawings and photographs can be taken to show the sequence of footprints. It is sometimes possible to follow a trail of a fox or deer for a number of miles, and to 'read' from this all sorts of stories and happenings.

Although plaster will faithfully reproduce the tracks of small animals such as mice, these are not often found on the ground, except in fine mud, dust or silt. To obtain such tracks, place a sheet of cardboard or a wooden board in a likely spot. On this place some suitable food as bait, and smear around this a ring of printer's ink. In reaching the food the animal should leave its footprints on the ink surface. In other places the ground can be prepared beforehand, by raking it over, or sprinkling some damp sand which is then smoothed over. A half mixture of modelling clay and silver sand mixed with water, then smeared onto a board, makes an excellent medium for small prints.

APPENDIX 11: Food niches occupied by vertebrate animals in Epping Forest observed between 1947 and 1955

Mammals

Fallow deer: grass, buds, acorns, crab apples, fungi, tree bark (holly). Feeds up to 4 feet.
Badger: ground feeder, largely omnivorous. (See Badger Project 1, p. 153).
Fox: mammals, birds, insects.
Grey squirrel: ground and tree feeder.Diurnal: acorns, nuts, buds, tree bark, fungi.
Woodmouse: night feeder on ground and in bushes, seeds, nuts, berries, insects. Keeps to hedgerows and woods.
Field vole: mainly grass. Night and day feeder in open glades.
Bank vole: seeds, nuts, berries in hedgerows and along banks and wood borders, nocturnal. Can climb.
Stoat: small mammals and birds, on and below ground.
Weasel: as above, but smaller prey. Seen more in hedgerows.
Rabbit: grass and herbs, in glades and fields. Much decimated by myxomatosis.
Hare: food as above—has increased, and often seen in wooded areas.
Hedgehog: nocturnal. Worms, insects.
Pipistrelle bat: nocturnal, aerial feeder at low level. Small insects.
Noctule bat: as above, at higher level. Larger insects, including moths and beetles.
Common shrew: feeds in leaf litter. Invertebrates.

Birds

Woodpeckers (note the different feeding 'levels')

Lesser spotted woodpecker: feeds in upper branches on insects, grubs.
Greater spotted woodpecker: feeds at lower levels on grubs, bark beetles, nuts and acorns.
Green woodpecker: feeds largely on ground, in glades and fields, especially on ants.

Leaf warblers (all three insect feeders but in slightly different niches)

Chiff-chaff: summer visitor. Insects mainly from trees.
Willow warbler: as above. Usually feeds closer to ground.
Wood warbler: as above, keeps more to light undergrowth.

Tits (another close sharing of food niches)

Blue tit: summer insects, winter seeds, in trees.
Great tit: as above but more among bushes than trees.
Coal tit: as above, tends to inhabit conifers and not very common in Forest.
Long-tailed tit: as above, tends to wander about more, in family parties.
Nuthatch: invertebrates in cracks in wood, also acorns and nuts.
Tree creeper: mainly invertebrates in cracks in bark.

Finches (similar close competition as for tits)

Chaffinch: seeds and insects from bushes and trees, more on ground during winter.
Goldfinch: as above but feeds on softer seeds such as thistle.
Greenfinch: as above, more numerous in bordering gardens near fruit.
Hawfinch: rare in the Forest.

Ground feeders (although these birds also feed above ground there is competition for food in the leaf litter of a woodland)

Woodpigeon: nuts, acorns, beechmast.
Jay: as above, also fruit, insects, birds' eggs. Will bury acorns.
Carrion crow: insects, carrion, eggs.
Pheasant: mainly escapes from a local reserve, seeds and insects.
Woodcock: mainly a winter visitor. Probes soft ground for worms, etc.

Birds-of-prey

Sparrow hawk: day hunter, small birds.
Kestrel: small mammals.
Tawny owl: night hunter, mice, voles and shrews.

Insects in flight

Swallow, house martin and swift: by day. Distribution depends largely on suitable nest sites.
Nightjar: by night. Very few nightjars now seen in Forest.

Reptiles

Grass snake: day hunter usually near water. Frogs, toads, newts, fish, voles. The last prey is somewhat unusual for the grass-snake, and was noted in a colony of grass snakes somewhat isolated from water, and living in heathland surroundings in competition with adders. Presumably there was enough food to go round.
Adder: day and night hunter (see Project 15, p. 191), voles, mice, shrews, viviparous lizards.
Slow-worm: earthworms, slugs and grubs above and below surface. Burrows frequently in leaf litter. Woodland borders.
Viviparous lizard: Open glades. Insects, spiders and earthworms.

Amphibians

Common frog: day hunter in damp places. Insects, earthworms, snails.
Common toad: night hunter. As above, especially ants.

APPENDIX 12: Life on the oak tree

During and after the death of a pollarded oak tree in Epping Forest, observations were made on the animals and plants which made use of the tree, either for shelter, for nesting purposes, or as a source of food. A selected list of animals seen at three different periods is given here:

1936. The tree still in apparent healthy condition.

Grey squirrel: feeding on acorns and building a drey; Fallow deer: feeding on acorns and using tree for shelter; Pipistrelle bat: roosting inside hollow of tree; Woodpigeon: acorns and nest; Carrion crow: seen removing egg from woodpigeon's nest; Jay: acorns (seen burying these in squirrel fashion); Nuthatch: searching for insects; Bluetit: passing through in winter flock; Leaf warblers: feeding on insects; Brambling: winter visitor; Great spotted woodpecker: acorns, also heard drumming on dead branch.

Polypody fern: epiphytic in pollarded crown at a time when these ferns were still a feature of the Forest; Ivy: Climbing up the trunk; Signs of bark beetles; Defoliating *Tortrix* moth caterpillars: a heavy infestation this year; Gall insects: three species producing the marble and currant galls and the oak-apple (see Fig. 69).

1948. The tree a year after it fell.

Grey squirrel: using fallen trunk as feeding platform; Woodmouse: nesting burrow under trunk; A single pipistrelle bat roosting under loose bark; Heavy activity of bark beetles; Various woodlice, centipedes and spiders under loose bark (species not identified); Fungi fruiting on rotting wood—Beefsteak (*Fistulina hepatica*); Oyster mushroom (*Pleurotus ostreatus*); *Polystictus versicolor*, and Sulphur tuft (*Hypholoma fasciculare*).

1955. The trunk in advanced state of decay. By this year many species of plants had germinated on the actual trunk, including the following:

Male fern; Wood sorrel; Wood violet; Young hornbeam; Young ivy; Also various fungi, including Myxomycetes, and various invertebrates of the litter fauna (see p. 64).

The above list is but a brief selection of a wide variety of animals and plants which become attracted to a tree during it long life of a hundred or more years. The importance of trees in the preservation of soil and the maintenance of life is discussed on page 74.

APPENDIX 13: Micro-habitats

The following lists are of animals and plants which occupied or visited three different kinds of micro-habitat examined by the author:

A log (in author's garden)

Woodlouse (3 species—*Oniscus*, *Porcellio* and *Armadillidium* spp.); earwig (*Forficula auricularia*); bark beetle species including *Platypus cylindrus*, longhorn beetle (*Clytus arietis*); centipedes (*Lithobius forficatus* and *Mecistocephalus punctifrons*); millepedes (*Geophilus longifrons* and *Julius sabulosus*); stag beetle (*Dorcus parallelipipedus*) which was introduced, see text; slug (*Agriolimax agrestis*); earthworms (*Lumbricus terrestris* and *Allolobophora* spp.); garden snail (*Helix aspersa*); queen wasp (found hibernating); all three British newts hibernating; toad (*Bufo bufo*) using log as retreat; wood mouse with nest; spotted woodpecker searching for grubs; blackbird, hedge sparrow and robin foraging among loose bark; fungi—a bracket (*Polystictus versicolor*), oyster mushroom (*Pleurotus ostreatus*), honey fungus (*Armillaria mellea*) and various slime fungi (Myxomycetes).

Carcase of a fallow deer

Black burying beetle (*Nechrophorus humator*); badger; fox; carrion crow; spotted flycatcher; tiger beetle (*Cicindela campestris*); violet ground beetle (*Carabus violaceus*); bluebottle (*Calliphora vomitoria*).

A cowpat

Tiger beetle (*Cicindela campestris*); dor beetle (*Geotropus stercorarius*); devil's coachhorse (*Staphylinus oleus*); dung fly (*Stomoxys calcibrans*); spotted flycatcher; swallow; hedgehog; toad; green woodpecker; various earthworms.

APPENDIX 14: A classified list of aquatic life recorded for Baldwin's Pond in Epping Forest

PROTOZOA. Non-cellular animals

1. RHIZOPODA (movement by means of pseudopods): *Amoeba* spp., *Arcella*, *Difflugia*, *Actinophrys*.
2. MASTIGOPHORA (movement with a flagellum): *Euglena*, *Ceratium*.
3. CILIOPHORA (movement with cilia): *Paramoecium*, *Spirostumum*, *Stentor*, *Vorticella*, *Opalina ranarum* (in rectum of frog).
4. SPOROZOA (parasitic protozoans with a spore forming stage): *Glugea* (cysts on carp), *Lankesterella* (red-leg disease in blood of frog—passed on by leech), *Ichthyopthyris* ('white-spot' disease found on stickleback).

PORIFERA. Sponges (mostly marine)
Pond sponge *Spongilla lacustris*.

COELENTERATA. Jellyfishes, anemones, etc., mainly marine)

1. HYDROZOA
Green Hydra *Chlorohydra viridissima* (= *Hydra viridis*), containing the green alga *Chlorella*, Brown Hydra *Hydra oligactis* (= *fusca*).

PLATYHELMINTHA. Flat-worms

1. TURBELLARIA. Planarians—free-swimming forms: *Polycelis nigra*, *Dugesia lugubris*, *Dendrocoelum lacteum*.
2. TREMATODA. Parasitic leaf-shaped flat-worms:
   a. Monogenea. *Gyrodactylus* (on gills of stickleback), *Diplozoon* (gills of carp), *Polystomum* (in frog's bladder).
   b. Digenea. *Cotylurus* (3 hosts—Ramshorn snail *Planorbis*, leech *Erpobdella* and a water fowl, unidentified).
   Haplometra (3 hosts—Water whelk *Limnea*, water beetle *Ilybius fuliginosus*, adult in lungs of frog), *Aspidogaster* (in mussel *Anodonta*).
3. CESTODA. Tapeworms:
   Schistocephalus gasterostei in stickleback (see p. 197).

NEMATODA. Roundworms.
Various specimens found but not identified. Some roundworms are free-living, others parasitic. None are easy to identify.

ROTIFERA. Wheel animalcules.
Proales gigantea (on eggs of snail, *Limnea*), *Floscularia ringens* (on *Elodea*), *Rotaria rotatoria* (= *R. vulgaris*), *Brachionus rubens* (on *Daphnia*), *Pleurotrocha daphnicola* (= *Proales daphnicola*) (on Daphnia), *Testudinella truncata* (on Water Louse *Asellus*), *Limnias ceratophylli* (on hornwort) *Stephanoceros fimbriatus* (= *eichornii*) (on *Elodea*), *Philodina roseola* (on *Polyzoa*).

BRYOZOA (POLYZOA). Moss animalcules
Cristatella (under floating leaves), *Lophopus* (on *Lemna*), *Plumatella* (on stones), *Fredericella* (roots of willow). Species not identified.

ANNELIDA. Segmented worms

1. CHAETOPODA. Bristle worms:
   a. Polychaeta (many bristles) marine.
   b. Oligochaeta (few bristles).
   Family Lumbricidae. *Eisenella tetrahedra* (in mud).
   Family Naiadidae. *Chaetogaster limnae* (in shell of *Limnaea*); *Naias* (in clump of algae).
   Family Tubificidae (sludge or river worms) *Tubifex* spp. (in mud).
   Family Enchytraeidae (White or Pot worms) *Enchytraeus* (in roots of *Potamogeton*).
2. HIRUDINEA. Leeches:
   a. Rhynchobdellae (leeches with a proboscis but no jaws).
   Family Ichthyobdellidae (fish leeches); *Piscicola geometra* (on carp).
   Family Glossiphoniidae (snail leeches); Helobdella (on *Planorbis*); *Glossiphonia* (free swimming).
   b. Arhynchobdellae (leeches with jaws).
   Family Hirudidae. *Haemopis sanguisuga*, the horse leech. The medicinal leech (*H. medicinalis*) was not found.
   Family Erpobdellidae *Erpobdella octoculata* (in mud).

CRUSTACEA

a. Branchiopoda (gill-footed crustaceans) the fairy shrimp (*Cheirocephalus diaphanus* (in a small Forest pool near Baldwin's Pond); *Apus* (= *Triops*) *cancriformis* (recorded for Epping Forest, but not on this survey).
b. Cladocera (water 'fleas'). *Daphnia* spp., *Simocephalus vetulus*, *Bosmina longirostris*, *Sida crystallina*.
c. Ostracoda (body enclosed in a hinged shell). *Cypris* spp.
d. Copepoda (pear-shaped body, no shell). *Cyclops* spp.
e. Branchiura (fish-lice, parasitic). *Argulus foliaceus*.
f. Isopoda (limbs of similar shape). *Asellus aquaticus*, the water louse or water slater.
g. Amphipoda (two kinds of thoracic limbs). *Gammarus pulex*, the freshwater shrimp.

INSECTA

1. APTERYGOTA. Wingless insects with incomplete metamorphosis:
   a. Collembola (Springtails) *Podura aquatica*.

2. PTERYGOTA. Mainly with two pairs of wings and a metamorphosis:

Exopterygota. An incomplete metamorphosis and no pupal stage:

b. Plecoptera (Stoneflies) Mainly in fast streams. None found.

c. Odonata (Dragonflies):
Anisoptera (Dragonfly hawks), Zygoptera (Damselflies).

d. Hemiptera (Bugs):
Heteroptera. All aquatic bugs belong to this group. *Hydrometra stagnorum*, the water measurer; *Vela caprai*, the water cricket; *Gerris najas*, the water strider; *Nepa cinerea*, the water scorpion; *Ranatra linearis*, the water stick-insect; *Ilyocoris* (= *Naucoris*) *cimicoides*, the saucer bug; *Notonecta glauca* spp. the backswimmer; *Corixa* spp. lesser water boatmen.

e. Ephemeroptera (Mayflies) *Cloeon* spp.

Endopterygota. A complete metamorphosis with pupal stage:

f. Megaloptera (Alder-flies) *Sialis lutaria*, the Alder-fly. The only species with an aquatic larva usually found in still-water.

g. Neuroptera (Lacewings) A few aquatic species as larvae. *Osmylus fulvicephalus* (in moss).

h. Trichoptera (Caddis flies) Larvae aquatic.

i. Lepidoptera (Butterflies and moths) Some larvae of the family Pyralidae aquatic. *Nymphula nympheata*, the China marks moth (under *Potamogeton* leaves).

j. Coleoptera (Beetles).
Family Hygrobiidae. *Hygrobius* (= *Pelobius*) *hermanni*, the screech beetle.
Family Dytiscidae. *Dytiscus marginalis*, the great diving beetle; *Acilius sulcatus; Ilybius fuliginosus*. Gyrinidae *Gyrinus* spp. Whirligig beetles.
Family Hydrophilidae. *Hydrophilus* (= *Hydrous*) *piceus*, the silver water beetle. One specimen found in ten years, probably a stray.
Family Chrysomelidae. *Donacia* spp. larva found among roots of *Potamogeton.*

k. Diptera (Flies).
Orthorrhapha. A free pupa, the adult emerging from a slit through the skin.
Nematocera. Adults with long and many jointed antennae.
Family Tipulidae (Craneflies). *Dicranota* (larvae found in wet mud).
Family Ptychopteridae. *Ptychoptera* (larva in mud in shallow water).
Family Psychodidae (Moth-flies).
Family Dixidae. *Dixa* spp. (larvae and pupae at surface).
Family Culicidae. *Culex pipiens*, the common mosquito or gnat. *Chaoborus* (= *Corethra*), the phantom midge.
Family Chironomidae (True midges. Larvae called 'blood-worms).
Family Ceratopogonidae (Biting midges). Species not identified.
Brachycera. Short horned flies.
Family Stratiomyidae (Soldier flies). *Stratiomys chameleon*, the chameleon fly.
Family Tabanidae (Horse flies and clegs). *Tabanus* spp.
Cyclorrhapha. Larva a maggot, pupa formed inside a hard case the puparium.
Family Syrphidae (Hover flies and Drone flies). *Tubifera* (= *Eristalis*) the rat-tailed maggot fly. Found in mud in shallow water.

ARACHNIDA. Water spiders

1. ARANEIDA. Spiders:
The water spider, *Argyroneta aquatica*; the wolf spider, *Pirata piscatorius*; the raft spider, *Dolomedes fimbriatus*.

2. HYDRACARINA. Water mites:
Hydracharina globosa (adult free-swimming, larvae on *Notonecta*); *Diplodontus despiciens*; *Hygrobates longipalpis* (among duckweed).

TARDIGRADA. Water 'bears'

Specimens seen in samples of damp moss, but not identified.

MOLLUSCA. Freshwater snails and mussels (univalves)

1. GASTEROPODA. Freshwater snails and limpets:
a. Pulmonata (water snails which take in air from the surface). The freshwater whelk, *Limnaea stagnalis*; dwarf pond snail, *Limnaea truncatula*; the wandering snail, *Limnaea peregra*; the marsh snail, *Limnaea palustris*; the bladder snail, *Physa fontinalis*; the lake limpet, *Ancylus* (= *Acroloxus lacustris*; the ram's-horn snails—*Planorbis corneus*, *P. complanatus*, *P. spiralis* and *P. carinatus*.

2. LAMELLIBRANCHIATA. Mussels and cockles (bivalves):
b. Unionacea. The painter's mussel, *Unio pictorum*; the swan mussel, *Anodonta cygnea*, the duck mussel, *Anodonta anatina*; the orb-shell cockle, *Sphaerium* spp.; the pea-shell cockle, *Pisidium* spp.

VERTEBRATA

1. PISCES. Fishes.
Family Cyprinidae. Common carp, *Cyprinus carpio*; crucian carp, *Carassius carassius*; gudgeon, *Gobio gobio*; rudd, *Scardinius eryophthalmus*.
Family Anguillidae. Common eel (*Anguilla anguilla*) Found only once (see p. 236).
Family Percidae. Perch, *Perca fluviatilis.*
Family Gasterosteidae. Three-spined stickleback, *Gasterosteus aculeatus.*
Family Esocidae. Pike, *Esox lucius.* A few small specimens.

1. Amphibia. Amphibians:

a. Caudata.

Family Salamandridae. Smooth newt, *Triturus vulgaris*; palmate newt, *Triturus helveticus*; gt. crested newt, *Triturus cristatus.*

b. Salientia.

Family Ranidae. Common frog, *Rana temporaria.*

Family Bufonidae. Common toad, *Bufo bufo.*

3. Reptilia. Reptiles.

Family Colubridae. Grass snake, *Natrix natrix.*

4. Aves. Birds (only water birds given in this record): Moorhen, coot, mallard, little grebe, kingfisher, heron, pochard, tufted duck, swallow, sedge warbler.

5. Mammalia. Mammals (species seen in or at the pondside):

Fallow deer, badger, grey squirrel, common shrew, water shrew, fox.

APPENDIX 15: pH

This symbol is commonly used in plant and animal studies to indicate the state of water and soil whose condition may be acid, alkaline or neutral. It is useful to know the pH of an area under survey, as this may have a bearing on the life which it supports. On chalk or limestone the soil is rich in mineral salts and said to be base saturated, especially with calcium (NaOH). It is strongly alkaline. In a place where there is a deficiency of mineral salts, and where plant decay is slow, e.g. on a moorland, an acid condition may arise with hydrochloric acid present (HCl).

Water (H_2O) in the soil, in whatever solution it occurs, contains numbers of minute electrically charged particles, called ions, which are either positively charged (the hydrogen ions H+) or negatively charged (the hydroxyl ions OH−). Their relative amounts will determine the alkalinity or acidity of the mixture.

At a temperature of 18°C there is one ten-millionth of a gramme of each kind of ion in a litre of pure water, which can be written as 1×10^{-7}. This gives a total of 10^{-14} ions which is always constant, so that if only one concentration is known the other can be determined. Thus, a figure of 10^{-8} hydrogen ions would mean that 10^{-6} hydroxyl ions are present in the solution.

In practice only the positive hydrogen ion content is measured, and for simplicity this is written as the symbol pH, followed by a figure from 1 to 14. pH stands for hydrogen and a German word *Partialdruck* (partial pressure). It represents the logarithm of the reciprocal of the H+ concentration, that is, $\log_{10} \frac{1}{\text{H ion.}}$ In a neutral solution, as in pure water, this would be 10^{-7}, and is written as pH 7. The positive and negative ions are in balance. In an acid solution the Hydrogen H+ ion concentration tends to increase, which means that the pH value (i.e. the reciprocal) falls. In an alkaline solution the reverse takes place as the Hydroxyl OH− ions increase. In other words acidity increases by pH 7 down to pH 1, and alkalinity rises from pH 7 to pH 14.

In natural waters these chemical changes are due mainly to the amount of carbon dioxide present, and to carbonates such as calcium. The former tends to make the water acid, and the carbonate to make it alkaline. In 'hard' water where carbonates are present, i.e. on a chalk soil, carbon dioxide is required in large amounts to change the pH, and the mixture tends to remain stable. In 'soft' water there is much more variation. Consequently a pH reading can give a fairly good idea of the amount of calcium which is present. This is very useful to know when working on water life, especially crustaceans and molluscs which depend upon calcium for shell growth. On land the pH figure acts as a guide to the whereabouts of certain chalk-loving plants (calcicoles) or chalk-haters (calcifuges).

During autumn, or when decomposition sets in (e.g. in a shaded woodland pool, a peat bog, or a base deficient soil) the pH will drop as the acidity rises. Readings taken at intervals can vary considerably as the CO_2 concentration rises and falls with the weather and season (see Fig. 47).

A simple soil test is to use diluted hydrochloric acid and pour a little on to the soil to be tested. If it bubbles the soil is calcareous. The acid reacts with the base to give off oxygen bubbles. No fizzing indicates a neutral or acid soil.

For more accurate results portable testing outfits, such as those supplied by the British Drug Houses Limited can be used, The B.D.H. 'Soil Indicator' Capillator set is useful for this work. It consists of liquid of neutral pH 7 containing indicators of different pH ranges which give different colours, also a trough and spoon. A small amount of the soil to be tested is placed in a clean trough at the larger end, covered with some indicator liquid, and allowed to mix. Surplus water is then drained over to the smaller end, and a colour comparison made with a chart. From neutral (no colour change) the solution will vary from bluish to yellow (slightly acid) through orange to red (highly acid).

Commercial instruments, called pH meters, give a highly accurate result. A liquid solution of the soil is used as part of the electrolyte in a battery. The battery voltage varies according to the concentration of H+ ions in the solution, so that it can be calibrated to give pH readings.

APPENDIX 16: The herbarium

A useful record of any field project is a herbarium compiled of plants found during the survey. Such a collection should be confined to the minimum number of specimens necessary, and should exclude any rarities which can just as well be recorded with a camera.

For collecting purposes the vasculum is perhaps a little outdated in these days of polythene, but still makes a useful container for specimens which are likely to damage in transit. It should be coloured white for easy recognition when put down somewhere, also to reflect the heat from the sun. Polythene bags can be carried inside for small and more delicate specimens which are likely to dry out, or from which the petals readily fall. A simple plant press may come in handy for such plants, and consists merely of two stiff cards containing drying-paper, and held together with elastic bands. Include some tie-on labels so that specimens can be identified on the spot.

At home the usual drying material available is newspaper, either placed under a heavy weight, the carpet, or in an old trousers' press. This is quite adequate, provided that the drying paper is changed frequently, at least once a day for the first week, so as to quickly dry out the specimen. Lay this in a natural position, showing upper and lower sides of leaves, between two sheets of thin paper. The plant remains inside this each time the drying paper is changed. In this it will dry in its permanent position without any upset.

Various refinements will present themselves with each collection. Fleshy stems can be gently crushed and flattened with a roller, and succulent leaves killed in boiling water. Tubular flowers are cut open and laid out flat. Thick stemmed portions are padded where necessary to give even pressure when drying, and so on. To hasten drying use two or three lattice-like frames made out of flat strips of wood, one placed at the bottom of the pile, one or two in the middle, and one on top under the weight. Dried specimens should be mounted on stiff white paper. Most serious collectors use the standard sheets as issued by the Kew Botanical Gardens Herbarium, measuring 11 inches by 7 inches. Specimens should not be pasted down, but held in place with strips of gummed paper. To avoid a bulge appearing in the herbarium sheets when stored together arrange the specimens a little off-centre. This will also leave room for more than one specimen per sheet.

Remember to give your specimens value by providing a history—the scientific name, locality of origin, county or vice-county, date and collector's name. Without this the plant is scientifically worthless.

A herbarium is usually compiled on systematic lines, either for the county or country, or for some kind of habitat. It shows the species which grow there, but may tell little else.

If a herbarium is to be a useful reference to field work, and one which supports the findings of a survey, then specimens should be collected accordingly. The grid sampling, transects and valence analyses which are carried out will tell which plants are dominant for the area and this should be shown in the herbarium, together with a selection of the associated plants. For example, an oak-wood community would show *Quercus robur* as dominant, with hazel (*Corylus*) and crab apple (*Malus*) in the shrub layer. Ground flora might include such herbs as *Scilla*, *Anemone*, *Arum*, *Nepeta*, and so on. As the collection grows it could be further subdivided into its more localised associations as indicated in the transect carried out for the Epping Forest Project 14 (p. 184). This shows an association for a beech wood, a bracken patch, footpath and ditch, each with its distinctive plants. In a herbarium collection for a sand-dune one might divide it up into the zones of foreshore, white and grey dune, dune slack and golf-links. In similar fashion the mountain, pond and salt-marsh have their zones and associations of plants. Each collection of specimens mounted on the herbarium sheets could have a separate colour folder for quick reference. In some cases the same species will turn up in different folders because of a wider tolerance to habitat. In such cases a single sheet could show the range of shape and size for the one species growing under these different conditions.

Apart from flowering plants one should not overlook the lower plants such as mosses, ferns, liverworts and fungi. In some cases these act as useful 'indicators' to the soil conditions of habitat (see Appendix 1, p. 212). Ferns are pressed in a similar manner to flowering plants, and because of their flatness can be mounted very neatly. Mosses and lichens can usually be air dried and kept in envelopes. Liverworts being more succulent need the drying paper treatment. Fungi, apart from the woody kinds, are best preserved in spirit, or cut into thin sections and rapidly dried under pressure. Perfectly preserved specimens can be produced by the freeze-drying technique now used in many museums, but this is probably beyond the scope and purse of the individual.

APPENDIX 17: Measuring angles and height in the field

To measure the angle of slope of ground, a protractor may be used to which a length of weighted cord is tied (Fig. 83). Sight the protractor as shown along its base towards a marked pole of the same height as the observer (along A-B). Then the angle between the weighted cord and the 90° line, i.e. COD will equal BOF, the angle of elevation, or GOA, the angle of depression.

To measure the height of a tree, cliff or building,

stand a distance AC from the object (say, 30 feet) and measure the angle between the line of sight to the treetop and the ground, i.e. BOF. Then the tree's height X = AC (30 feet) × (tan BOF + height of protractor from ground). Should the ground be sloping, take an average figure of the two angles measured from the same distance on opposite sides of the tree.

An alternative method is based on similar triangles. From a point A which is 30 feet from the base of the tree C, sight to the top of the tree B. While doing so get someone to place a 6-foot rod DE in such a position that D is in the line of sight (see Fig. 83). Triangles ABC and ADE are now of similar shape. Assume that AE works out at 10 feet. Then the tree's height should be $\frac{AC}{AE} \times DE$ or $\frac{30}{10} \times 6 = 18$ feet. This method will only give good results if the ground is level.

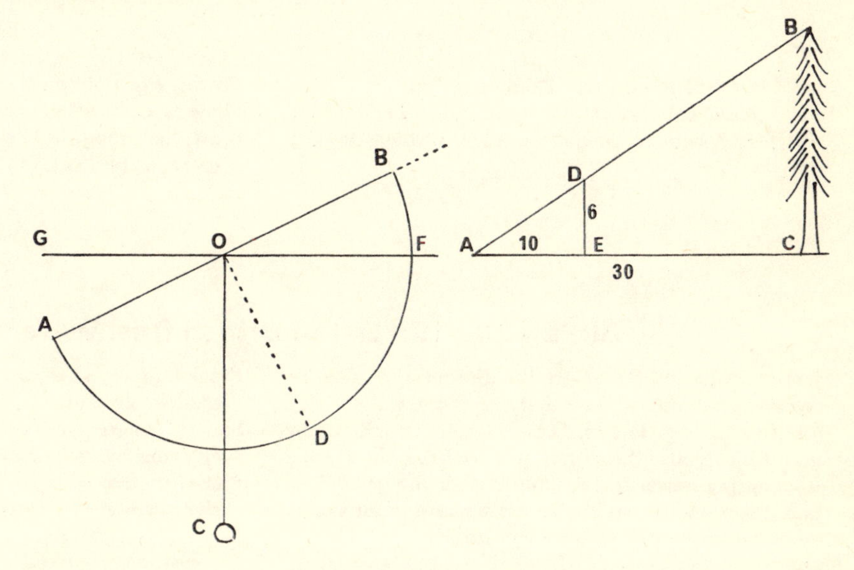

FIG. 83. Two methods of measuring the height of a tree, etc. in the field—with a protractor (left) and right, with a measuring pole (see text page 190).

APPENDIX 18: Main habitat regions in Britain, with a short list of typical flora and fauna

Chalk

S.E. downlands—North and south downs, Chilterns, Yorkshire and Lincolnshire wolds, Salisbury Plain.

Beech, dogwood; cowslip, harebell, pyramidal orchid; common hare; cuckoo, skylark, kestrel; marbled white, burnet moth.

Limestone

Yorkshire wolds, Mendips, Cotswolds, Westmorland, southern Peak District, Gower (South Wales), Central Ireland, Burren.

Ash, hawthorn; rockrose, carline thistle, spotted orchid; badger; jackdaw, wheatear; brown argus.

Clay

South and midland farmland—Weald, East Anglia, Thames valley.

Pedunculate oak, hazel; meadow buttercup, red clover, early purple orchid; fox, stoat; rook, tawny owl, nightingale; meadow brown.

Inland sands and gravels

Southern heathlands—New Forest, Ashdown Forest, Surrey commons, Dorset heaths, Breckland.

Silver birch (pine introduced by Forestry), gorse; bell heather, tormentil, sundew; roe deer; adder; willow warbler, nightjar; silver studded blue, emperor moth.

Acid rocks

Millstone grits and granites on mountains and moorland—South Wales hills, Snowdonia, Peak District, Pennines, Lake District, Highlands, West Country moorlands.

Scots Pine, common birch; bilberry, ling, cotton grass; red deer, wild cat, pine marten; viviparous lizard; meadow pipit, red grouse, peregrine; small heath butterfly.

Rocky coast

N.W. Scotland, Ireland, Wales, West Country.

Sea buckthorn; sea pink, sea campion, wall pennywort; grey seal; kittiwake, puffin, herring gull.

Coastal sand and shingle

Braunton Burrows (Devon), Blakeney (Norfolk) Newborough Warren (Anglesey), Culbin Sands (Morayshire), Chesil Bank (Dorset), Dorset and Lancashire sand dunes.

Few trees (Corsican Pine introduced); marram grass, sea rocket, sea holly; rabbit, common seal; common tern, ringed plover; grayling butterfly.

Rivers

Avon, Dart, Shannon, Wye, Thames.

Willow, sessile oak; water iris, great willow herb; otter, water vole, salmon; kingfisher, sedge warbler, dipper; mayfly.

Lakes and reservoirs

Scottish lochs, Lake District.

Willow, rowan; bullrush; great crested grebe, diver, trout.

Ponds and marsh

Norfolk Broads.

Alder; reed, water lily; coypu; common frog; bittern; bearded tit, moorhen; swallow tail butterfly, dragon fly.

Built-up areas

Towns, e.g. London.

Plane, sycamore; Oxford ragwort, rosebay, dandelion; rat, hedgehog; feral pigeon, black-headed gull, house sparrow; peacock butterfly; house-fly.

APPENDIX 19: List of British freshwater fishes

Salmon (*Salmo salar*). 40-50 lb. Enters the cleaner rivers (anadromous) especially in Scotland, Wales and Ireland (e.g. Tweed, Tees, Wye, Severn, Shannon and Liffey), also English rivers such as the Avon, from spring onwards, breeding late in the year. The parr takes about two years to reach the smolt stage, when return migration commences.

Brown trout (*Salmo trutta*). In clear streams and lakes, varying greatly in size and markings, from tiny brook trout to large lake specimens. The sea trout is migratory, entering rivers to spawn.

Char (*Salvelinus alpinus*). Up to 1 foot recorded. In some Scottish lochs, the Lakes and Ireland.

Eel (*Anguilla anguilla*). Almost any freshwater, with autumnal migration towards sea. Summer arrival of elvers. Up to 3 feet long in female, 12-18 inches in male.

Barbel (*Barbus barbus*). Up to 30 lb. Deep rivers such as the Thames, Trent and Yorkshire Ouse.

Allis shad (*Alosa alosa*). Up to 2 feet long. A member of the herring family, rare in Britain except in Severn and Irish Shannon, where it enters in spring to breed. Up to 8 lb. recorded. The Twaite shad is a smaller species.

Grayling (*Thymallus thymallus*). Up to 3 lb. Fast streams, locally abundant in England and Wales, introduced into Scotland. A freshwater salmonid.

Bass. (*Morone labrax*). Up to 30 lb. Mainly on south coast, entering estuaries in summer.

Burbot (*Lota lota*). Up to 8 lb. A freshwater member of the cod family in rivers entering North Sea.

River lamprey (*Lampetra fluviatilis*). Up to 15 inches. Ascends clean rivers to spawn, such as the Severn, Trent, Ouse and Dee. A parasitic animal which is a primitive survivor of a group which preceded true fishes.

Brook lamprey (*Lampetra planeri*). Up to 6 inches. Clear shallow streams.

Smelt (*Osmerus eperlanus*). Up to 12 inches. Marine, entering rivers to spawn, especially in Scotland.

Pike (*Esox lucius*). Up to 50 lb. Weedy lakes and sluggish rivers. Large specimens in Ireland.

Perch (*Perca fluviatilis*). Up to 5 lb. Rivers, lakes and ponds, among weeds.

Ruffe or pope (*Acerina cernua*). Up to 8 inches. Slow rivers and canals, usually on bottom, in Midlands and South England, especially East Anglia.

Common carp (*Cyprinus carpio*). Up to 50 lb. Introduced from Eastern Asia by the monks. Still waters of lakes, gravel pits, stew ponds, etc.

Crucian or bronze carp (*Carassius carassius*). Up to 3½ lb. Introduced. Small specimens commonly found in ponds.

Tench (*Tinca tinca*). Up to 8 lb. Still, weedy waters, mainly on bottom.

Chub (*Squalius cephalus*). Up to 8 lb. Larger rivers, absent Ireland, West Country and North Scotland.

Bream (*Abramis brama*). Up to 12 lb. Lakes, slow rivers and canals, mainly England and Ireland, especially in East Anglia.

Dace (*Leuciscus leuciscus*). Up to 1½ lb. Slow-moving rivers, mainly England and Wales. Absent Ireland, except in River Blackwater.

Roach (*Rutilus rutilus*). Up to 3 lbs. Widespread in lakes, rivers and canals, but rare in West Country, absent parts of Wales, North Scotland and Ireland.

Rudd or red-eye (*Scardinius eryophthalmus*). Up to 2 lb. Local England and Wales, absent Scotland, common in Ireland, in weedy lakes and sluggish rivers.

Gudgeon (*Gobio gobio*). Up to 8 inches. Shallow water, still or moving, on gravel or sand. Widespread, absent Scotland.

Minnow (*Phoxinus phoxinus*). Up to 4 inches. Very common in streams and moving shallows, on gravel or sand. Gregarious.

Bleak (*Alburnus alburnus*). Up to 8 inches. Widespread, in shallow swift water. Absent Scotland and Ireland.

Stone loach (*Nemacheilus barbatula*). Up to 5 inches. Widespread in shallow, clear streams among stones, usually hidden.

Spiny loach (*Cobitis taenia*). A little smaller than Stone loach. More local, mainly eastern England.

Bullhead (*Cottus gobio*). See Stone loach. Up to 4 inches.

Three-spined stickleback (*Gasterosteus aculeatus*). 3-4 inches. Common almost everywhere, including brackish, even salt, waters.

Ten-spined stickleback (*Pygosteus pungitus*). Locally widespread, rare in Ireland, confined more to freshwater.

APPENDIX 20: List of British reptiles and amphibians

Reptiles

Adder or northern viper (*Vipera berus*). Quiet places, usually in dry surroundings, in woods, heaths, moorland, cliffs, quarries, etc. The most widespread snake. Absent Ireland.

Grass or ringed snake (*Natrix natrix*). Generally in vicinity of water, and an excellent swimmer. Absent Ireland, most of Scotland.

Smooth snake (*Coronella austriaca*). Rare and localised to heaths in southern counties, especially Hampshire and Dorset.

Common or viviparous lizard (*Lacerta vivipara*). Widespread and common over most of Britain, including parts of Ireland, in glades, hedgerows, heaths and moors, waste ground, cliffs and mountains.

Sand lizard (*Lacerta agilis*). Very local and becoming rare, in sand dunes and heaths, mainly in Surrey, Dorset, Hampshire and Lancashire. Threatened by expansion of the holiday industry.

Slow-worm (*Anguis fragilis*). Widespread along woodland borders, hedgerows, quarries, waste ground, cliffs, and undisturbed places such as rubbish dumps, churchyards and railway embankments. Absent Ireland.

Amphibians

Common frog (*Rana temporaria*). Widespread in most counties, including much of Ireland, usually in grassy places or undergrowth near water, sometimes high in mountains. Is disappearing in many places due to loss of ponds.

Edible frog (*Rana esculenta*). An introduction, as far back as 1837. Once very common in East Anglian fens and broads, now practically extinct except for a few odd colonies reintroduced around London. Occasionally seen as an escape.

Marsh frog (*Rana ridibunda*). Introduced from Hungary into the Romney marsh, Kent, in 1934. Has spread all over the Kent marshes where it is still common in the dykes and canals.

Common toad. (*Bufo bufo*). Common and widespread, often far from water but centred on certain breeding ponds. Absent Ireland.

Natterjack (*Bufo calamita*), Very localised, becoming rare, in similar places to the Sand lizard, and threatened for similar reasons, also from loss of breeding ponds. A further colony in South West Ireland (Kerry).

Smooth newt (*Triturus vulgaris*). Widespread in most areas, including much of Ireland, but avoiding mountains and moorlands.

Palmated newt (*Triturus helveticus*). Widespread, absent Ireland. Often the only newt found on acid soils. A montane species.

Crested or great warty newt (*Triturus cristatus*). Widespread but more localised, absent Ireland. Usually missing in built-up areas due to collectors.

APPENDIX 21: Animals in the post

When sending live animals through the post, care should be taken to see that they are not damaged or killed in transit through jolting, and that they have sufficient air and moisture. In most cases the small animals, such as invertebrates, also small specimens of reptiles and amphibians, can be packed in dampish moss placed inside a perforated tin or box with firmly secured lid. Use sufficient material, i.e. moss, grass, leaves, etc., so that the animals can move about but at the same time are not shaken up too much.

Holes through metal should be punched from the inside, and the jagged edges flattened with a hammer. Aquatic animals such as insects and their larvae, molluscs, worms and crayfish travel best in wet moss or water weed, rather than in a water container.

Water, of course, will be required for sending fish, and the container will have to travel by rail or coach.

Dead specimens which may be dry and brittle require extra care in packing. Insects which are already pinned, or those placed in a folded paper (i.e. before setting) should be put in a strong box. If pinned, see that this is well secured in the box in its cork holder. If in a wrapper then use enough packing material to make it secure from movement.

Liquid-preserved specimens should be saturated in 70% alcohol or 5% formalin according to the species (see Hints to Collectors in the Bibliography) and wrapped in polythene with sufficient packing material around it. Small specimens, e.g. water insects, can be placed in a strong glass or metal tube with firm cork stopper or screw cap, containing the correct preservative, and also some cotton wool or tissue paper to hold the specimen still.

Mammals and birds obviously travel as postal packages, but are sent as goods by rail, coach or air. The dispatch office will advise on what can, and what cannot, be sent. There may be certain regulations to follow and forms to fill out. For long journeys by sea or air there is usually a trained person in charge of any livestock, whose duty it is to watch over the animals during transit. At air terminals and other places there may be rest centres or animal hospitals staffed by air or rail personnel, or run by welfare organisations such as the R.S.P.C.A., so that there should be little danger these days of an animal not reaching its destination in good condition. It is in the interests of the transport service as well as the recipient, and in particular the animal(s), that each item should be clearly marked LIVESTOCK, also giving the name of the animal.

APPENDIX 22: Cleaning a skeleton

The bones, skulls and teeth which turn up from time to time during field work may require identification. These can occur in bird pellets, as stomach remains or in decaying carcases. It is useful for reference purposes to have a collection of skeletons or skulls to which one can refer.

Obtain if possible freshly killed animals, which should then be skinned or plucked carefully and the carcase gently boiled in water. Use an alkaline water (adding a pinch of washing soda or bleach if necessary). To this is added a digestive enzyme *pancreatin* (from a chemist) to hasten the process.

Wash the bones and skull clear of flesh (the brain will probably run out) and whiten in a solution of 20 vol. hydrogen peroxide by further gentle boiling. Warning—keep the material away from the sides of the pan as contact may caused a stain to appear on the bone. Suspend the bones, skull etc. in a gauze bag or on a shelf of perforated zinc.

The author has tried experiments in bone cleaning by using animals such as the wood ant (*Formica rufa*). A small dead animal placed on the nest will soon end up as a skeleton, but must be watched as the bones get buried or become scattered. Delicate creatures such as shrews may be difficult to reassemble.

Another agent worth trying is tadpoles. They once did a perfect job on a dead bat suspended in their bowl, by cleaning away the skin and flesh, but leaving the skeleton intact as the ligaments were untouched. Tadpoles are ravenous carrion feeders during the late stage before metamorphosis, and have been known to feed on sick fish which cannot escape. It is a pity that these obliging 'laboratory assistants' are only available for a limited period during late spring.

APPENDIX 23: The grid reference

The National Grid is a system which enables a point on the landscape of Britain to be identified or recorded from the O.S. map. It consists of a grid of squares overlying the map which is based on the Transverse Mercator plan, whose central meridian lies in a straight N.-S. line 2° west of Greenwich.

The point of origin for the grid is taken as zero or 0, and it lies on this line a little to the south-west of Ireland. The position of any spot to be taken becomes the distance in metres east and west of this zero.

The National Grid is thus a series of lines parallel and at right angles to the 2° W. line just mentioned, formed into squares. According to the scale of the O.S. map these squares have sides of 10 km., 1 km. or 100 metres. For field work a one-inch map is generally used, where the grid lines are spaced at 1 km. intervals. The large numbers in the margin of the map are the 1 km. intervals spaced from the zero off Ireland, and read from west to east and from south to north. From these squares it is possible to give the position on the map of any place, by quoting the co-ordinates on which it lies. They are the distances read off the map margin east and north of zero. and called Eastings and Northings.

A full reference comes in six figures, thus E459321, N625437 (these numbers have been chosen at random as an example). Since Eastings always precede Northings, the letters may be dropped. Also, if a position need only be given to the nearest 100 metres, as is usual with field recording, then the last two figures representing tens and units of metres, can also be dropped. The reference then reads 4593.6254. Should the locality be well known, e.g. Epping Forest, then the first figure representing hundreds of metres is not necessary. The reference now reads 593.254. This is called the Normal National Grid Reference.

Such a figure will repeat itself on the grid at every 100 km. interval and if the place is not well known it may be necessary to give the Full National Grid Reference. For this purpose each of the 100 km. squares covering Britain has been numbered, and the figure indicating the square in which the place lies, e.g. 12, is added. The Full National Grid Reference then reads 12/4593.6254. Numbered squares are given on a small map in the bottom margin of the O.S. map. On the inside covers of the inch and 2-inch maps information is given on the grid system. See also O.S. booklet No. 1/45, published by H.M.S.O. at 4d.

Bibliography

GENERAL WORKS

Bates, Marston: *The Forest and the Sea* (Museum Press 1961).
Carson, Rachael: *Silent Spring* (Penguin 1965).
Cave, B. V.: *Advance of Life* (Pergamon Press 1966).
Cave, R. P.: *Elementary Map Reading* (Methuen 1943).
Coward, T. A.: *Life of the Wayside and Woodland* (Warne 1963).
Darlington, C. L.: *The Microscope* (Museum Press).
Darwin, Charles: *Darwin on Humus and the Earthworm* (Faber 1966); *The Origin of Species* (reprint of 6th edition with foreword by Gavin de Beer) (O.U.P. 1967).
Farb, Peter: *Living Earth* (Constable 1960).
Fisher, James: *Shell Nature Lover's Atlas* (Michael Joseph 1966).
Fitter, R. S. R.: *Wildlife in Britain* (Penguin 1963); *The Ark in our Midst* (Collins 1964).
Heywood, V. H.: *Plant Taxonomy* (Edward Arnold 1967).
Leutscher, Alfred: *Animal Camouflage* (Daily Mail Publn.).
Library Association: *Readers Guide to Books on Natural History* (No. 73), from the Publications Officer, Lindsey and Holland County Library, 45 Newlands, Lincoln.
Portman, A.: *Animal Camouflage* (Univ. of Michigan Press 1950).
Reade, Winwood and R. M. Stuttard (eds.): *A Handbook for Naturalists* (Evans 1968).
Reid, Leslie: *Earth's Company* (John Murray 1958); also published as *The Sociology of Nature* (Penguin 1962).
Savory, T. H.: *The World of Small Animals* (U.L.P. 1955).
Storey, J. H.: *The Web of Life* (Vincent Stuart n. d.).
Watson, G. G.: *The Junior Naturalist's Handbook* (A. & C. Black 1962).
White, Gilbert: *The Natural History of Selborne*, with introduction by R. B. M. Lockley (Everyman-Dent 1966).
Wilson, R. A.: *An Introduction to Parasitology* (Edward Arnold 1967).

EVOLUTION

de Beer, Gavin: *Story of Evolution* (Rathbone 1967); *A Handbook on Evolution* (British Museum, Natural History 1964); *Atlas of Evolution* (Nelson 1964).
Carter, G. S.: *Animal Evolution* (Sidgwick and Jackson 1954).
Darwin, Charles and Wallace, Alfred: *Evolution by Natural Selection* (C.U.P. 1958).
Dowdeswell, W. H.: *The Mechanism of Evolution* (Heinemann 1963).
Huxley, Julian: *Evolution, the Modern Synthesis* (Allen and Unwin 1963); *The Wonderful World of Life—the Story of Evolution* (Garden City Books, N. York 1958).
Ross, H. H.: *Understanding Evolution* (Prentice Hall 1967).
Sheppard, P. M.: *Natural Selection and Heredity* (Hutchinson 1959).

BIOLOGY

Burton M.: *Animal senses* (Routledge 1961).
Butler, J. A. V.: *The Life of the Cell* (Allen and Unwin 1964).
Clare, June: *The Staff of Life* (Phoenix House 1964).
Fisher *et al.*: *Nature—Earth, Plants and Animals* (Macdonald 1960).
Hurst, Rona: *Loom of Life* (Barrie and Rockliffe 1964).
Mittwoch, Ursula: *Sex Chromosomes* (Academic Press 1967).
Vines, A. E. and Rees, N.: *Plant and Animal Biology* (Pitman 1964), 2 vols.
Waddington, C. H.: *Introduction to Modern Genetics* (Allen and Unwin 1959).

BEHAVIOUR

Ardrey, Robert: *The Territorial Imperative* (Collins 1967).
Burton, M.: *Animal Courtship* (Hutchison 1953).
Barnett, A.: *Instinct and Intelligence* (MacGibbon and Kee).
Carthy, J. D.: *Animal Navigation* (Allen and Unwin 1956); *Animal Behaviour* (Aldus Books 1965); *The Study of Behaviour* (Edward Arnold 1966).
Fox, H. M.: *Personality in Animals* (Penguin 1952).
Howard, Len: *Birds as Individuals* (Collins 1952).
Lorenz, Konrad: *King Solomon's Ring* (Methuen 1955); *On Aggression* (Methuen 1966).
Scott, J. P.: *Animal Behaviour* (C.U.P. 1958).
Thorpe, W. H.: *Learning and Instinct in Animals* (Methuen 1963).

Thorpe, W. H. and Zangwill, O. L.: *Current Problems in Animal Behaviour* (C.U.P. 1961).
Tinbergen, N.: *Social Behaviour in Animals* (Methuen 1953); *The Herring Gull's World* (Collins New Naturalist 1953).

GEOLOGY

Ager, R. V.: *Introducing Geology* (Faber 1961).
Davies, G. M.: *A Student's Introduction to Geology* (Allen and Unwin 1949).
Evans, I. O.: *The Observer's Book of Geology* (Warne 1957).
Frazer R.: *The Habitable Earth* (Hodder and Stoughton 1964).
Holmes, A.: *Principles of Physical Geology* (Nelson 1944).
Read, H. H. and Watson, Janet: *Beginning Geology* (Macmillan 1966).
Stamp, L. Dudley: *Britain's Structure and Scenery* (Collins New Naturalist 1946).
Steers, J.: *The Sea Coast* (Collins New Naturalist 1953).

CLIMATE AND WEATHER

Johnson, C. G. and Smith, L. P.: *The Biological Significance of Climate Changes in Britain* (Academic Press 1965).
Manley, G.: *Climate and the British Scene* (Collins New Naturalist 1952).
Smith, L. P.: *Weathercraft* (Blandford 1960).

DICTIONARIES

Abercrombie M. *et al.*: *A Dictionary of Biology* (Penguin 1966).
Burton M. and Day N. C.: *Oxford Junior Encyclopaedia*, Vol. 2. Natural History (O.U.P. 1949).
Fitter, R. S. R. and M.: *Dictionary of British Natural History* (Penguin 1967).
Hanson, H. C.: *Dictionary of Ecology* (Owen 1962).
Henderson, I. F. and W. D.: *A Dictionary of Scientific Terms* (Oliver and Boyd 1960).
Lysaght, A.: *Dictionary of Natural History and Other Field Study Societies in Gt. Britain* (British Assn. for the Advancement of Science 1959).

TRACKS, SIGNS AND MOVEMENT

Ennion, E. A. R. and Tinbergen, N.: *Tracks* (O.U.P. 1967).
Gray, J.: *How Animals Move* (Penguin 1959).
Lawrence, M. J. and Brown, R. W.: *Mammals of Britain—their Tracks, Trails and Signs* (Blandford Press).
Leutscher, Alfred: *Tracks and Signs of British Animals* (Cleaver-Hume 1960).
Speakman, F.: *Tracks, Trails and Signs* (Bell 1954).

ROCKS AND THE SOIL

Brade-Berks, S. G.: *Good Soil* (English U. P. 1949).
Hall, A. D.: *The Soil* (John Murray 1947).
Jackson, R. M.: *Life in the Soil* (Edward Arnold 1966).
Kevan, D. K. McE.: *Soil Zoology* (Butterworth 1955).
Page, L. W.: *Rocks and Minerals* (U.L.P. 1966).
Russell, Sir E. J.: *The World of the Soil* (Collins 1961).
Vanstone, E.: *The Soil and the Plant* (Macmillan 1947).

CARE OF ANIMALS

Boulenger, E. G., rev. by A. Leutscher: *Keep an Aquarium* (Ward Lock 1956).
Keeping Pets—a pamphlet giving a list of books on pet care (1961) from the Publicity Officer, Lindsey and Holland County Library, 45 Newland, Lincoln.
Leutscher, Alfred: *Vivarium Life* (Cleaver-Hume 1961).
Sposzynska, Joy O.: *The Aquarium; Vivarium and Terrariums* (both Nelson 1967).
Suffolk Nat. Hist. Society: *Keeping Wild Animals in Schools* (Mammal Society of the B. Isles).
Universities Federation for Animal Welfare: *The UFAW Handbook on the Care and Management of Laboratory Animals* (Livingstone 1967).

ECOLOGY

Arvill, Robert: *Man and Environment* (Pelican 1967).
Ashby, M.: *Introduction to Plant Ecology* (Macmillan 1961).
Benton, A. H. and Werner W. E.: *Principles of Field Biology and Ecology* (McGraw-Hill 1958).
Cloudsley-Thompson, J. L.: *Microecology* (Arnold 1967).
Dowdeswell, W. H.: *Animal Ecology* (Methuen 1952).
Elton, C.: *Animal Ecology* (Sidgwick and Jackson 1927).
Elton, C.: *The Ecology of Animals* (Methuen 1950).
Farb, Peter: *Ecology* (Life Nature Library, Time Inc. 1965).
Friedlander, C. P.: *Heathland Ecology* (Heinemann 1960).
Hanson, H. C.: *Dictionary of Ecology* (Owen 1962).
Instructions for Collectors: *Fishes* (No. 619); *Insects* (No. 395a); *Invertebrates, Animals other than Insects* (No. 225); *Plants* (No. 560) (British Museum, Natural History).
Kendligh, S. C.: *Animal Ecology* (Prentice Hall 1961).
Leach, W.: *Plant Ecology* (Methuen 1949).
Lewis, T. and Taylor, L. R.: *Introduction to Experimental Ecology* (Academic Press 1967).
Macfadyen, A.: *Animal Ecology* (Pitman 1963).
McLean, R. C. and Cook, W. R.: *Practical Field Ecology* (Allen and Unwin 1950) with useful keys.
Miles, D. M. and H. B.: *Freshwater Ecology* (Hulton Ed. Publn. 1967); *Seashore Ecology* (Hulton Ed. Publn. 1967).
Neal, Ernest: *Woodland Ecology* (Heinemann 1958).
Prime, C. T.: *The Young Botanist* (Nelson 1963).
Sankey, J. H. P.: *A Guide to Field Biology* (Longmans Green 1958).
Tait, R. V.: *Elements of Marine Ecology* (Butterworth).
Tansley, A. G.: *The British Isles and their vegetation* (C.U.P. 1953); *Introduction to Plant Ecology* (Allen and Unwin 1946).

CONSERVATION

Arvill, Robert: *Man and Environment* (Pelican 1967).
Carson, Rachel: *Silent Spring* (Penguin 1965).
Christian, Garth: *Tomorrow's Countryside* (John Murray 1966).
While some Trees Stand (G. Newnes 1963); *A Place for Animals* (Lutterworth 1958).
Crowe, Sylvia: *Forests and the Landscape* (H. M. Stationery Office).
Fitter, R. S. R., ed.: *Kingfisher* (monthly, 6 Gombards Street, St. Albans, Herts.). A news sheet on conservation matters.

Hadfield, Miles: *One Man's Garden* (Phoenix House 1966).
Marchant, R. A.: *Man and Beast* (Bell 1966).
Mellanby, Keith: *Pesticides and Pollution* (Collins New Naturalist 1967).
Newman, L. Hugh: *Creating a Butterfly Garden* (John Baker 1967).
Nicholson, E. M.: *Britain's Nature Reserves* (Country Life 1957).
Stamp, L. Dudley: *Man and the Land* (Collins New Naturalist 1964); *Nature Conservation in Britain* (Collins New Naturalist 1966).
Tansley, A. G.: *Britain's Green Mantle* (Allen and Unwin 1949); *Our Heritage of Wild Nature* (C.U.P. 1945).
Wheeler, L. R.: *Harmony of Nature* (Ed. Arnold 1947).

MAMMALS

Animals of Britain Series (Sunday Times Publications 1962-3), edited by L. Harrison Matthews: No. 1, *Badgers*—E. G. Neal; No. 2, *Horseshoe Bats*—John Hooper; No. 3, *Hedgehogs*—Maxwell Knight; No. 4, *Water Voles*—S. R. Ryder; No. 5, *Grey Squirrels*—Monica Shorten; No. 6, *Red Squirrels*—Monica Shorten; No. 7, *Grey Seals*—H. R. Hewer; No. 8, *Otters*—E. G. Neal; No. 9, *Foxes*—H. G. Hurrell; No. 10, *Dormice*—Elaine Hurrell; No. 11, *Fallow Deer*—F. J. Taylor Page; No. 12, *Roe Deer*—F. J. Taylor Page; No. 13, *Red Deer*—F. J. Taylor Page; No. 14, *Weasels*—Ian Linn; No. 15, *Moles*—Gillian Godfrey; No. 16, *Brown Rats*—Colin Matheson; No. 17, *Shrews*—Peter Crowcroft; No.18, *Bats, (Noctule, Leisler's and Serotine)*—Michael Blackmore; No. 19, *Harvest Mice*—Maxwell Knight; No. 20, *Ancient White Cattle*—Kenneth Whitehead; No. 21, *Rabbits*—Harry V. Thompson; No. 22, *Pine Marten*—H. G. Hurrell; No. 23, *Mountain Hares*—A. Watson and R. Hewson; No. 24, *Baleen Whales*—L. Harrison Matthews.
Blackmore, M.: *Mammals in Britain* (Collins 1948).
Brink, H. van den: *A Field Guide to Mammals of the British Isles and Europe* (Collins 1966).
Burton, M.: *Wild Animals of the British Isles, based on Animal Life of the British Isles* (Warne 1960).
Cadman, W. A.: *Roe Deer* and *Fallow Deer* (H.M. Stationery Office); *Dawn, Dusk and Deer* (Country Life 1966).
Corbet, G. B.: *The Identification of British Mammals* (British Museum Natural History 1964); *The Terrestrial Mammals of Western Europe* (Foulis 1967).
Crowcroft, Peter: *Mice All Over* (Foulis 1966).
Fitzgerald, B. Vesey: *British Bats* (Methuen 1949).
Godfrey, G. and Crowcroft, W. P.: *The Life of the Shrew* (Museum Press 1960).
Knight, Maxwell: *How to Observe our Wild Mammals* (Routledge and Kegan Paul 1957).
Lancum, F. H.: *Wild Mammals of the Land* (H.M.S.O. Min. of Agriculture 1957).
Lockley, R. M.: *Grey Seal, Common Seal* (Deutsch 1966).
Matthews, L. Harrison: *British Mammals* (Collins New Naturalist 1952).
Neal, Ernest: *The Badger* (Collins New Naturalist 1948).
Page, F. J. Taylor: *Field Guide to the British Deer* (Mammal Soc. 1966).
Shorten, Monica: *Squirrels* (Collins New Naturalist 1954).
Southern, N. ed. *The Handbook of British Mammals* (Mammal Soc. 1964).
Stehl, G. and Brohmer. P., ed. and trans. by Alfred Leutscher: *The Young Specialist Looks at Animals (Mammals)* (Burke 1965).
Street, Philip: *Mammals of the British Isles* (Hale 1961).
Thompson, H. and Worden. A.: *The Rabbit* (Collins New Naturalist 1956).
Council for Nature: *Predatory Mammals in Britain* (*Council for Nature* 1966).

BIRDS

Armstrong, E. A.: *The Wren* (Collins New Naturalist 1955).
Atkinson-Willis, G. L., ed.: *Wildfowl in Great Britain* (H.M.S.O.).
Baker, J. A.: *The Peregrine* (Collins New Naturalist 1967).
Benson, S. V.: *The Observer's Book of Birds* (Warne 1960).
Brown, P. and Waterson G.: *Return of the Osprey* (Collins 1962).
Buxton, J.: *The Redstart* (Collins New Naturalist 1950).
Campbell, Bruce: *The Oxford Book of Birds* (O.U.P.).
Colquhoun, M. K.: *The Woodpigeon in Britain* (H.M.S.O. 1951).
Dorst, J.: *The Migration of Birds* (Heinemann 1962).
Ennion, E. A.: *The Lapwing* (Methuen 1949).
Fisher, J.: *The Fulmar* (Collins New Naturalist 1952).
Fitter, R. S. R.: *Collins Guide to Bird Watching* (Collins 1963).
Fitter, R. S. R. and Richardson, R. A.: *Collins Pocket Guide to British Birds* (Collins 1966); *Collins Pocket Guide to Nests and Eggs* (Collins 1954).
Gordon, S.: *The Golden Eagle* (Collins 1955).
Hillstead, A. F.: *The Blackbird* (Faber 1945).
Hollom, P.A.D.: *The Popular Handbook of British Birds* (Witherby 1955); *The Popular Handbook of Rarer British Birds* (Witherby 1960).
Hosking, E. and Newberry, C.: *The Swallow* (Collins 1946).
Lack, D.: *Life of the Robin* (Witherby 1946); *Swifts in a Tower* (Methuen 1956).
Lockley, R. M.: *Puffins* (Dent 1953); *Shearwaters* (Dent 1942).
Lowe, F. A.: *The Heron* (Collins New Naturalist 1954).
Macleod, R. D.: *Key to the Names of British Birds* (Pitman 1954).
Matthews, G. V.: *Bird Navigation* (C.U.P. 1955).
Mountfort, G.: *The Hawfinch* (Collins New Naturalist 1957).
Peterson, R., Mountfort, G. and Hollom, P.: *A Field Guide to the Birds of Britain and Europe* (Collins 1954).
Richmond, W. K.: *British Birds of Prey* (Lutterworth 1959).
Scott, Peter: *Wildfowl of the British Isles* (Country Life 1957).
Smith, S.: *The Yellow Wagtail* (Collins New Naturalist 1950).
Smith, S. B.: *British Waders in their Haunts* (Bell 1950).
Summers-Smith, D.: *The House Sparrow* (Collins New Naturalist 1963).
Thompson, D. Nethersole: *The Greenshank* (Collins New Naturalist 1951).
Ticehurst, N. F.: *The Mute Swan in England* (Cleaver-Hume 1957).

Wenzel, F.: *The Buzzard* (Allen and Unwin 1959).
Yeates, G. K.: *The Life of the Rook* (P. Allan 1934).

MIGRATION

Christian, Garth: *Down the Long Wind* (Newnes 1961).
Dorst, J.: *The Migration of Birds* (Heinemann 1962).
Hollom P. A. and Brownlow, H. G.: *Trapping Methods for Bird Ringers* (Br. Trust for Ornithology, Oxford 1958).
Lockley. R. M. and Russell, R.: *Bird Ringing* (Crosby Lockwood 1953).
Matthews, G. V. *Bird Navigation* (C.U.P. 1955).
Simms, E.: *Bird Migrants* (Cleaver-Hume 1952).

REPTILES AND AMPHIBIANS

Hellmich, Walter, ed. Alfred Leutscher: *Reptiles and Amphibians of Europe* (Blandford Press 1956).
Knight, Maxwell: *Frogs, Toads and Newts in Britain* (1962); *Reptiles in Britain* (1965) (Brockhampton).
Leutscher, Alfred, ed. and trans.: *The Young Specialist looks at Reptiles* (Burke 1966).
Savage, R. M.: *The Ecology and Life History of the Common Frog* (Pitman 1961).
Smith, Malcolm: *The British amphibians and Reptiles* (Collins New Naturalist 1954).

FISHES

Frost, W. E. and Brown, M. E.: *The Trout* (Collins New Naturalist 1967).
Jenkins, J. T.: *The Fishes of the British Isles* (Warne 1936).
Jones, J. W.: *The Salmon* (Collins New Naturalist 1959).
Macleod, R. D.: *Key to the Names of British Fishes, Amphibians and Reptiles* (Pitman 1956).
Taylor, F. J.: *Fishes of Rivers, Lakes and Ponds* (Blandford 1961).
Wells, A. L.: *The Observer's Book of Freshwater Fishes* (Warne 1957).

INSECTS

Burr, M.: *British Grasshoppers and their Allies* (P. Allan 1936).
Butler, C. G.: *The World of the Honeybee* (Collins New Naturalist 1954).
Coyler, C. N. and Hammond, C. O.: *Flies of the British Isles* (Warne 1951).
Ford, E. B.: *Butterflies* (Collins New Naturalist 1957); *Moths* (Collins N. Naturalist 1955).
Ford, R. L. E.: *Practical Entomology* (Warne 1963).
Free, J. B., and Butler C. G.: *Bumblebees* (Collins New Naturalist 1959).
Hahnewald, Edgar, ed. N. D. Riley: Insects in Colour (Blandford Press 1963).
Hickin, N. E.: *Caddis* (Methuen 1952).
Imms, A.: *Insect Natural History* (Collins New Naturalist 1947).
Joy, N. N.: *Practical Handbook of British Beetles* (Witherby 1932).
Linssen, E. F.: *Beetles of the British Isles* (Warne 1959), 2 vols.
Longfield, Cynthia: *The Dragonflies of the British Isles* (Warne 1949).
Lucas, W. J: *Monograph of the British Orthoptera* (Ray Society 1920)
Macan, T. T.: *A Revised Key to the British Water Bugs* (Freshwater Biological Assn. 1956).
Macleod, R. D.: *A Key to the Names of British Butterflies and Moths* (Pitman 1959).
Marshall J. F.: *The British Mosquitos* (British Museum Natural History 1938).
Morley, D. W.: *Ants* (Collins New Naturalist 1953).
Newman, L. H.: *Looking at Butterflies* (Collins 1959); *Complete British Butterflies in Colour* (EburyPress 1967).
Nixon, G. E.: *The World of Bees* (Hutchinson 1959).
Oldroyd, H.: *Insects and their World* (British Museum, Natural History 1966); *Collecting, Preserving and Studying Insects* (Hutchinson, London 1958).
Ragge, David: *Grasshoppers, Crickets and Cockroaches of the British Isles* (Warner 1965).
Sandars, E.: *An Insect Book for the Pocket* (O.U.P. 1946).
South, R.: *The Butterflies of the British Isles* (Warne 1941); *The Moths of the British Isles, rev.* by H. L. Edelsten (Warne 1961), 2 vols.
Southwood, T. R. E. and Leston, D.: *Land and Water Bugs of the British Isles* (Warne 1959).
Step, E.: *Bees, Wasps, Ants and Allied Insects of the British Isles* (Warne 1932).
Stokoe, W. J.: *Butterflies and Moths of the Wayside and Woodland* (Warne 1952).
Stokoe, W. J. and Stovin, G. H. T.: *The Caterpillars of the British Butterflies* (Warne 1944); *The Caterpillars of the British Moths* (Warne 1959), 2 vols.

OTHER INVERTEBRATES

Bristowe, W. S.: *The World of Spiders* (Collins New Naturalist 1958).
Cernosvitov, L. and Evans, A. C.: *Lumbricidae* (*Annelida*) (Linnean Soc. monograph No. 6, 1947).
Clark, Ailsa: *Starfishes and their Relations* (British Museum Natural History 1962).
Cloudsley-Thompson, J. L.: *Spiders, Scorpions, Centipedes and Mites: the Ecology and Natural History of Woodlice, Myriapos and Arachnids* (Pergamon Press 1958).
Cloudsley-Thompson, J. L. and Sankey, J.: *Land invertebrates* (Methuen 1961).
Donner, J.: *Rotifers* (Warne 1966).
Eason, E. H.: *Centipedes of the British Isles* (Warne 1964).
Edney, E. B.: *British Woodlice* (Synopsis of the British fauna, Linnean Soc. No. 9. 1954).
Macan, T. T.: *A Guide to the Freshwater Invertebrate Animals* (Longmans Green 1959); *A Key to the British Fresh and Brackish Water Gastropods, with Notes on their Ecology* (Freshwater Biological Assn. 1960).
Morton, J. E.: *Molluscs* (Hutchinson 1958).
Quick, H. E.: *British Slugs* (British Museum Natural History 1960).
Savory, T. H.: *The Spiders and Allied Orders of the British Isles* (Warne 1945); *Arachnida* (Academic Press 1964).
Step, E.: *Shell Life* (Warne 1937).
Street, P.: *Shell Life on the Seashore* (Faber 1961).

PLANTS, GENERAL

Bastin, H.: *Plants Without Flowers* (Hutchinson 1955).
Brimble, L. J. F.: *Intermediate Botany* (Macmillan 1953).

Elliott, J. H.: *Teach Yourself Botany* (E.U.P. 1958).
Lowson, J. M.: *Textbook of Botany* (University Tutorial Press 1962).
Ramsbottom, J.: *Popular Book of Botany* (Ward Lock 1953).
Turrill, W. B.: *British Plant Life* (Collins New Naturalist 1948).
Woodward, Marcus: *Gerard's Herball* (Spring Books 1964).

FLOWERING PLANTS

Brimble L. J. F.: *The Floral Year* (Macmillan 1949).
Clapham, A. R. and others: *Flora of the British Isles* (C.U.P. 1962); *Flora of the British Isles* (illustrations C.U.P. 1957 etc.) in parts.
Gilmour, J. and Walters, M.: *Wildflowers* (Collins N. Naturalist 1954).
Hepburn, Ian: *Flowers of the Coast* (Collins New Naturalist 1952).
Martin, W. Keble: *The Concise British Flora in Colour* (Ebury Press 1965).
McClintock, D. and Fitter, R. S. R.: *Collins Pocket Guide to Wildflowers* (Collins 1956)—with supplement by D. McClintock, 1957—(C. M. Rob, Thirsk).
Melderis, A. and Bangerter, E. B.: *A Handbook of British Flowering Plants* (Ward Lock 1955).
Perring F. H. and Walters, S. M.: *Atlas of the British Flora* (Nelson, B.S.B.I. 1962).
Prime, C. T. and Deacock, R. I.: *The Shorter British Flora* (Methuen 1953).
Raven, J. and Walters, S. M.: *Mountain Flowers* (Collins New Naturalist 1956).
Salisbury, E. J.: *Flowers of the Woods* (Penguin 1946).
Scott, D. H. and Brooks: *Flowering Plants* (Black 1948).
Step, E.: *Wayside and Woodland Blossoms* (Warne 1963), 3 vols.
Summerhayes, V. S.: *Wild Orchids of Britain* (Collins New Naturalist 1951).

FLOWERLESS PLANTS

Chapman, V. J.: *The Algae* (Macmillan 1962).
Dickinson, C. I.: *British Seaweeds* (Eyre and Spottiswoode 1963).
Duncan, U. K.: *A Guide to the Study of Lichens* (Arbroath, Buncle 1959).
Hodgson, N. B.: *Grasses, Sedges, Rushes and Ferns* (Eyre and Spottiswoode 1949).
Hubbard, C. E.: *Grasses* (Penguin 1954).
Jewell, A. L.: *The Observer's Book of Mosses and Liverworts* (Warne 1955).
Kershaw, K. A. and Alwin, K. L.: *The Observer's Book of Lichens* (Warne 1963).
Newton, L.: *A Handbook of the British Seaweeds* (British Museum Natural History 1931).
Smith, A. L.: *Handbook of the British Lichens* (British Museum, Natural History 1921).
Step. E.: *Wayside and Woodland Ferns; a Guide to the British Ferns, Horsetails and Clubmosses* (Warne 1945).
Stokoe, W. K.: *The Observer's Book of Grasses, Sedges and Rushes* (Warne 1942); *The Observer's Book of Ferns* (Warne 1950); *The Observer's Book of Grasses* (Warne 1942).
Taylor, P.: *British Ferns and Mosses* (Eyre and Spottiswoode 1960).
Watson, E. V.: *British Mosses and Liverworts* (C.U.P. 1955).
Watson, H.: *Woodland Mosses* (H.M.S.O. Forestry Commission 1947).

FUNGI

Hvass, E. and H.: *Mushrooms and Toadstools in Colour* (Blandford 1961).
Lange, M. and Hora, F. B.: *Collin's Guide to Mushrooms and Toadstools* (Collins 1963).
Pilat, A. and Usak, O.: *A Handbook of Mushrooms* (Spring Books 1959).
Ramsbotton, J.: *Mushrooms and Toadstools* (Collins New Naturalist 1953); *A Handbook of the Larger Fungi* (British Museum Natural History 1944).
Rayner, M. C.: *Trees and Toadstools* (Faber 1955).
Wakefield, E. M.: *The Observer's Book of Common Fungi* (Warne 1954).

TREES

Brimble, L. J. F.: *Trees in Britain* (Macmillan 1946).
Edlin, H. L.: *Trees, Woods and Man* (Collins New Naturalist 1956); *Know your Conifers* (H.M.S.O. 1960).
Holbrook, A. W.: *Pocket Guide to Trees* (Country Life).
Jay, B. A.: *Conifers in Britain* (Black 1952).
Meikle, R. D.: *British Trees and Shrubs* (Eyre and Spottiswoode 1958).
Step, E.: *Wayside and Woodland Trees* (Warne 1960).
Stokoe, W. J.: *The Observer's Book of Trees* (Warne 1960).
Vedel, Helge and Lange, Johan: *Trees and Bushes in Wood and Hedgerow* (Methuen 1960).

WOODLAND

Anderson, Mark: *The Natural Woodlands of Britain and Ireland* (Department of Forestry, Oxford. Holywell Press 1932).
Barth, G. Mandahl: *Woodland Life* (Blandford Press, 1966).
Barclay Smith, Phyllis: *Woodland Birds* (Penguin 1955).
Brimble, J. A.: *London's Epping Forest* (Country Life 1950).
Chrystal, R. N.: *Insects of the British Woodlands* (Warne 1937).
Edlin, H. L.: *Trees, Woods and Man* (Collins New Naturalist 1956); *Wildlife of Wood and Forest* (Hutchinson 1960); *England's Forests* (Faber 1958).
Neal, Ernest: *Woodland Ecology* (Heinemann 1958).
Qvist, Alan: *Epping Forest* (Corporation of London 1958).
Sage, Brian: *Northaw Great Wood; its History and Natural History* (Education Dept, Hertfordshire County Council 1966).
Salisbury, E. *Flowers of the Woods* (Penguin 1946).
Wooldridge, S. W. and Goldring, F.: *The Weald* (Collins New Naturalist 1953).
Yapp, W. B.: *Birds and Woods* (O.U.P. 1962).

FIELDS, COMMONS AND HEDGEROWS

Hoskins, H. G. and Stamp, L. Dudley: *The Common Lands of England and Wales* (Collins New Naturalist 1959).

Lancum, E. H.: *Wild Birds of the Land* (H.M.S.O. Min. of Agriculture 1951).
Locke, G. M.: *A Sample Survey of Field and other Boundaries in Great Britain* (*Quarterly Journal of Forestry*, Vol. LVI, No. 2, April 1962).
Moore, I.: *Grass and Grasslands* (Collins New Naturalist 1966).

CHALKLAND AND DUNES

Lousley, J. E.: *Wildflowers of Chalk and Limestone* (Collins New Naturalist 1950).
Salisbury, E. J.: *Downs and Dunes* (Bell 1952).
Sankey, J.: *Chalkland Ecology* (Heinemann 1966).

BUILT UP AREAS AND TOWNS

Barclay-Smith, Phyllis: *Garden Birds* (Penguin 1952).
Cohen, E. and Campbell, B.: *Nest Boxes* (British Trust for Ornithology 1957).
Cohen, Julius B. and Ruston A. G.: *Smoke: A Study of Town Air* (Arnold 1925).
Economic Series, British Museum (Natural History): 6, *Ants*; 32, *Blowflies*, 376, *Fleas*; 409, *Woodboring Beetles*; 545, *Carpet Beetles*; 547, *Silverfish and Firebrat*; 552, *Bed Bug*; 550, *Lice*; 551, *The Housefly*; 554, *Clothes Moths*; 564, *The Cockroach*; 565, *Insects of Stored Food Products*; 608, *Crickets*; 626, *Common Earwig*.
Fitter, R. S. R.: *London's Natural History* (Collins New Naturalist 1945).
Holmes, R. C. *et al.*: *The Birds of the London Area since 1900* (London Nat. hist. soc., Collins 1957).
Kirkpatrick, R.: *The Biology of Waterworks* (British Museum Natural History 1924).
Knight, Maxwell: *Bird Gardening* (Routledge 1954).
Lousely, J. E.: *Natural History of the City* (Corporation of London).
Ministry of Works: *Bird Life in the Royal Parks* (H.M.S.O. 1947) annually.
Newman, L. Hugh: *Create a Butterfly Garden* (John Baker 1967).
Salisbury, E. J.: *Weeds and Aliens* (Collins New Naturalist 1961).
Staples, C. P.: *Birds in a Garden Sanctuary* (Warne 1946).

MOUNTAINS AND MOORLAND

Condry, William: *The Snowdonia National Park* (Collins New Naturalist 1966).
Coombs, R.: *Mountain Birds* (Penguin 1952).
Darling, F. F. and Boyd, T. M.: *The Highlands and Islands* (Collins New Naturalist 1947).
Edwards, K. C. *et al.*: *The Peak District* (Collins New Naturalist 1962).
Friedlander, C. P.: *Heathland Ecology* (Heinemann 1960).
Harvey, L. A. and St Leger-Gordon: *Dartmoor* (Collins New Naturalist 1962).
Pearsall, W. H.: *Mountains and Moorlands* (Collins New Naturalist 1950); *Ben Lawers and its Alpine Flowers* (National Trust for Scotland 1954).
Raven, J. and Walters. M.: *Mountain Flowers* (Collins New Naturalist 1956).
Guide Books to the National Parks (H.M.S.O.)—Dartmoor, Brecon Beacons, Snowdonia, Peak District, N. Yorks moors, Glen More, the Border.

PONDS, LAKES AND RIVERS

Brown, E. S.: *Life in Freshwater* (O.U.P. 1955).
Carpenter, K.: *Life in Inland Waters* (Sidgwick and Jackson 1928).
Clegg, John, ed.: *Pond and Stream Life of Europe* (Blandford 1963); *Freshwater Life of the British Isles* (Warne 1965); *The Observer's Book of Pond Life* (Warne 1967).
Ellis, E. A.: *The Broads* (Collins New Naturalist 1965).
Fisher, R., ed. W. Engelhardt and H. Merxmuller: *The Young Specialist looks at Pond Life* (Burke 1964).
Lowe-McConnell, R. H.: *Man-made Lakes* (Academic Press 1966).
Macan, T. T.: *A Guide to the Freshwater Invertebrate Animals* (Longmans Green 1959).
Macan, T. T. and Worthington, E. B.: *Life in Lakes and Rivers* (Collins New Naturalist 1951).
Mellanby, Helen: *Animal Life in Freshwater* (Methuen).
Miles, D. A. and H. B.: *Freshwater Ecology* (Hulton Ed. Pub. 1967).
Ward, H. B. and Whipple, G. C, ed. W. Edmondson: *Freshwater Biology*, (Wiley, New York and Chapman and Hall, London 1959).

THE SEA SHORE

Barrett, J. H. and Yonge, C. M.: *Collins Pocket Guide to the Seashore* (Collins 1958).
Eales, N. B.: *The Littoral Fauna of the British Isles* (C.U.P. 1961).
Evans, I. O.: *The Observer's Book of Sea and Seashore* (Warne 1962).
Fisher, J. and Lockley, R. M.: *Sea Birds* (Collins New Naturalist 1954).
Flattely, F. W. and Walton, C. L.: *The Biology of the Seashore* (Sidgwick and Jackson 1946).
Hepburn, Ian: *Flowers of the Coast* (Collins New Naturalist 1952).
Hill, C. A. Gibson: *Birds of the Coast* (Witherby 1949).
Miles, D. H. and H. B.: *Seashore Ecology* (Hulton Ed. Pub. 1967).
Southward, A. J.: *Life on the Seashore* (Warne).
Street, P.: *Between the Tides* (U.L.P. 1952).
Vevers, H. G.: *The British Seashore* (Routledge 1954).
Wilson, D. P.: *Life of the Shore and Shallow Sea* (Nicholson and Watson 1935).
Yonge, C. M.: *The Seashore* (Collins New Naturalist 1949).

Some useful addresses and sources of information

Only permanent addresses are given in this list. Many addresses change with an alteration in the secretary. The latest address can usually be obtained from the Council for Nature (q.v.).

Amateur Entomologists' Society, 1 West Ham Lane, London E. 15. Founded 1935.

Association of Applied Biologists. Publication—*Annals of Applied Biology*.

Association of School Natural History Societies. Founded 1947. Seeks to bring school societies into closer contact with one another, and with senior bodies. An annual exhibition. Journal—*Starfish*.

B.B.C. Natural History Unit, Broadcasting House, Whiteladies Road, Bristol 8. Founded 1957. Has a very comprehensive library of sounds and films of British wildlife, some for hire.

Botanical Society of the British Isles, c/o Department of Botany, British Museum (Natural History), London, S.W. 7. Founded 1836, originally the Botanical Society of London. Publication—*Watsonia*.

British Association for the Advancement of Science, 3 Sanctuary Buildings, Great Smith Street, London S.W. 1. Publication—*The Advancement of Science*.

British Bryological Society (Mosses and Liverworts). Founded 1896 as the Moss Exchange Club. Publication—*Transactions of the B.B.S.*

British Ecological Society. Founded 1904 as the British Vegetation Committee. Publication—*Journal of Ecology*.

British Lichen Society. Founded 1958. Publication—the *Lichenologist*.

British Herpetological Society (Reptiles and Amphibia). Founded 1947. Publication—*British Journal of Herpetology*.

British Mycological Society (Fungi). Founded 1896. Publication—the *Transactions*.

British Naturalists' Association (with branches throughout the country). Founded 1905 as British Empire Naturalists' Association. Journal—*Countryside*.

British Ornithologists' Union, c/o Bird Room, British Museum (Natural History), London, S.W. 7. Founded 1858. Journal—the *Ibis*.

British Trust for Ornithology, Beech Grove, Tring, Hertfordshire. Founded 1932. Carries out field studies and bird observations. Is the main ringing centre for Britain. Numbered rings carry the address of the British Museum (Nat. Hist.). Publication—*Bird Study*.

British Phycological Society (Algae). Founded 1952. Publication—*British Phycological Bulletin*.

British Pteridological Society (Ferns). Founded 1891. Journal—*British Fern Gazette*.

Commons, Open Spaces and Footpaths Preservation Society, 166 Shaftesbury Avenue, London, W.C. 2. Founded 1865 to protect commons and other open public spaces, such as Epping Forest. Publication—*Journal of the C.O.S.F.P.S.*

Conchological Society of Great Britain and Ireland c/o British Museum (Natural History), London, S.W. 7. Founded 1872. Publication—*Journal of Conchology*.

Council for Nature, c/o Zoological Society of London, Regent's Park, London, N.W. 1. Founded 1958. The central body and clearing house for natural history interests in Britain. Has a large following in membership among local societies and other bodies. Runs an Intelligence Unit which offers advice and answers to enquiries, also a Conservation Corps movement which carries out useful and periodical conservation measures in the field where and when asked for. Publication—*News for Naturalists and Habitat*. Organises National Nature Week.

County Naturalists' Trusts. Main Headquarters—the Manor House, Alford, Lincolnshire. Addresses of individual County Trusts available from headquarters or from the Society for the Promotion of Nature Reserves (q.v.).

Council for the Preservation of Rural England, 4 Hobart Place, London, S.W. 1. Founded 1926. The same address for Wales (Founded 1928).

Fauna Preservation Society, c/o Zoological Society of London. Regent's Park, London, N.W. 1. Founded 1903. Publication—*Oryx*.

Field Studies Council, Devereux Buildings, Devereux St., London, W.C. 2. Founded 1943. Runs a fine set of field centres (see under Educational Facilities).

Forestry Commission, 22 Saville Row, London, W.1. A pioneer Government department in conservation, which offers facilities to naturalists and schools. Contact the appropriate Conservator of the Forest to be visited.

Geological Association. Founded 1858. Publication—the *Proceedings*.

Freshwater Biological Association, the Ferry House, Far Sawrey, Ambleside, Westmorland. A research centre into freshwater biology on the shores of lake Windermere. Founded 1929.

Institute of Biology, 41 Queen's Gate, London, S.W. 7. Founded 1950. Offers useful advice in careers and studies of a biological nature. Publication—*Journal of the Institute of Biology*.

Linnean Society of London, Burlington House, London, Piccadilly, W. 1. Founded 1788. Publishes a *Journal* and *Proceedings*.

Mammal Society, c/o Zoological Society of London, N.W. 1. Founded 1954. Publishes a *Bulletin*.

Marine Biological Association of the United Kingdom, the Laboratory, Citadel Hill, Plymouth, Devon. Founded 1884. Publication—*Journal of the M.B.A. of the U.K.*

Men of the Trees. Devoted to the preservation and appreciation of trees.

National Trust, 42 Queen Anne's Gate, London, S.W. 1, also National Trust for Scotland, 5 Charlotte Square, Edinburgh, 2. Both these Trusts are owners of nature reserves and places of natural beauty available to the general public.

Nature Conservancy, 19 Belgrave Square, London, S.W. 1. Placed under Royal Charter in 1949. This is the Government's conservation organisation which manages some 150 nature reserves. Address in Scotland: 12 Hope Terrace, Edinburgh 9 and Wales: Pemrhos Road, Bangor, Caernarvonshire. Publication—an *Annual Report*.

Queckett Microscopical Club, c/o the Royal Society, Burlington House, Piccadilly, London, W. 1. Founded 1865. Publication—*Journal of the Q.M.C.*

Ramblers' Association, 48 Park Road, Baker Street, London, N.W. 1.

Royal Entomological Society of London, 41 Queen's Gate, London, S.W. 7. Founded 1833. Publications—*Transactions and Proceedings.*

Romany Society. Founded 1945 to perpetuate the memory of B.B.C.'s Romany (S. Bramwell Evans) and to further interest in outdoor life. Publication—the *Romany Magazine.*

Royal Society, 6 Carlton House Terrace, London, S.W. 1. Founded 1663.

Royal Society for the Protection of Birds, the Lodge, Sandy, Bedfordshire, also 21 Regent Terrace, Edinburgh, 7. Founded 1889. Owns and manages some two dozen bird reserves. Publication—*Bird Notes.*

School Natural Science Society, formerly School Nature Study Union. Caters for science teachers.

Scottish Field Studies Association, 141 Bath Street, Glasgow, C. 2. Runs field centres (see under Educational Facilities).

Scottish Marine Biological Association, Marine Station, Millport, Isle of Cumbrae, Scotland.

Scottish Ornithologists' Club, 21 Regent Terrace, Edinburgh 7.

Selborne Society. Founded 1885 to perpetuate the memory of Gilbert White. Publication—*Bulletin and Selborne Magazine.*

Society for the Promotion of Nature Reserves, c/o British Museum (Natural History), London, S.W. 7. Founded 1912. Owns many reserves. Assists and co-ordinates the work of the County Trusts (see County Naturalists' Trusts). Publication—*Handbook.*

Systematics Association, c/o British Museum (Natural History), London, S.W. 7. Founded 1937. Publication—*Bibliography of Key Works for the identification of the British Fauna and Flora.*

Universities Federation for Animal Welfare, formerly University of London Animal Welfare Society. Founded 1926 to promote a humane attitude towards wild and domesticated animals. Publications—*U.F.A.W. Courier* and many pamphlets on the care of animals.

William Pengelly Cave Studies Association. Founded 1962. Has a cave study centre at Buckfastleigh, South Devon. Journal—*Studies in Speleology.*

Wildfowl Trust, formerly the Severn Wildfowl Trust, New Grounds, Slimbridge, Gloucestershire. Founded 1946. A valuable sanctuary and centre for the study and conservation of wildfowl. Publication—*Annual Report.*

World Wildlife Fund, an International body founded in 1961 to collect and distribute funds for world-wide conservation measures. The British National Appeal (B.N.A.) was formed in 1962 under the Presidency of the Duke of Edinburgh. Address 7-8 Plumtree Court, London, E.C. 4.

Wildlife Youth Service, junior branch of the World Wildlife Fund, 120A London Road, Morden, Surrey. Founded 1963.

Youth Hostels Association, 29 John Adam Street, London, W.C. 2.

Zoological Society of London, Regent's Park, London, N.W. 1. Founded 1826, under Royal Charter 1829. Maintains the London Zoological Gardens at Regent's Park, and the Whipsnade Zoological Park near Dunstable. Publications—*Journal of Zoology* (proceedings), the *Zoological Record* and *Nomenclator Zoologicus.*

Educational facilities

The following bodies provide sparetime educational facilities for amateur naturalists, either as lectures or as field studies:

Extra Mural Department, London University, 7 Ridgmount Street, London, W.C. 1. This and many other similar departments attached to various universities, run a large number of evening courses, including natural history subjects. For copies of the syllabus apply to the appropriate university department.

L.E.A. courses. Here again, many suitable subjects as evening courses are offered by local education authorities. Apply to the local or county council education department.

Field Studies Council, 9 Devereux Court, Strand, London, W.C. 2. The Council manages the following excellent centres, each of which provides accommodation for residential courses in natural history. Apply in each case to the Warden:

- Dale Fort Field Centre, near Haverford West, Pembrokeshire. Also arranges visits to Skokholm Island for bird study.
- Flatford Mill Field Centre, East Bergholt, Colchester, Essex.
- Juniper Hall Field Centre, Box Hill, near Dorking, Surrey.
- Malham Tarn Field Centre, near Settle, Yorkshire.
- Orielton Field Centre, near Pembroke, Pembrokeshire.
- Preston Mountford Field Centre, near Shrewsbury.
- Slapton Ley Field Centre, near Kingsbridge, South Devon.
- The Draper's Field Centre, Rhyd-y-Creuau, Bettws-y-Coed, Caerns.
- The Leonard Wills Field Centre, Nettlecombe Court, Williton, Taunton, Somerset.

The Scottish Field Studies Association, 141 Bath Street, Glasgow, C.2. Similar to above. Has a centre at Garth Memorial Youth Hostel, Fortingall, Aberfeldy, Perthshire.

NOTE. In addition to the above, useful information of a local nature, such as lectures, film shows, exhibitions etc., can usually be obtained from the public library or newspaper.

Index

For main subject entries, see Contents List. This index is not intended to be an exhaustive guide to all species mentioned.